Generalized Quantum Calculus with Applications

Generalized Quantum Calculus with Applications

Svetlin G. Georgiev

Sanket Tikare

ACADEMIC PRESS

An imprint of Elsevier

Academic Press is an imprint of Elsevier
125 London Wall, London EC2Y 5AS, United Kingdom
525 B Street, Suite 1650, San Diego, CA 92101, United States
50 Hampshire Street, 5th Floor, Cambridge, MA 02139, United States

ISBN: 978-0-443-32804-6

For information on all Academic Press publications
visit our website at https://www.elsevier.com/books-and-journals

Publisher: Mara Conner
Acquisitions Editor: Stephanie Cohen
Editorial Project Manager: Mason Malloy
Production Project Manager: Erragounta Saibabu Rao
Cover Designer: Victoria Pearson Esser

Typeset by VTeX

Printed in the United Kingdom

Last digit is the print number: 9 8 7 6 5 4 3 2 1

Contents

Preface

Quantum calculus is the modern name for the investigation of calculus without limits. The quantum calculus or q-calculus began with FH Jackson in the early twentieth century, but this kind of calculus had already been worked out by Euler and Jacobi. Recently, it arose interest due to high demand of mathematics that models quantum computing. The q-calculus appeared as a connection between mathematics and physics. It has a lot of applications in different mathematical areas such as number theory, combinatorics, orthogonal polynomials, basic hypergeometric functions, and other sciences such as quantum theory, mechanics, and the theory of relativity. The book by Kac and Cheung [16] covers many of the fundamental aspects of quantum calculus. Recently, the general quantum difference operators that generalize the quantum calculus are defined.

The present book is devoted to the qualitative theory of general quantum calculus and its applications to general quantum differential equations and inequalities. The book is intended for senior undergraduate students and beginning graduate students of engineering and science courses. There are eight chapters in this book. Each chapter contains a good percentage of new material reflecting the advances in the subject. The chapters in the book are pedagogically organized. Each chapter concludes with a section with practical problems.

In Chapter 1, we define first- and second-kind generalized quantum operators. We introduce the concept of β-differentiation. The main properties and main rules of β-differentiation are deduced. Moreover, the chain rules, mean values theorems, and the Leibniz formula are presented.

Chapter 2 deals with β-integral and some of its properties. The basic rules for β-integration are deduced. The β-Taylor formula is also presented. Improper integrals of the first and the second kind are introduced and studied.

Chapter 3 is devoted on the study of some β-elementary functions. Some elementary functions such as β-exponential, β-trigonometric, and β-hyperbolic functions are defined, and their basic properties are given.

In Chapter 4, we define the generalized quantum Laplace transform and prove some of its properties. Some particular classes of functions ψ_k, $k \in \mathbb{N}_0$, are defined. The generalized quantum Laplace transforms for some elementary generalized quantum functions are presented.

Chapter 5 investigates homogeneous and nonhomogeneous initial value problems for first-order β-differential equations. The integral representation of their solutions is given. The method of β-Laplace transform for solving initial value problems is presented.

In Chapter 6, we investigate homogeneous and nonhomogeneous second-order linear β-differential equations. The concept of β-Wronskian is introduced, and some of their properties are given. The β-analogue of the Abel theorem is proved. A fundamental system of solutions is defined, and using it, some representations of general solutions of the considered classes β-differential equations are given. The β-Euler–Cauchy equations are investigated. The annihilator method and the β-Laplace transform method are presented.

Chapter 7 deals with the β-differential systems and their structures. The β-exponential matrix function is defined, and some of its properties are deduced. The β-analogue of the classical Liouville theorem is proved. The β-Putzer algorithm for finding the β-exponential matrix function in the case where the matrix is constant is introduced. The existence and uniqueness of solutions to nonlinear β-differential systems are also investigated.

In Chapter 8, some linear β-integral inequalities such as Gronwall-type inequalities, Bellman-type inequalities, Volterra-type β-integral inequalities, Gamidov-type inequalities, and Pachpatte-type inequalities are obtained.

This book is addressed to a wide audience of specialists such as mathematicians, physicists, engineers, and biologists. It can be used as a textbook at the graduate level and as a reference book for several disciplines.

For further reading, the readers can refer [1–23].

Sanket Tikare acknowledges the valuable and continued support from Dr. Usha Mukundan, Academic Director of Hindi Vidya Prachar Samiti, and Dr. Himanshu Dawda, Principal of Ramniranjan Jhunjhunwala College, during the preparation of this book. Sanket extends his sincere gratitude to his coauthor Svetlin G. Georgiev for the lively and productive discussions, which provided deep insight into the subject. Further, Sanket extends thanks to his wife Shaila and daughter Manasvini for their moral support and unflinching encouragement. Sanket has dedicated this book to his lifetime teacher Late Shri Manohar Hari Risbud.

The authors wish to express their gratitude to their families and friends for their constant support and encouragement. Moreover, the authors are grateful to the production staff at Elsevier.

Instructor and student resources

Helpful ancillaries have been prepared to aid learning and teaching.

Please visit the student companion site for more details: https://www.elsevier.com/books-and-journals/book-companion/9780443328046

For qualified professors, additional, instructor-only teaching material can be requested here: https://educate.elsevier.com/9780443328046

Paris, France   *Svetlin G. Georgiev*
Mumbai, India   *Sanket Tikare*
                                                                  December 2023

CHAPTER

General quantum differentiation[⊛]

1

In this chapter, we define first- and second-kind general quantum operators, and using them, we define the β-derivative and deduct some of its properties. The main rules for β-differentiation are given. The chain rules, the Rolle theorem, and the Leibniz formula for β-derivarives are also deduced and proved.

Throughout this chapter, we suppose that $I \subset \mathbb{R}$ and $s_0 \in I$.

1.1 The β-operators

Definition 1.1.1. We say that $\beta\colon I \to \mathbb{R}$ is a first-kind general quantum operator provided it is a strictly increasing continuous function on I with $\beta(t) \in I$ for every $t \in I$, s_0 is the unique fixed point of β, and

$$(t - s_0)(\beta(t) - t) \le 0, \quad t \in I.$$

Definition 1.1.2. We say that $\beta\colon I \to \mathbb{R}$ is a second-kind general quantum operator provided it is a strictly increasing continuous function on I with $\beta(t) \in I$ for every $t \in I$, s_0 is the unique fixed point of β, and

$$(t - s_0)(\beta(t) - t) \ge 0, \quad t \in I.$$

Note 1.1.3. When we say that $\beta\colon I \to \mathbb{R}$ is a general quantum operator, this means that it is a first- or second-kind general quantum operator.

Example 1.1.4. Let $I = [3, \infty)$ and $s_0 = 3$. Define

$$\beta(t) = 2\log\left(\frac{t}{3}\right) + 3, \quad t \in I.$$

Note that for each $t \in I$,

$$\log\left(\frac{t}{3}\right) \ge 0,$$

and hence $\beta(t) \in I$ for every $t \in I$. Also,

$$\beta'(t) = \frac{2}{t}$$
$$> 0, \quad t \in I,$$

⊛ This book has a companion website hosting complementary materials. Visit this URL to access it: https://www.elsevier.com/books-and-journals/book-companion/9780443328046.

Generalized Quantum Calculus with Applications. https://doi.org/10.1016/B978-0-44-332804-6.00006-6

1

that is, β is strictly increasing on I. Moreover,

$$\beta(3) = 2\log\left(\frac{3}{3}\right) + 3$$
$$= 2\log 1 + 3$$
$$= 3,$$

that is, 3 is the unique fixed point of β. Set

$$h(t) = 2\log\left(\frac{t}{3}\right) + 3 - t, \quad t \in I,$$

and

$$g(t) = (t - 3)h(t), \quad t \in I.$$

We have

$$h'(t) = \frac{2}{t} - 1$$
$$= \frac{2 - t}{t}$$
$$< 0, \quad t \in I.$$

Thus h is strictly decreasing on I, and then

$$h(t) \le h(3)$$
$$= 2\log\left(\frac{3}{3}\right) + 3 - 3$$
$$= 2\log 1$$
$$= 0.$$

Hence, since $t - 3 \ge 0$, $t \in I$, we obtain

$$g(t) \le 0, \quad t \in I.$$

Therefore, β is a first-kind general quantum operator.

Example 1.1.5. Let $I = \mathbb{R}$, $q \in (0, 1)$, and $\omega \ge 0$. Define

$$\beta(t) = qt \pm \omega, \quad t \in I.$$

Then $\beta(t) \in I$ for every $t \in I$, and

$$\beta'(t) = q$$
$$> 0, \quad t \in I.$$

Thus, β is a strictly increasing continuous function on I. Consider the equation

$$\beta(s_0) = s_0.$$

This gives

$$q s_0 \pm \omega = s_0,$$

or

$$(1 - q)s_0 = \pm\omega,$$

whereupon

$$s_0 = \pm \frac{\omega}{1 - q}$$

is the unique fixed point of β. Next, consider the function

$$g(t) = \left(t \mp \frac{\omega}{1 - q} \right) ((q - 1)t \pm \omega), \quad t \in I.$$

1. Let $t \geq \pm\dfrac{\omega}{1 - q}$. Then

$$t \mp \frac{\omega}{1 - q} \geq 0,$$

and

$$(q - 1)t \pm \omega = -(1 - q)t \pm \omega$$
$$= -(1 - q)\left(t \mp \frac{\omega}{1 - q} \right)$$
$$\leq 0.$$

Hence, $g(t) \leq 0$.

2. Let $t \leq \pm\dfrac{\omega}{1 - q}$. Then

$$t \mp \frac{\omega}{1 - q} \leq 0,$$

and

$$(q - 1)t \pm \omega = -(1 - q)\left(t \mp \frac{\omega}{1 - q} \right)$$
$$\geq 0.$$

Hence, $g(t) \leq 0$.

Therefore, β is a first-kind general quantum operator.

Example 1.1.6. Let $I = \mathbb{R}$, $q > 1$, and $\omega \geq 0$, and let β, g, and s_0 be as in Example 1.1.5. Then β is a strictly increasing function on I with unique fixed point s_0. Moreover,

1. if $t \geq \pm\dfrac{\omega}{1 - q}$, then we have

$$(q - 1)t \pm \omega \geq 0$$

and $g(t) \geq 0$;

2. if $t \leq \pm \dfrac{\omega}{1-q}$, then

$$(q-1)t \pm \omega \leq 0$$

and $g(t) \geq 0$.

Thus, β is a second-kind general quantum operator.

Example 1.1.7. Let $q \in (0, 1)$, $n \in 2\mathbb{N}+1$, and $I = \left(-q^{\frac{1}{1-n}}, q^{\frac{1}{1-n}}\right)$. Define

$$\beta(t) = qt^n, \quad t \in I.$$

Then

$$\beta'(t) = qnt^{n-1}$$
$$> 0, \quad t \in I, \quad t \neq 0,$$

and $\beta'(0) = 0$. Thus β is a strictly increasing continuous function on I. Hence, using the fact that $\beta(0) = 0$, we conclude that $s_0 = 0$ is the unique fixed point of β. Next,

$$t\left(qt^n - t\right) = t^2\left(qt^{n-1} - 1\right)$$
$$\leq t^2\left(q\left(q^{\frac{1}{1-n}}\right)^{n-1} - 1\right)$$
$$= 0, \quad t \in I.$$

Thus, β is a first-kind general quantum operator.

Exercise 1.1.8. Let $I = \left[1, \dfrac{3}{2}\right]$ and define

$$\beta(t) = \frac{1}{2}t + \frac{3}{4}, \quad t \in I.$$

Then prove that β is a first-kind general quantum operator.

Definition 1.1.9. Let β be a general quantum operator. For $k \in \mathbb{N}_0$, define

$$\beta^k(t) = \underbrace{\beta \circ \beta \circ \cdots \circ \beta(t)}_{k \text{ times}},$$
$$\beta^{-k}(t) = \underbrace{\beta^{-1} \circ \beta^{-1} \circ \cdots \circ \beta^{-1}(t)}_{k \text{ times}}, \quad \text{and}$$
$$\beta^0(t) = t, \quad t \in I.$$

Definition 1.1.10. For $q > 0$, $q \neq 1$, and $n \in \mathbb{N}_0$, define

$$[n]_q = \frac{1-q^n}{1-q}.$$

Example 1.1.11. Let I, β, q, ω, and s_0 be as in Example 1.1.5. Then

$$qt = \beta(t) \mp \omega, \quad t \in I,$$

whereupon

$$t = \frac{\beta(t) \mp \omega}{q}, \quad t \in I.$$

Hence,

$$\beta^{-1}(t) = \frac{t \mp \omega}{q}, \quad t \in I.$$

Moreover, we have

$$\begin{aligned}
\beta^2(t) &= \beta \circ \beta(t) \\
&= q\beta(t) \pm \omega \\
&= q(qt \pm \omega) \pm \omega \\
&= q^2 t \pm q\omega \pm \omega \\
&= q^2 t \pm (q+1)\omega \\
&= q^2 t \pm \frac{1-q^2}{1-q}\omega \\
&= q^2 t \pm [2]_q \omega, \quad t \in I,
\end{aligned}$$

and

$$\begin{aligned}
\beta^3(t) &= \beta \circ \beta \circ \beta(t) \\
&= \beta \circ \beta^2(t) \\
&= \beta\left(\beta^2(t)\right) \\
&= q\beta^2(t) \pm \omega \\
&= q\left(q^2 t \pm (1+q)\omega\right) \pm \omega \\
&= q^3 t \pm (q + q^2)\omega \pm \omega \\
&= q^3 t \pm (1 + q + q^2)\omega \\
&= q^3 t \pm \frac{1-q^3}{1-q}\omega \\
&= q^3 t \pm [3]_q \omega, \quad t \in I.
\end{aligned}$$

Assume that

$$\beta^n(t) = q^n t \pm [n]_q \omega, \quad t \in I,$$

for some $n \in \mathbb{N}$. We will prove that

$$\beta^{n+1}(t) = q^{n+1} t \pm [n+1]_q \omega, \quad t \in I.$$

Indeed, we have

$$
\begin{aligned}
\beta^{n+1}(t) &= \beta \circ \beta^n(t) \\
&= \beta\left(\beta^n(t)\right) \\
&= q\beta^n(t) \pm \omega \\
&= q\left(q^n t \pm (1 + q + \cdots + q^{n-1})\omega\right) \pm \omega \\
&= q^{n+1}t \pm \left(q + q^2 + \cdots + q^n\right)\omega \pm \omega \\
&= q^{n+1}t \pm \left(1 + q + q^2 + \cdots + q^n\right)\omega \\
&= q^{n+1}t \pm \frac{1 - q^{n+1}}{1 - q}\omega \\
&= q^{n+1}t \pm [n+1]_q\omega, \quad t \in I.
\end{aligned}
$$

Hence, applying the principle of mathematical induction, we conclude that

$$
\beta^n(t) = q^n t \pm [n]_q\omega, \quad t \in I,
$$

for each $n \in \mathbb{N}_0$. Next,

$$
\begin{aligned}
\beta^{-2}(t) &= \beta^{-1} \circ \beta^{-1}(t) \\
&= \beta^{-1}\left(\beta^{-1}(t)\right) \\
&= \frac{\beta^{-1}(t) \mp \omega}{q} \\
&= \frac{\frac{t \mp \omega}{q} \mp \omega}{q} \\
&= \frac{t \mp \omega \mp q\omega}{q^2} \\
&= \frac{t \mp (1 + q)\omega}{q^2} \\
&= \frac{t \mp \frac{1 - q^2}{1 - q}\omega}{q^2} \\
&= \frac{t \mp [2]_q\omega}{q^2}, \quad t \in I,
\end{aligned}
$$

and

$$
\begin{aligned}
\beta^{-3}(t) &= \beta^{-1} \circ \beta^{-2}(t) \\
&= \beta^{-1}\left(\beta^{-2}(t)\right) \\
&= \frac{\beta^{-2}(t) \mp \omega}{q}
\end{aligned}
$$

$$= \frac{\frac{t \mp (1+q)\omega}{q^2} \mp \omega}{q}$$

$$= \frac{t \mp (1+q)\omega \mp q^2\omega}{q^3}$$

$$= \frac{t \mp (1+q+q^2)\omega}{q^3}$$

$$= \frac{t \mp \frac{1-q^3}{1-q}\omega}{q^3}$$

$$= \frac{t \mp [3]_q \omega}{q^3}, \quad t \in I.$$

Assume that

$$\beta^{-n}(t) = \frac{t \mp [n]_q \omega}{q^n}, \quad t \in I,$$

for some $n \in \mathbb{N}$. We will prove that

$$\beta^{-(n+1)}(t) = \frac{t \mp [n+1]_q \omega}{q^{n+1}}, \quad t \in I.$$

Indeed,

$$\beta^{-(n+1)}(t) = \beta^{-1} \circ \beta^{-n}(t)$$

$$= \beta^{-1}\left(\beta^{-n}(t)\right)$$

$$= \frac{\beta^{-n}(t) \mp \omega}{q}$$

$$= \frac{\frac{t \mp (1+q+\cdots+q^{n-1})\omega}{q^n} \mp \omega}{q}$$

$$= \frac{t \mp (1+q+\cdots+q^{n-1})\omega \mp q^n\omega}{q^{n+1}}$$

$$= \frac{t \mp (1+q+\cdots+q^{n-1}+q^n)\omega}{q^{n+1}}$$

$$= \frac{t \mp \frac{1-q^{n+1}}{1-q}\omega}{q^{n+1}}$$

$$= \frac{t \mp [n+1]_q \omega}{q^{n+1}}, \quad t \in I.$$

Hence, applying the principle of mathematical induction, we conclude that

$$\beta^{-n}(t) = \frac{t \mp [n]_q \omega}{q^n}, \quad t \in I,$$

for each $n \in \mathbb{N}_0$.

Example 1.1.12. Let I, β, q, and n be as in Example 1.1.7. Then

$$t^n = \frac{\beta(t)}{q}, \quad t \in I,$$

and

$$t = \left(\frac{\beta(t)}{q}\right)^{\frac{1}{n}}, \quad t \in I.$$

Therefore,

$$\beta^{-1}(t) = \left(\frac{t}{q}\right)^{\frac{1}{n}}, \quad t \in I.$$

Moreover,

$$\begin{aligned}
\beta^2(t) &= \beta \circ \beta(t) \\
&= \beta(\beta(t)) \\
&= q(\beta(t))^n \\
&= q\left(qt^n\right)^n \\
&= qq^n t^{n^2} \\
&= q^{1+n} t^{n^2} \\
&= q^{\frac{1-n^2}{1-n}} t^{n^2} \\
&= q^{[2]_n} t^{n^2}, \quad t \in I.
\end{aligned}$$

Assume that

$$\beta^l(t) = q^{[l]} t^{n^l}, \quad t \in I,$$

for some $l \in \mathbb{N}$. We will prove that

$$\beta^{l+1}(t) = q^{[l+1]_n} t^{n^{l+1}}, \quad t \in I.$$

Indeed, we have

$$\begin{aligned}
\beta^{l+1}(t) &= \beta \circ \beta^l(t) \\
&= \beta\left(\beta^l(t)\right) \\
&= q\left(\beta^l(t)\right)^n \\
&= q\left(q^{1+n+\cdots+n^{l-1}} t^{n^l}\right)^n \\
&= qq^{n+n^2+\cdots+n^l} t^{n^{l+1}} \\
&= q^{1+n+\cdots+n^l} t^{n^{l+1}} \\
&= q^{\frac{1-n^{l+1}}{1-n}} t^{n^{l+1}}
\end{aligned}$$

$$= q^{[l+1]_n} t^{n^{l+1}}, \quad t \in I.$$

Hence, applying the principle of mathematical induction, we get

$$\beta^k(t) = q^{[k]_n} t^{n^k}, \quad t \in I,$$

for each $k \in \mathbb{N}_0$. Next,

$$\beta^{-2}(t) = \beta^{-1} \circ \beta^{-1}(t)$$

$$= \beta^{-1}\left(\beta^{-1}(t)\right)$$

$$= \left(\frac{\beta^{-1}(t)}{q}\right)^{\frac{1}{n}}$$

$$= \left(\frac{\left(\frac{t}{q}\right)^{\frac{1}{n}}}{q}\right)^{\frac{1}{n}}$$

$$= \left(\frac{\frac{t^{\frac{1}{n}}}{q^{\frac{1}{n}}}}{q}\right)^{\frac{1}{n}}$$

$$= \left(\frac{t^{\frac{1}{n}}}{q^{\frac{n+1}{n}}}\right)^{\frac{1}{n}}$$

$$= \frac{t^{\frac{1}{n^2}}}{q^{\frac{n+1}{n^2}}}$$

$$= \frac{t^{\frac{1}{n^2}}}{q^{\frac{1}{n^2} \frac{1-n^2}{1-n}}}$$

$$= \frac{t^{\frac{1}{n^2}}}{q^{\frac{[2]_n}{n^2}}}, \quad t \in I.$$

Assume that

$$\beta^{-l}(t) = \frac{t^{\frac{1}{n^l}}}{q^{\frac{[l]_n}{n^l}}}, \quad t \in I,$$

for some $l \in \mathbb{N}$. We will prove that

$$\beta^{-(l+1)}(t) = \frac{t^{\frac{1}{n^{l+1}}}}{q^{\frac{[l+1]_n}{n^{l+1}}}}, \quad t \in I.$$

Indeed,

$$
\begin{aligned}
\beta^{-(l+1)}(t) &= \beta^{-1} \circ \beta^{-l}(t) \\
&= \beta^{-1}\left(\beta^{-l}(t)\right) \\
&= \left(\frac{\beta^{-l}(t)}{q}\right)^{\frac{1}{n}} \\
&= \left(\frac{t^{\frac{1}{n^l}}}{\dfrac{q^{\frac{[l]_n}{n^l}}}{q}}\right)^{\frac{1}{n}} \\
&= \left(\frac{t^{\frac{1}{n^l}}}{\dfrac{q^{\frac{1+n+\cdots+n^{l-1}}{n^l}}}{q}}\right)^{\frac{1}{n}} \\
&= \left(\frac{t^{\frac{1}{n^l}}}{q^{\frac{1+n+\cdots+n^{l-1}+n^l}{n^l}}}\right)^{\frac{1}{n}} \\
&= \frac{t^{\frac{1}{n^{l+1}}}}{q^{\frac{\frac{1-n^{l+1}}{1-n}}{n^{l+1}}}} \\
&= \frac{t^{\frac{1}{n^{l+1}}}}{q^{\frac{[l+1]_n}{n^{l+1}}}}, \quad t \in I.
\end{aligned}
$$

Hence, applying the principle of the mathematical induction, we get

$$
\beta^{-k}(t) = \frac{t^{\frac{1}{n^k}}}{q^{\frac{[k]_n}{n^k}}}, \quad t \in I,
$$

for each $k \in \mathbb{N}_0$.

Exercise 1.1.13. Let I and β be as in Exercise 1.1.8. Then find

1. $\beta^{-1}(t), t \in I$;
2. $\beta^{k}(t), k \in \mathbb{N}_0, t \in I$;
3. $\beta^{-k}(t), t \in I, k \in \mathbb{N}$.

Theorem 1.1.14. *Let $\beta \colon I \to \mathbb{R}$ be a general quantum operator. Then we have the following:*

1. *The sequence $\{\beta^k\}_{k \in \mathbb{N}_0}$ uniformly converges to s_0 on any compact interval $J \subseteq I$ containing s_0.*

2. *The series*

$$\sum_{k=0}^{\infty} |\beta^k(t) - \beta^{k+1}(t)|$$

uniformly converges to $|t - s_0|$ *on any compact interval* $J \subset I$ *containing* s_0.

Proof. **1.** Let $J = [a, b]$ and $s_0 \in [a, b]$.
    (i) Suppose that $t \in [s_0, b]$. Then

$$t - s_0 \geq 0$$

and

$$\beta(t) \leq t.$$

Hence, since β is strictly increasing on I, we get

$$\beta^2(t) \leq \beta(t),$$
$$\beta^3(t) \leq \beta^2(t),$$
$$\vdots$$
$$\beta^{k+1}(t) \leq \beta^k(t), \quad k \in \mathbb{N}_0.$$

Then $\{\beta^k\}_{k \in \mathbb{N}_0}$ is decreasing to s_0. Now, applying the Dini test, we find that the sequence $\{\beta^k\}_{k \in \mathbb{N}_0}$ uniformly converges to s_0 on $[s_0, b]$.
    (ii) Suppose that $t \in [a, s_0]$. Then

$$t - s_0 \leq 0$$

and

$$\beta(t) \geq t.$$

Hence, since β is strictly increasing on I, we get

$$\beta^2(t) \geq \beta(t),$$
$$\beta^3(t) \geq \beta^2(t),$$
$$\vdots$$
$$\beta^{k+1}(t) \geq \beta^k(t), \quad k \in \mathbb{N}_0.$$

Then $\{\beta^k\}_{k \in \mathbb{N}_0}$ is increasing to s_0. Now, applying the Dini test, we find that the sequence $\{\beta^k\}_{k \in \mathbb{N}_0}$ uniformly converges to s_0 on $[a, s_0]$.
From (i) and (ii) the conclusion follows.
2. Consider the sequence $\{S_n\}_{n \in \mathbb{N}}$ defined by

$$S_n(t) = \sum_{k=0}^{n} |\beta^k(t) - \beta^{k+1}(t)|, \quad t \in J.$$

(i)   Let $t \in [s_0, b]$. Then

$$S_n(t) = t - \beta(t) + \beta(t) - \beta^2(t) + \cdots + \beta^n(t) - \beta^{n+1}(t)$$
$$= t - \beta^{n+1}(t).$$

Now from the above discussion, it follows that $t - \beta^{n+1}(t)$ converges uniformly to $t - s_0$.

(ii)   Let $t \in [a, s_0]$. Then

$$S_n(t) = \sum_{k=0}^{n} \left(\beta^{k+1}(t) - \beta^k(t) \right)$$
$$= \beta^{n+1}(t) - \beta^n(t) + \cdots + \beta(t) - t$$
$$= \beta^{n+1}(t) - t.$$

Now from the above discussion, it follows that $\beta^{n+1}(t) - t$ converges uniformly to $s_0 - t$. This completes the proof. $\square$

We further suppose that X is a Banach space with norm $\| \cdot \|$.

Corollary 1.1.15. *Let β be a first-kind general quantum operator. If $f : I \rightarrow X$ is continuous at s_0, then the sequence $\{ f(\beta^k(t)) \}_{k \in \mathbb{N}}$ converges uniformly to $f(s_0)$ on every compact interval $J \subseteq I$ containing s_0.*

Theorem 1.1.16. *Let β be a first-kind general quantum operator. If $f : I \rightarrow X$ is continuous at s_0, then the series*

$$\sum_{k=0}^{\infty} \left| \beta^k(t) - \beta^{k+1}(t) \right| \| f(\beta^k(t)) \| \tag{1.1}$$

is uniformly convergent on every compact interval $J \subset I$ containing s_0.

Proof. Let $J \subset I$ be a compact interval. By Corollary 1.1.15, we have that the sequence $\{ f(\beta^k(t)) \}_{k \in \mathbb{N}_0}$ converges uniformly to $f(s_0)$ on J. Hence, there exists $k_0 \in \mathbb{N}$ such that

$$\| f(\beta^k(t)) - f(s_0) \| < 1, \quad k \geq k_0, \quad t \in J.$$

Then

$$\| f(\beta^k(t)) \| < 1 + \| f(s_0) \|, \quad k \geq k_0, \quad t \in J.$$

Therefore, we have

$$|(\beta^k(t) - \beta^{k+1}(t))| \| f(\beta^k(t)) \| \leq |\beta^k(t) - \beta^{k+1}(t)|(1 + \| f(s_0) \|), \quad k \geq k_0, \quad t \in J.$$

Set

$$D_n(t) = \sum_{k=0}^{n} |\beta^k(t) - \beta^{k+1}(t)| \| f(\beta^k(t)) \|, \quad t \in J, \quad n \in \mathbb{N}_0,$$

and

$$C_n(t) = \sum_{k=0}^{n} |\beta^k(t) - \beta^{k+1}(t)|(1 + \|f(s_0)\|), \quad t \in J, \quad n \in \mathbb{N}_0.$$

Then by Theorem 1.1.14, we see that the sequence $\{C_n(t)\}$ converges uniformly to $|t - s_0|(1 + \|f(s_0)\|)$, $t \in J$. Hence, by the Cauchy test, it follows that for every $\varepsilon > 0$, there is $m_0 \in \mathbb{N}_0$ such that

$$\|C_n(t) - C_m(t)\| < \varepsilon, \quad t \in J, \quad n, m \geq m_0.$$

Then

$$\|D_n(t) - D_m(t)\| = \left\| \sum_{k=0}^{n} |\beta^k(t) - \beta^{k+1}(t)| \|f(\beta^k(t))\| - \sum_{k=0}^{m} |\beta^k(t) - \beta^{k+1}(t)| \|f(\beta^k(t))\| \right\|$$

$$= \left\| \sum_{k=m+1}^{n} |\beta^k(t) - \beta^{k+1}(t)| \|f(\beta^k(t))\| \right\|$$

$$= \sum_{k=m+1}^{n} |\beta^k(t) - \beta^{k+1}(t)|(1 + \|f(s_0)\|)$$

$$= \|C_n(t) - C_m(t)\|$$

$$< \varepsilon, \quad t \in J, \quad n > m \geq \max\{k_0, m_0\}.$$

Thus, (1.1) is uniformly convergent on J. This completes the proof. $\qquad\square$

Example 1.1.17. Let I, β, q, and ω be as in Example 1.1.5. Then

$$\beta^k(t) = q^k t \pm \frac{1 - q^k}{1 - q}\omega,$$

$$\beta^{-k}(t) = \frac{t \mp \frac{1-q^k}{1-q}\omega}{q^k}, \quad t \in I, \quad k \in \mathbb{N}.$$

Therefore

$$\lim_{k \to \infty} \beta^k(t) = \lim_{k \to \infty} \left(q^k t \pm \frac{1 - q^k}{1 - q}\omega \right)$$

$$= \pm \frac{\omega}{1 - q}$$

$$= s_0, \quad t \in I,$$

and

$$\lim_{k \to \infty} \beta^{-k}(t) = \lim_{k \to \infty} \left(\frac{t \mp \frac{1-q^k}{1-q}\omega}{q^k} \right)$$

$$= \begin{cases} \infty & \text{if } t > s_0, \\ -\infty & \text{if } t < s_0. \end{cases}$$

Example 1.1.18. Let I, β, q, and n be as in Example 1.1.7. Then

$$\beta^k(t) = q^{\frac{1-n^k}{1-n}} t^{n^k},$$

$$\beta^{-k}(t) = \frac{t^{\frac{1}{n^k}}}{q^{\frac{1-n^k}{n^k(1-n)}}}, \qquad t \in I, \quad k \in \mathbb{N}.$$

Thus

$$\lim_{k \to \infty} \beta^k(t) = \lim_{k \to \infty} \left(q^{\frac{1-n^k}{1-n}} t^{n^k} \right)$$
$$= 0, \quad t \in I,$$

and

$$\lim_{k \to \infty} \beta^{-k}(t) = \lim_{k \to \infty} \left(\frac{t^{\frac{1}{n^k}}}{q^{\frac{1-n^k}{n^k(1-n)}}} \right)$$

$$= \begin{cases} q^{\frac{1}{1-n}} & \text{if } t > 0, \\ 0 & \text{if } t = 0, \\ -q^{\frac{1}{1-n}} & \text{if } t < 0. \end{cases}$$

Exercise 1.1.19. Let $n \in 2\mathbb{N} + 1$ and $I = (-1, 1)$. Define

$$\beta(t) = t^n, \quad t \in I.$$

1. Prove that β is a first-kind general quantum operator.
2. Prove that

$$\beta^k(t) = t^{n^k}, \quad t \in I, \quad k \in \mathbb{N}.$$

3. Prove that

$$\beta^{-k}(t) = t^{-n^k}, \quad t \in I, \quad k \in \mathbb{N}.$$

4. Find

$$\lim_{k \to \infty} \beta^k(t), \quad t \in I.$$

5. Find

$$\lim_{k \to \infty} \beta^{-k}(t), \quad t \in I.$$

Remark 1.1.20. Note that Theorem 1.1.14, Corollary 1.1.15, and Theorem 1.1.16 are valid in the case when β is a second-kind general quantum operator.

1.2 Definition of β-derivative and examples

Throughout this book, we suppose that $I \subset \mathbb{R}$, X is a Banach space with norm $\|\cdot\|$, and $\beta : I \to \mathbb{R}$ is a general quantum operator.

Definition 1.2.1. For a function $f : I \to X$, define the β-difference operator of f as follows:

$$
D_\beta f(t) =
\begin{cases}
\dfrac{f(\beta(t)) - f(t)}{\beta(t) - t} & \text{if } t \neq s_0, \\[2ex]
f'(s_0) & \text{if } t = s_0,
\end{cases}
$$

provided that $f'(s_0)$ exists. In this case, we say that $D_\beta f(t)$ is the β-derivative of f at t. We say that f is β-differentiable on I if $D_\beta f(t)$ exists for every $t \in I$.

Example 1.2.2. Let $I = \left[1, \dfrac{3}{2} \right]$ and $X = \mathbb{R}$. Define

$$
\beta(t) = \frac{1}{2}t + \frac{3}{4} \quad \text{and}
$$
$$
f(t) = 4t^2 - 9t, \quad t \in I.
$$

We will find $D_\beta f(t)$, $t \in I$. Here $s_0 = \dfrac{3}{2}$. Then

$$
f'(t) = 8t - 9, \quad t \in I,
$$

and

$$
D_\beta f(t) =
\begin{cases}
\dfrac{f(\beta(t)) - f(t)}{\beta(t) - t} & \text{if } t \neq \dfrac{3}{2}, \\[2ex]
f'\left(\dfrac{3}{2}\right) & \text{if } t = \dfrac{3}{2},
\end{cases}
$$

$$
=
\begin{cases}
\dfrac{4\left(\frac{1}{2}t + \frac{3}{4}\right)^2 - \left(\frac{1}{2}t + \frac{3}{4}\right)}{\frac{1}{2}t + \frac{3}{4} - t} & \text{if } t \neq \dfrac{3}{2}, \\[2ex]
8\dfrac{3}{2} - 9 & \text{if } t = \dfrac{3}{2},
\end{cases}
$$

$$
=
\begin{cases}
\dfrac{4\left(\frac{1}{4}t^2 + \frac{3}{4}t + \frac{9}{16}\right) - \frac{1}{2}t - \frac{3}{4}}{\frac{3}{4} - \frac{1}{2}t} & \text{if } t \neq \dfrac{3}{2}, \\[2ex]
12 - 9 & \text{if } t = \dfrac{3}{2},
\end{cases}
$$

$$
= \begin{cases} \dfrac{t^2 + 3t + \frac{9}{4} - \frac{1}{2}t - \frac{3}{4}}{\frac{3-2t}{4}} & \text{if } t \neq \dfrac{3}{2}, \\[2em] 3 & \text{if } t = \dfrac{3}{2} \end{cases}
$$

$$
= \begin{cases} \dfrac{t^2 + \frac{9}{2}t + \frac{3}{2}}{\frac{3-2t}{4}} & \text{if } t \neq \dfrac{3}{2}, \\[2em] 3 & \text{if } t = \dfrac{3}{2} \end{cases}
$$

$$
= \begin{cases} \dfrac{2(2t^2 + 9t + 3)}{3 - 2t} & \text{if } t \neq \dfrac{3}{2}, \\[1.5em] 3 & \text{if } t = \dfrac{3}{2}. \end{cases}
$$

Example 1.2.3. Let $I = [-1, 1]$ and $X = \mathbb{R}$. Define

$$
\beta(t) = \frac{1}{4}t + \frac{1}{4}, \quad t \in I, \quad \text{and}
$$

$$
f(t) = \begin{cases} t & \text{if } t \in (-1, 0), \\ -t & \text{if } t \in (0, 1), \\ 1 & \text{if } t = 0, \\ 0 & \text{if } t = 1, -1. \end{cases}
$$

We will find $D_\beta f(t)$, $t \in I$. Here $s_0 = \dfrac{1}{3}$. Set

$$
g(t) = -t, \quad t \in I.
$$

Then

$$
g'(t) = -1,
$$
$$
\beta(0) = \frac{1}{4},
$$
$$
\beta(1) = \frac{1}{2},
$$
$$
\beta(-1) = 0.
$$

Hence,

$$
D_\beta f(t) = \begin{cases} \dfrac{f(\beta(t)) - f(t)}{\beta(t) - t} & \text{if } t \neq \dfrac{1}{3}, \\[1.5em] f'\left(\dfrac{1}{3}\right) & \text{if } t = \dfrac{1}{3}. \end{cases}
$$

$$
= \begin{cases}
\dfrac{\beta(t) - t}{\beta(t) - t} & \text{if } t \in (-1, 0), \\[2mm]
-\dfrac{\beta(t) - t}{\beta(t) - t} & \text{if } t \in \left(0, \dfrac{1}{3}\right) \cup \left(\dfrac{1}{3}, 1\right), \\[2mm]
g'\left(\dfrac{1}{3}\right) & \text{if } t = \dfrac{1}{3}, \\[2mm]
\dfrac{f(\beta(0)) - f(0)}{\beta(0) - 0} & \text{if } t = 0, \\[2mm]
\dfrac{f(\beta(-1)) - f(-1)}{\beta(-1) + 1} & \text{if } t = -1, \\[2mm]
\dfrac{f(\beta(1)) - f(1)}{\beta(1) - 1} & \text{if } t = 1
\end{cases}
$$

$$
= \begin{cases}
1 & \text{if } t \in (-1, 0), \\[2mm]
-1 & \text{if } t \in (0, 1), \\[2mm]
\dfrac{f\left(\frac{1}{4}\right) - f(0)}{\frac{1}{4} - 0} & \text{if } t = 0, \\[2mm]
\dfrac{f(0) - f(-1)}{0 + 1} & \text{if } t = -1, \\[2mm]
\dfrac{f\left(\frac{1}{2}\right) - f(1)}{\frac{1}{2} - 1} & \text{if } t = 1
\end{cases}
$$

$$
= \begin{cases}
1 & \text{if } t \in (-1, 0), \\[2mm]
-1 & \text{if } t \in (0, 1), \\[2mm]
\dfrac{-\frac{1}{4} - 1}{\frac{1}{4} - 0} & \text{if } t = 0, \\[2mm]
\dfrac{1 - 0}{1} & \text{if } t = -1, \\[2mm]
\dfrac{-\frac{1}{2} - 0}{-\frac{1}{2}} & \text{if } t = 1
\end{cases}
$$

$$
= \begin{cases}
1 & \text{if } t \in (-1, 0), \\[1mm]
-1 & \text{if } t \in (0, 1), \\[1mm]
-5 & \text{if } t = 0, \\[1mm]
1 & \text{if } t = 1, -1.
\end{cases}
$$

Example 1.2.4. Let $I = \mathbb{R}$ and $X = \mathbb{R}$. Define

$$
\beta(t) = 2t + 3 \quad \text{and}
$$

$$f(t) = t^{10}, \quad t \in I.$$

We will find $D_\beta f(t), t \in I$. Here $s_0 = -3$ and

$$f'(t) = 10t^9, \quad t \in I.$$

Then

$$D_\beta f(t) = \begin{cases} \dfrac{f(2t+3) - f(t)}{2t+3-t} & \text{if} \quad t \neq -3, \\[2ex] f'(-3) & \text{if} \quad t = -3 \end{cases}$$

$$= \begin{cases} \dfrac{(2t+3)^{10} - t^{10}}{t+3} & \text{if} \quad t \neq -3, \\[2ex] 10(-3)^9 & \text{if} \quad t = -3 \end{cases}$$

$$= \begin{cases} \dfrac{(2t+3-t)((2t+3)^9 + t(2t+3)^8 + \cdots + t^9)}{t+3} & \text{if} \quad t \neq -3, \\[2ex] -10(3)^9 & \text{if} \quad t = -3 \end{cases}$$

$$= \begin{cases} (2t+3)^9 + t(2t+3)^8 + \cdots + t^9 & \text{if} \quad t \neq -3, \\[2ex] -10(3)^9 & \text{if} \quad t = -3. \end{cases}$$

Example 1.2.5. Let $I = \mathbb{R}$ and $X = \mathbb{R}^2$. Define

$$\beta(t) = \frac{1}{2}t + 1, \quad t \in I,$$

and $f : I \to \mathbb{R}^2$ by

$$f(t) = (t^2, 2t), \quad t \in I.$$

We will find $D_\beta f(t), t \in I$. Here $s_0 = 2$. Set

$$f_1(t) = t^2,$$
$$f_2(t) = 2t, \quad t \in I.$$

Then

$$f_1'(t) = 2t,$$
$$f_2'(t) = 2, \quad t \in I,$$

and

$$D_\beta f_1(t) = \begin{cases} \dfrac{f_1\left(\frac{1}{2}t + 1\right) - f_1(t)}{\frac{1}{2}t + 1 - t} & \text{if} \quad t \neq 2, \\[2ex] f_1'(2) & \text{if} \quad t = 2 \end{cases}$$

$$
= \begin{cases} \dfrac{\left(\frac{1}{2}t+1\right)^2 - t^2}{1 - \frac{1}{2}t} & \text{if } t \neq 2, \\[2ex] 2 \cdot 2 & \text{if } t = 2 \end{cases}
$$

$$
= \begin{cases} \dfrac{\left(\frac{1}{2}t+1-t\right)\left(\frac{1}{2}t+1+t\right)}{1 - \frac{1}{2}t} & \text{if } t \neq 2, \\[2ex] 4 & \text{if } t = 2 \end{cases}
$$

$$
= \begin{cases} 1 + \dfrac{3}{2}t & \text{if } t \neq 2, \\[2ex] 4 & \text{if } t = 2, \end{cases}
$$

and

$$
D_\beta f_2(t) = \begin{cases} \dfrac{f_2\left(\frac{1}{2}t+1\right) - f_2(t)}{\frac{1}{2}t + 1 - t} & \text{if } t \neq 2, \\[2ex] f_2'(2) & \text{if } t = 2 \end{cases}
$$

$$
= \begin{cases} \dfrac{2\left(\frac{1}{2}t+1\right) - 2t}{1 - \frac{1}{2}t} & \text{if } t \neq 2, \\[2ex] 2 & \text{if } t = 2 \end{cases}
$$

$$
= \begin{cases} \dfrac{2(t+2-2t)}{2-t} & \text{if } t \neq 2, \\[2ex] 2 & \text{if } t = 2 \end{cases}
$$

$$
= \begin{cases} 2 & \text{if } t \neq 2, \\[1ex] 2 & \text{if } t = 2. \end{cases}
$$

Therefore,

$$
D_\beta f(t) = (D_\beta f_1(t), D_\beta f_2(t))
$$

$$
= \begin{cases} \left(1 + \dfrac{3}{2}t, 2\right) & \text{if } t \neq 2, \\[2ex] (4, 2) & \text{if } t = 2 \end{cases}
$$

$$
= \left(1 + \dfrac{3}{2}t, 2\right), \quad t \in I.
$$

Example 1.2.6. Let $I = \mathbb{R}$, and let $X = \mathscr{M}_{2\times 2}$, the Banach space of 2×2 real matrices. Define

$$
\beta(t) = \frac{1}{4}t, \quad t \in I,
$$

and $f : I \to \mathscr{M}_{2 \times 2}$ by

$$f(t) = \begin{pmatrix} t^3 & 1 \\ t & t^2 \end{pmatrix}, \quad t \in I.$$

We will find $D_\beta f(t), t \in I$. Set

$$
\begin{aligned}
f_1(t) &= t^3, \\
f_2(t) &= 1, \\
f_3(t) &= t, \\
f_4(t) &= t^2, \quad t \in I.
\end{aligned}
$$

Here $s_0 = 0$, and

$$
\begin{aligned}
f_1'(t) &= 3t^2, \\
f_2'(t) &= 0, \\
f_3'(t) &= 1, \\
f_4'(t) &= 2t, \quad t \in I.
\end{aligned}
$$

Then

$$
D_\beta f_1(t) =
\begin{cases}
\dfrac{f_1\left(\frac{1}{4}t\right) - f_1(t)}{\frac{1}{4}t - t} & \text{if} \quad t \neq 0, \\[3mm]
f_1'(0) & \text{if} \quad t = 0
\end{cases}
$$

$$
=
\begin{cases}
\dfrac{\left(\frac{1}{4}t\right)^3 - t^3}{-\frac{3}{4}t} & \text{if} \quad t \neq 0, \\[3mm]
3 \cdot 0^2 & \text{if} \quad t = 0
\end{cases}
$$

$$
=
\begin{cases}
\dfrac{\left(\frac{1}{4}t - t\right)\left(\frac{1}{16}t^2 + \frac{t^2}{4} + t^2\right)}{-\frac{3}{4}t} & \text{if} \quad t \neq 0, \\[3mm]
0 & \text{if} \quad t = 0
\end{cases}
$$

$$
=
\begin{cases}
\dfrac{21}{16}t^2 & \text{if} \quad t \neq 0, \\[3mm]
0 & \text{if} \quad t = 0,
\end{cases}
$$

$$
D_\beta f_2(t) =
\begin{cases}
\dfrac{f_2\left(\frac{1}{4}t\right) - f_2(t)}{\frac{1}{4}t - t} & \text{if} \quad t \neq 0, \\[3mm]
f_2'(0) & \text{if} \quad t = 0
\end{cases}
$$

$$
=
\begin{cases}
\dfrac{1 - 1}{-\frac{3}{4}t} & \text{if} \quad t \neq 0, \\[3mm]
0 & \text{if} \quad t = 0
\end{cases}
$$

$$= \begin{cases} 0 & \text{if } t \neq 0, \\ 0 & \text{if } t = 0, \end{cases}$$

$$D_\beta f_3(t) = \begin{cases} \dfrac{f_3\left(\frac{1}{4}t\right) - f_3(t)}{\frac{1}{4}t - t} & \text{if } t \neq 0, \\ f_3'(0) & \text{if } t = 0 \end{cases}$$

$$= \begin{cases} \dfrac{\frac{1}{4}t - t}{-\frac{3}{4}t} & \text{if } t \neq 0, \\ 1 & \text{if } t = 0 \end{cases}$$

$$= \begin{cases} 1 & \text{if } t \neq 0, \\ 1 & \text{if } t = 0, \end{cases}$$

and

$$D_\beta f_4(t) = \begin{cases} \dfrac{f_4\left(\frac{1}{4}t\right) - f_4(t)}{\frac{1}{4}t - t} & \text{if } t \neq 0, \\ f_4'(0) & \text{if } t = 0 \end{cases}$$

$$= \begin{cases} \dfrac{\left(\frac{1}{4}t\right)^2 - t^2}{\frac{1}{4}t - t} & \text{if } t \neq 0, \\ 2 \cdot 0 & \text{if } t = 0 \end{cases}$$

$$= \begin{cases} \dfrac{\left(\frac{1}{4}t - t\right)\left(\frac{1}{4}t + t\right)}{-\frac{3}{4}t} & \text{if } t \neq 0, \\ 0 & \text{if } t = 0 \end{cases}$$

$$= \begin{cases} \dfrac{5}{4}t & \text{if } t \neq 0, \\ 0 & \text{if } t = 0. \end{cases}$$

Therefore,

$$D_\beta f(t) = \begin{pmatrix} D_\beta f_1(t) & D_\beta f_2(t) \\ D_\beta f_3(t) & D_\beta f_4(t) \end{pmatrix}$$

$$= \begin{cases} \begin{pmatrix} \dfrac{21}{16}t^2 & 0 \\ 1 & \dfrac{5}{4}t \end{pmatrix} & \text{if } t \neq 0, \\[2em] \begin{pmatrix} 0 & 0 \\ 1 & 0 \end{pmatrix} & \text{if } t = 0 \end{cases}$$

$$= \begin{pmatrix} \dfrac{21}{16}t^2 & 0 \\ 1 & \dfrac{5}{4}t \end{pmatrix}, \quad t \in I.$$

Exercise 1.2.7. Let $I = \mathbb{R}$ and $X = \mathbb{R}$. Define

$$\beta(t) = \frac{1}{8}t + \frac{7}{8} \quad \text{and}$$
$$f(t) = t^2 - 2t + 2, \quad t \in I.$$

1. Prove that β is a first-kind general quantum operator.
2. Find $D_\beta f(t), t \in I$.

1.3 Properties of the β-derivative

In this section, we prove some basic properties of the β-derivative.

Theorem 1.3.1. *The β-difference operator is a linear operator.*

Proof. Let $f, g \colon I \to X$, and let $a \in \mathbb{C}$. Then

$$D_\beta(f+g)(t) = \begin{cases} \dfrac{(f+g)(\beta(t)) - (f+g)(t)}{\beta(t) - t} & \text{if } t \neq s_0, \\[2mm] (f+g)'(s_0) & \text{if } t = s_0 \end{cases}$$

$$= \begin{cases} \dfrac{f(\beta(t)) + g(\beta(t)) - f(t) - g(t)}{\beta(t) - t} & \text{if } t \neq s_0, \\[2mm] f'(s_0) + g'(s_0) & \text{if } t = s_0 \end{cases}$$

$$= \begin{cases} \dfrac{f(\beta(t)) - f(t)}{\beta(t) - t} + \dfrac{g(\beta(t)) - g(t)}{\beta(t) - t} & \text{if } t \neq s_0, \\[2mm] f'(s_0) + g'(s_0) & \text{if } t = s_0 \end{cases}$$

$$= \begin{cases} \dfrac{f(\beta(t)) - f(t)}{\beta(t) - t} & \text{if } t \neq s_0, \\[2mm] f'(s_0) & \text{if } t = s_0 \end{cases}$$
$$+ \begin{cases} \dfrac{g(\beta(t)) - g(t)}{\beta(t) - t} & \text{if } t \neq s_0, \\[2mm] g'(s_0) & \text{if } t = s_0 \end{cases}$$
$$= D_\beta f(t) + D_\beta g(t), \quad t \in I,$$

and

$$D_\beta(af)(t) = \begin{cases} \dfrac{(af)(\beta(t)) - (af)(t)}{\beta(t) - t} & \text{if } t \neq s_0, \\[2mm] (af)'(s_0) & \text{if } t = s_0 \end{cases}$$

$$= \begin{cases} \dfrac{af(\beta(t)) - af(t)}{\beta(t) - t} & \text{if } t \neq s_0, \\[2mm] af'(s_0) & \text{if } t = s_0 \end{cases}$$

$$= a \begin{cases} \dfrac{f(\beta(t)) - f(t)}{\beta(t) - t} & \text{if } t \neq s_0, \\[2mm] f'(s_0) & \text{if } t = s_0 \end{cases}$$

$$= aD_\beta f(t), \quad t \in I.$$

Thus, D_β is a linear operator. This completes the proof. $\square$

Now we present the so-called simple useful formula for β-differential calculus.

Theorem 1.3.2. *If f is β-differentiable on I, then*

$$f(\beta(t)) = f(t) + (\beta(t) - t)D_\beta f(t), \quad t \in I. \tag{1.2}$$

Proof. We will consider the following two cases.

(i) Let $t \neq s_0$. Then

$$D_\beta f(t) = \frac{f(\beta(t)) - f(t)}{\beta(t) - t},$$

whereupon we get (1.2).

(ii) Let $t = s_0$. Then

$$\beta(s_0) = s_0$$

and

$$f(\beta(s_0)) = f(s_0)$$
$$= f(s_0) + (\beta(s_0) - s_0)D_\beta f(s_0).$$

This completes the proof. $\square$

Theorem 1.3.3. *Let f be β-differentiable on I. Then f is continuous at s_0.*

Proof. Since $f'(s_0)$ exists, we have that f is continuous at s_0. $\square$

1.4 Rules for β-differentiation

In this section, we prove the basic rules for β-differentiation.

Theorem 1.4.1. *Let $f : I \to X$ and $g : I \to \mathbb{R}$ be β-differentiable functions at $t \in I$. Then we have the following:*

1. *The product $fg : I \to X$ is β-differentiable at t, and*

$$D_\beta(fg)(t) = (D_\beta f(t))g(t) + f(\beta(t))D_\beta g(t)$$
$$= (D_\beta f(t))g(\beta(t)) + f(t)D_\beta g(t), \quad t \in I.$$

2. *The quotient $\dfrac{f}{g}$ is β-differentiable at t, and*

$$D_\beta\left(\frac{f}{g}\right)(t) = \frac{(D_\beta f(t))g(t) - f(t)D_\beta g(t)}{g(t)g(\beta(t))}, \qquad t \in I.$$

Proof. **1.** We have the following two cases.

    (i) Let $t = s_0$. Then

$$\beta(s_0) = s_0,$$

and

$$
\begin{aligned}
(fg)'(s_0) &= f'(s_0)g(s_0) + f(s_0)g'(s_0)\\
&= f'(s_0)g(\beta(s_0)) + f(s_0)g'(s_0)\\
&= D_\beta f(s_0)g(s_0) + f(\beta(s_0))D_\beta g(s_0).
\end{aligned}
$$

    (ii) Let $t \neq s_0$. Then

$$
\begin{aligned}
D_\beta(fg)(t) &= \frac{(fg)(\beta(t)) - (fg)(t)}{\beta(t) - t}\\
&= \frac{f(\beta(t))g(\beta(t)) - f(t)g(t)}{\beta(t) - t}\\
&= \frac{f(\beta(t))(g(\beta(t)) - g(t)) + (f(\beta(t)) - f(t))g(t)}{\beta(t) - t}\\
&= f(\beta(t))\frac{g(\beta(t)) - g(t)}{\beta(t) - t} + \frac{f(\beta(t)) - f(t)}{\beta(t) - t}g(t)\\
&= f(\beta(t))D_\beta g(t) + (D_\beta f(t))g(t),
\end{aligned}
$$

and

$$
\begin{aligned}
D_\beta(fg)(t) &= \frac{(fg)(\beta(t)) - (fg)(t)}{\beta(t) - t}\\
&= \frac{(f(\beta(t)) - f(t))g(\beta(t)) + f(t)(g(\beta(t)) - g(t))}{\beta(t) - t}\\
&= \frac{f(\beta(t)) - f(t)}{\beta(t) - t}g(\beta(t)) + f(t)\frac{g(\beta(t)) - g(t)}{\beta(t) - t}\\
&= (D_\beta f(t))g(\beta(t)) + f(t)D_\beta g(t).
\end{aligned}
$$

2. We will consider the following two cases.

    (i) Let $t = s_0$ and $g'(s_0) \neq 0$. Then

$$g(s_0) = g(\beta(s_0)),$$

and

$$
\begin{aligned}
\left(\frac{f}{g}\right)'(s_0) &= \frac{f'(s_0)g(s_0) - f(s_0)g'(s_0)}{(g(s_0))^2}\\
&= \frac{(D_\beta f(s_0))g(s_0) - f(s_0)D_\beta g(s_0)}{g(s_0)g(\beta(s_0))}.
\end{aligned}
$$

(ii)   Let $t \neq s_0$. Then

$$D_\beta \left(\frac{f}{g} \right)(t) = \frac{\left(\frac{f}{g} \right)(\beta(t)) - \left(\frac{f}{g} \right)(t)}{\beta(t) - t}$$

$$= \frac{\frac{f(\beta(t))}{g(\beta(t))} - \frac{f(t)}{g(t)}}{\beta(t) - t}$$

$$= \frac{f(\beta(t))g(t) - f(t)g(\beta(t))}{(\beta(t) - t)g(t)g(\beta(t))}$$

$$= \frac{(f(\beta(t)) - f(t))g(t) - f(t)(g(\beta(t)) - g(t))}{(\beta(t) - t)g(t)g(\beta(t))}$$

$$= \frac{\frac{f(\beta(t)) - f(t)}{\beta(t) - t}g(t) - f(t)\frac{g(\beta(t)) - g(t)}{\beta(t) - t}}{g(t)g(\beta(t))}$$

$$= \frac{(D_\beta f(t))g(t) - f(t)D_\beta g(t)}{g(t)g(\beta(t))}.$$

This completes the proof. $\square$

Example 1.4.2. Let $I = \mathbb{R}$ and define

$$\beta(t) = \frac{1}{2}t + 1 \quad \text{and}$$
$$h(t) = t^3 + t^2 + t, \quad t \in I.$$

Note that

$$\beta'(t) = \frac{1}{2}, \quad t \in I,$$

that is, $\beta \colon I \to I$ is a strictly increasing continuous function on I. Next, consider the equation

$$\beta(t) = t, \quad t \in I.$$

This gives

$$t = 2,$$

which is the unique fixed point of β, that is, $s_0 = 2$. Moreover,

$$(t - s_0)(\beta(t) - t) = (t - 2)\left(\frac{1}{2}t + 1 - t \right)$$
$$= \frac{(t - 2)(2 - t)}{2}$$
$$= -\frac{(t - 2)^2}{2}$$
$$\leq 0, \quad t \in I.$$

Therefore, β is a first-kind general quantum operator on I. We will find $D_\beta h(t)$, $t \in I$. We have

$$h'(t) = 3t^2 + 2t + 1, \quad t \in I,$$

and

$$\begin{aligned}
h'(2) &= 3 \cdot 2^2 + 2 \cdot 2 + 1 \\
&= 3 \cdot 4 + 4 + 1 \\
&= 12 + 5 \\
&= 17 \\
&= D_\beta h(2).
\end{aligned}$$

Next,

$$\begin{aligned}
D_\beta h(t) &= \frac{h(\beta(t)) - h(t)}{\beta(t) - t} \\
&= \frac{(\beta(t))^3 + (\beta(t))^2 + \beta(t) - t^3 - t^2 - t}{\beta(t) - t} \\
&= \frac{(\beta(t) - t)((\beta(t))^2 + t\beta(t) + t^2) + (\beta(t) - t)(\beta(t) + t) + \beta(t) - t}{\beta(t) - t} \\
&= (\beta(t))^2 + t\beta(t) + t^2 + \beta(t) + t + 1 \\
&= (\beta(t))^2 + (t + 1)\beta(t) + t^2 + t + 1 \\
&= \left(\frac{1}{2}t + 1\right)^2 + (t + 1)\left(\frac{1}{2}t + 1\right) + t^2 + t + 1 \\
&= \frac{1}{4}t^2 + t + 1 + \frac{1}{2}t^2 + t + \frac{1}{2}t + 1 + t^2 + t + 1 \\
&= \frac{7}{4}t^2 + \frac{7}{2}t + 3, \quad t \neq 2.
\end{aligned}$$

Observe that

$$h(t) = t(t^2 + t + 1), \quad t \in I.$$

Then we write

$$h(t) = f(t)g(t), \quad t \in I,$$

where

$$\begin{aligned}
f(t) &= t \quad \text{and} \\
g(t) &= t^2 + t + 1, \quad t \in I.
\end{aligned}$$

Then

$$\begin{aligned}
f'(t) &= 1, \\
g'(t) &= 2t + 1, \quad t \in I,
\end{aligned}$$

$$g(2) = 2^2 + 2 + 1$$
$$= 4 + 3$$
$$= 7,$$
$$g'(2) = 2 \cdot 2 + 1$$
$$= 5,$$

and

$$f(\beta(t)) = \beta(t)$$
$$= \frac{1}{2}t + 1,$$
$$g(\beta(t)) = (\beta(t))^2 + \beta(t) + 1$$
$$= \left(\frac{1}{2}t + 1\right)^2 + \frac{1}{2}t + 1 + 1$$
$$= \frac{1}{4}t^2 + t + 1 + \frac{1}{2}t + 2$$
$$= \frac{1}{4}t^2 + \frac{3}{2}t + 3, \quad t \in I.$$

Next, we have

$$D_\beta f(t) = 1, \quad t \in I,$$

and

$$D_\beta g(t) = \frac{g(\beta(t)) - g(t)}{\beta(t) - t}$$
$$= \frac{(\beta(t))^2 + \beta(t) + 1 - t^2 - t - 1}{\beta(t) - t}$$
$$= \frac{(\beta(t) - t)(\beta(t) + t) + \beta(t) - t}{\beta(t) - t}$$
$$= \beta(t) + t + 1$$
$$= \frac{1}{2}t + 1 + t + 1$$
$$= \frac{3}{2}t + 2, \quad t \in I, \quad t \neq 2,$$

and

$$D_\beta g(2) = g'(2)$$
$$= 5.$$

Therefore

$$D_\beta h(t) = D_\beta (fg)(t)$$
$$= (D_\beta f(t))g(t) + f(\beta(t))D_\beta g(t)$$

$$= 1 \cdot (t^2 + t + 1) + \left(\frac{1}{2}t + 1\right)\left(\frac{3}{2}t + 2\right)$$

$$= t^2 + t + 1 + \frac{3}{4}t^2 + t + \frac{3}{2}t + 2$$

$$= \frac{7}{4}t^2 + \frac{7}{2}t + 3, \quad t \neq 2,$$

$$D_\beta h(t) = D_\beta(fg)(t)$$
$$= (D_\beta f(t))g(\beta(t)) + f(t)D_\beta g(t)$$
$$= 1 \cdot \left(\frac{1}{4}t^2 + \frac{3}{2}t + 3\right) + t\left(\frac{3}{2}t + 2\right)$$
$$= \frac{1}{4}t^2 + \frac{3}{2}t + 3 + \frac{3}{2}t^2 + 2t$$
$$= \frac{7}{4}t^2 + \frac{7}{2}t + 3, \quad t \in I, \quad t \neq 2,$$

and

$$D_\beta h(2) = (D_\beta f(2))g(2) + f(2)(D_\beta g(2))$$
$$= (1)(7) + (2)(5)$$
$$= 7 + 10$$
$$= 17.$$

Example 1.4.3. Let $I = (-\sqrt{2}, \sqrt{2})$. Define

$$\beta(t) = \frac{1}{2}t^3 \quad \text{and}$$

$$h(t) = \frac{t^2 - 4t + 1}{t^2 + t + 1}, \quad t \in I.$$

From Example 1.1.7, it follows that β is a first-kind general quantum operator. Here $s_0 = 0$. We will find $D_\beta h(t)$, $t \in I$. Let

$$f(t) = t^2 - 4t + 1,$$
$$g(t) = t^2 + t + 1, \quad t \in I.$$

Then

$$g(\beta(t)) = g\left(\frac{1}{2}t^3\right)$$
$$= \left(\frac{1}{2}t^3\right)^2 + \frac{1}{2}t^3 + 1$$
$$= \frac{1}{4}t^6 + \frac{1}{2}t^3 + 1,$$
$$f'(t) = 2t - 4, \quad \text{and}$$
$$g'(t) = 2t + 1, \quad t \in I,$$

which gives

$$f(0) = 1,$$
$$g(0) = 1,$$
$$f'(0) = -4,$$
$$g'(0) = 1.$$

Hence,

$$
\begin{aligned}
D_\beta h(0) &= \frac{f'(0)g(0) - f(0)g'(0)}{(g(0))^2} \\
&= \frac{(-4)(1) - (1)(1)}{1^2} \\
&= -4 - 1 \\
&= -5.
\end{aligned}
$$

Now, let $t \in (-\sqrt{2}, 0) \cup (0, \sqrt{2})$. Then

$$
\begin{aligned}
D_\beta f(t) &= \frac{f(\beta(t)) - f(t)}{\beta(t) - t} \\
&= \frac{(\beta(t))^2 - 4\beta(t) + 1 - t^2 + 4t - 1}{\beta(t) - t} \\
&= \frac{(\beta(t) - t)(\beta(t) + t) - 4(\beta(t) - t)}{\beta(t) - t} \\
&= \beta(t) + t - 4 \\
&= \frac{1}{2}t^3 + t - 4,
\end{aligned}
$$

$$
\begin{aligned}
D_\beta g(t) &= \frac{g(\beta(t)) - g(t)}{\beta(t) - t} \\
&= \frac{(\beta(t))^2 + \beta(t) + 1 - t^2 - t - 1}{\beta(t) - 1} \\
&= \frac{(\beta(t) - t)(\beta(t) + t) + \beta(t) - t}{\beta(t) - t} \\
&= \beta(t) + t + 1 \\
&= \frac{1}{2}t^3 + t + 1,
\end{aligned}
$$

and

$$
\begin{aligned}
D_\beta h(t) &= \frac{(D_\beta f(t))g(t) - f(t)D_\beta g(t)}{g(t)g(\beta(t))} \\
&= \frac{\left(\frac{1}{2}t^3 + t - 4\right)(t^2 + t + 1) - (t^2 - 4t + 1)\left(\frac{1}{2}t^3 + t + 1\right)}{(t^2 + t + 1)\left(\frac{1}{4}t^6 + \frac{1}{2}t^3 + 1\right)}
\end{aligned}
$$

$$= \frac{1}{(t^2+t+1)\left(\frac{1}{4}t^6+\frac{1}{2}t^3+1\right)}\Big(\frac{1}{2}t^5+\frac{1}{2}t^4+\frac{1}{2}t^3+t^3+t^2+t$$

$$-4t^2-4t-4-\frac{1}{2}t^5-t^3-t^2+2t^4+4t^2+4t-\frac{1}{2}t^3-t-1\Big)$$

$$= \frac{\frac{5}{2}t^4-5}{(t^2+t+1)\left(\frac{1}{4}t^6+\frac{1}{2}t^3+1\right)}$$

$$= \frac{10(t^4-2)}{(t^2+t+1)(t^6+2t^3+4)}.$$

Example 1.4.4. Let α be a constant, and let $m \in \mathbb{N}$. Define

$$f(t) = (t-\alpha)^m \quad \text{and}$$

$$g(t) = \frac{1}{(t-\alpha)^m}, \quad t \in I.$$

Then we have

$$D_\beta f(s_0) = m(s_0-\alpha)^{m-1}$$

and

$$D_\beta f(t) = \frac{f(\beta(t))-f(t)}{\beta(t)-t}$$

$$= \frac{(\beta(t)-\alpha)^m-(t-\alpha)^m}{\beta(t)-t}$$

$$= \frac{1}{\beta(t)-t}(\beta(t)-t)$$

$$\times \left((\beta(t)-\alpha)^{m-1}+(t-\alpha)(\beta(t)-\alpha)^{m-2}+\cdots+(t-\alpha)^{m-1}\right)$$

$$= \sum_{k=0}^{m-1}(\beta(t)-\alpha)^{m-1-k}(t-\alpha)^k, \quad t \in I, \quad t \neq s_0.$$

Next,

$$D_\beta g(s_0) = -m(s_0-\alpha)^{m-1},$$

and

$$D_\beta g(t) = D_\beta \frac{1}{f(t)}$$

$$= \frac{(D_\beta 1)f(t)-D_\beta f(t)}{f(t)f(\beta(t))}$$

$$= -\frac{\displaystyle\sum_{k=0}^{m-1}(\beta(t)-\alpha)^{m-1-k}(t-\alpha)^k}{(t-\alpha)^m(\beta(t)-\alpha)^m}$$

$$= -\sum_{k=0}^{m-1} \frac{(\beta(t)-\alpha)^{m-1-k}(t-\alpha)^k}{(t-\alpha)^m(\beta(t)-\alpha)^m}$$

$$= -\sum_{k=0}^{m-1} \frac{1}{(\beta(t)-\alpha)^{k+1}(t-\alpha)^{m-k}}, \quad t \in I, \quad t \neq s_0.$$

Exercise 1.4.5. Let $I = \mathbb{R}$. Define

$$\beta(t) = \frac{2}{3}t + 1 \quad \text{and}$$

$$f(t) = \frac{t+1}{t^2+4}, \quad t \in I.$$

Find $D_\beta f(t), t \in I$.

1.5 Properties of β-differentiable functions

In this section, we prove some properties of β-differentiable functions.

Theorem 1.5.1. *Let $f : I \to X$ be a β-differentiable function such that*

$$D_\beta f(t) = 0, \quad t \in I. \tag{1.3}$$

Then

$$f(t) = f(s_0), \quad t \in I.$$

Proof. From (1.3), we get that

$$f(\beta(t)) = f(t), \quad t \in I, \quad t \neq s_0.$$

Hence, since $\beta(t) \in I$ for every $t \in I$, we find

$$\begin{aligned} f(t) &= f(\beta(t)) \\ &= f(\beta^2(t)) \\ &\ \ \vdots \\ &= f(\beta^k(t)), \quad t \in I, \quad t \neq s_0, \quad k \in \mathbb{N}. \end{aligned} \tag{1.4}$$

By Corollary 1.1.15, we obtain

$$\lim_{k \to \infty} f(\beta^k(t)) = f(s_0), \quad t \in I.$$

Hence, in view of (1.4), we arrive at

$$f(t) = \lim_{k \to \infty} f(t)$$

$$= \lim_{k\to\infty} f\left(\beta^k(t)\right)$$
$$= f(s_0), \quad t \in I.$$

This completes the proof. $\qquad\qquad\square$

Corollary 1.5.2. *Let $f, g\colon I \to X$ be β-differentiable functions such that*

$$D_\beta(f - g)(t) = 0, \quad t \in I. \tag{1.5}$$

Then

$$f(t) - f(s_0) = g(t) - g(s_0), \quad t \in I.$$

Proof. Since f and g are β-differentiable functions, $f - g$ is also a β-differentiable function. Now applying Theorem 1.5.1, we find

$$(f - g)(t) = (f - g)(s_0), \quad t \in I,$$

that is,

$$f(t) - g(t) = f(s_0) - g(s_0), \quad t \in I.$$

This completes the proof. $\qquad\qquad\square$

Definition 1.5.3. Let $s_0 \in [a, b] \subseteq I$. We define the β-interval

$$[a, b]_\beta = \{\beta^k(a)\colon k \in \mathbb{N}_0\} \cup \{\beta^k(b)\colon k \in \mathbb{N}_0\} \cup \{s_0\}$$

and for $c \in I$, the class

$$[c]_\beta = \{\beta^k(c)\colon k \in \mathbb{N}_0\} \cup \{s_0\}.$$

For a set $A \subseteq \mathbb{R}$, define

$$A^* = A\setminus\{s_0\}.$$

Example 1.5.4. Let $I = (-\sqrt{3}, \sqrt{3})$ and define

$$\beta(t) = \frac{1}{3}t^3, \quad t \in I.$$

Take

$$[a, b] = [-1, 1].$$

Then

$$[-1, 1]^* = [-1, 0) \cup (0, 1].$$

Next,

$$\beta(-1) = -\frac{1}{3},$$

$$\beta^2(-1) = \beta(\beta(-1))$$
$$= \beta\left(-\frac{1}{3}\right)$$
$$= \frac{1}{3}\left(-\frac{1}{3}\right)^3$$
$$= -\frac{1}{3^4},$$
$$\beta^3(-1) = \beta(\beta^2(-1))$$
$$= \beta\left(-\frac{1}{3^4}\right)$$
$$= \frac{1}{3}\left(-\frac{1}{3^4}\right)^3$$
$$= -\frac{1}{3^{13}},$$
$$\beta^4(-1) = \beta(\beta^3(-1))$$
$$= \beta\left(-\frac{1}{3^{13}}\right)$$
$$= \frac{1}{3}\left(-\frac{1}{3^{13}}\right)^3$$
$$= -\frac{1}{3^{40}},$$
$$\vdots,$$

and

$$\beta(1) = \frac{1}{3},$$
$$\beta^2(1) = \frac{1}{3^4},$$
$$\beta^3(1) = \frac{1}{3^{13}},$$
$$\beta^4(1) = \frac{1}{3^{40}},$$
$$\vdots$$

Thus,

$$[-1, 1]_{\frac{1}{3}t^3} = \left\{-1, -\frac{1}{3}, -\frac{1}{3^4}, -\frac{1}{3^{13}}, -\frac{1}{3^{40}}, \cdots\right\}$$
$$\cup \left\{1, \frac{1}{3}, \frac{1}{3^4}, \frac{1}{3^{13}}, \frac{1}{3^{40}}, \cdots\right\} \cup \{0\}.$$

Moreover,

$$\beta\left(\frac{1}{2}\right) = \frac{1}{3 \cdot 2^3},$$

$$\beta^2\left(\frac{1}{2}\right) = \beta\left(\beta\left(\frac{1}{2}\right)\right)$$

$$= \beta\left(\frac{1}{3 \cdot 2^3}\right)^3$$

$$= \frac{1}{3^4 \cdot 2^9},$$

$$\beta^3\left(\frac{1}{2}\right) = \beta\left(\beta^2\left(\frac{1}{2}\right)\right)$$

$$= \beta\left(\frac{1}{3^4 \cdot 2^9}\right)$$

$$= \frac{1}{3}\left(\frac{1}{3^4 \cdot 2^9}\right)^3$$

$$= \frac{1}{3^{13} \cdot 2^{27}},$$

$$\vdots$$

Therefore,

$$\left[\frac{1}{2}\right]_{\frac{1}{3}t^3} = \left\{\frac{1}{2}, \frac{1}{3 \cdot 2^3}, \frac{1}{3^4 \cdot 2^9}, \frac{1}{3^{13} \cdot 2^{27}}, \dots\right\}.$$

Exercise 1.5.5. Let $I = \mathbb{R}$ and define

$$\beta(t) = \frac{1}{2}t + 1, \quad t \in I.$$

Find

1. $[0, 2]_\beta$;
2. $[1]_\beta$.

Theorem 1.5.6. *Let $[a, b] \subseteq I$, and let $f : [a, b] \to \mathbb{R}$ be continuous at s_0. Then f is strictly increasing on $[a, b]_\beta$ if and only if*

$$D_\beta f(t) > 0, \quad t \in [a, b]_\beta^*. \tag{1.6}$$

Proof. Suppose that (1.6) holds and β is a first-kind general quantum operator. Without loss of generality, suppose that $s_0 \notin \{a, b\}$. Since

$$(a - s_0)(\beta(a) - a) \leq a,$$

we get

$$a < \beta(a).$$

Now, since β is strictly increasing on $[a, b]$, we obtain

$$a < \beta(a) < \beta^2(a) < \cdots < \beta^k(a) < \cdots < s_0, \quad k \in \mathbb{N}.$$

Next, we have

$$(b - s_0)(\beta(b) - b) \leq 0,$$

and thus

$$\beta(b) < b.$$

Hence,

$$s_0 < \cdots < \beta^m(b) < \cdots < \beta^2(b) < \beta(b) < b, \quad m \in \mathbb{N}.$$

Now, applying Corollary 1.1.15, we arrive at

$$f(a) < f(\beta(a)) < \cdots < f(\beta^k(a)) < \cdots < f(s_0)$$

and

$$f(s_0) < \cdots < f(\beta^m(b)) < \cdots < f(\beta(b)) < f(b),$$

$k, m \in \mathbb{N}$, that is, f is strictly increasing on $[a, b]_\beta$.

Conversely, suppose that f is strictly increasing on $[a, b]_\beta$. Let $t \in [a, b]_\beta^*$. If

$$\beta(t) > t,$$

then

$$f(\beta(t)) > f(t),$$

and thus (1.6) holds. Also, if

$$\beta(t) < t,$$

then

$$f(\beta(t)) < f(t),$$

and thus (1.6) holds. This completes the proof. $\qquad\qquad \square$

Example 1.5.7. Let $I = \left[1, \dfrac{3}{2}\right]$. Define

$$\beta(t) = \frac{1}{2}t + \frac{3}{4} \quad \text{and}$$
$$f(t) = 4t^2 - 9t, \quad t \in I.$$

Here $s_0 = \dfrac{3}{2}$, and

$$
\begin{aligned}
D_\beta f(t) &= \frac{f(\beta(t)) - f(t)}{\beta(t) - t} \\
&= \frac{4(\beta(t))^2 - 9\beta(t) - 4t^2 + 9t}{\beta(t) - t} \\
&= \frac{4(\beta(t) - t)(\beta(t) + t) - 9(\beta(t) - t)}{\beta(t) - t} \\
&= 4(\beta(t) + t) - 9 \\
&= 4\left(\frac{1}{2}t + \frac{3}{4} + t\right) - 9 \\
&= 6t + 3 - 9 \\
&= 6t - 6 \\
&\geq 0, \quad t \in I.
\end{aligned}
$$

On the other hand,

$$
\begin{aligned}
f(1) &= 4 \cdot 1^2 - 9 \cdot 1 \\
&= 4 - 9 \\
&= -5.
\end{aligned}
$$

Next,

$$
\begin{aligned}
f(1.1) &= 4 \cdot (1.1)^2 - 9 \cdot 1.1 \\
&= 4.84 - 9.9 \\
&= -5.06.
\end{aligned}
$$

Therefore, f is not strictly increasing on $\left[1, \dfrac{3}{2}\right]$.

Similarly to Theorem 1.5.6, we can prove the following result.

Theorem 1.5.8. *Let $[a, b] \subseteq I$, and let $f : [a, b] \to \mathbb{R}$ be continuous at s_0. Then f is strictly decreasing on $[a, b]_\beta$ if and only if*

$$
D_\beta f(t) < 0, \quad t \in [a, b]_\beta^*.
$$

1.6 Chain rules

In this section, we discuss some chain rules for β-differential calculus.

Theorem 1.6.1. *Let $g : I \to \mathbb{R}$ be a continuous and β-differentiable function, and let $f : I \to X$ be a continuously differentiable function. Then there exists a point c between t and $\beta(t)$ such that*

$$
D_\beta(f \circ g)(t) = f'(g(c))D_\beta g(t), \quad t \in I. \tag{1.7}
$$

Proof. If $t = s_0$, then the statement is obvious. Suppose $t \neq s_0$. Then

$$D_\beta(f \circ g)(t) = \frac{f(g(\beta(t))) - f(g(t))}{\beta(t) - t}$$

$$= \frac{f(g(\beta(t))) - f(g(t))}{g(\beta(t)) - g(t)} \frac{g(\beta(t)) - g(t)}{\beta(t) - t}$$

$$= f'(\eta) D_\beta g(t),$$

where η is between $g(\beta(t))$ and $g(t)$. Since g is continuous, there is a point c between t and $\beta(t)$ such that

$$\eta = g(c).$$

Consequently, we get (1.7). This completes the proof. $\qquad\square$

Example 1.6.2. Let $I = \mathbb{R}$ and define

$$\beta(t) = \frac{4}{5}t + 1,$$

$$f(t) = t^3, \quad \text{and}$$

$$g(t) = t^2, \quad t \in I.$$

We will find $c \in \mathbb{R}$ such that

$$D_\beta(f \circ g)(0) = f'(g(c)) D_\beta g(0). \tag{1.8}$$

Here $s_0 = 5$. Then

$$f'(t) = 3t^2,$$

$$f'(g(t)) = 3(t^2)^2$$

$$= 3t^4, \quad t \in I,$$

$$f'(g(c)) = 3c^4,$$

and

$$D_\beta g(0) = \frac{g(\beta(0)) - g(0)}{\beta(0) - 0}$$

$$= \frac{(\beta(0))^2}{\beta(0)}$$

$$= \beta(0)$$

$$= 1,$$

$$f(g(t)) = (t^2)^3$$

$$= t^6, \quad t \in I,$$

and

$$D_\beta(f \circ g)(0) = \frac{(\beta(0))^6 - 0^6}{\beta(0) - 0}$$
$$= (\beta(0))^5 + (\beta(0))^4 + (\beta(0))^3 + (\beta(0))^2 + \beta(0) + 1$$
$$= 1 + 1 + 1 + 1 + 1 + 1$$
$$= 6.$$

Therefore, (1.8) gives

$$6 = 3c^4,$$

that is,

$$c = \pm\sqrt[4]{2},$$

which is the required c.

Exercise 1.6.3. Let $I = \mathbb{R}$ and define

$$\beta(t) = \frac{1}{2}t + 1,$$
$$f(t) = t^2, \quad \text{and}$$
$$g(t) = t^2, \quad t \in I.$$

Find $c \in \mathbb{R}$ such that

$$D_\beta(f \circ g)(0) = f'(g(c))D_\beta g(0).$$

Theorem 1.6.4. *Let $f: \mathbb{R} \to \mathbb{R}$ be continuously differentiable, and let $g: I \to \mathbb{R}$ be β-differentiable. Then $f \circ g: I \to \mathbb{R}$ is β-differentiable, and*

$$D_\beta(f \circ g)(t) = \left(\int_0^1 f'\big(g(t) + h(\beta(t) - t)D_\beta g(t)\big) dh \right) D_\beta g(t), \quad t \in I. \tag{1.9}$$

Proof. For $t = s_0$, the statement is obvious. Suppose that $t \neq s_0$. Then we have

$$f(g(\beta(t))) - f(g(t)) = \int_{g(t)}^{g(\beta(t))} f'(y)dy$$
$$= \left(\int_0^1 f'\big(g(t) + h(g(\beta(t)) - g(t))\big) dh \right) (g(\beta(t)) - g(t)),$$

whereupon

$$\frac{f(g(\beta(t))) - f(g(t))}{\beta(t) - t} = \left(\int_0^1 f'\left(g(t) + h(\beta(t) - t)\frac{g(\beta(t)) - g(t)}{\beta(t) - t}\right) dh \right) \left(\frac{g(\beta(t)) - g(t)}{\beta(t) - t} \right),$$

and subsequently, we get (1.9). This completes the proof. $\qquad\square$

Example 1.6.5. Let $I = \mathbb{R}$ and define

$$\beta(t) = \frac{7}{8}t + 1,$$

$$f(t) = \frac{1}{1+t^2}, \quad \text{and}$$

$$g(t) = 1 + t, \quad t \in I.$$

Then

$$f'(t) = -\frac{2t}{(1+t^2)^2},$$

$$D_\beta g(t) = 1, \quad t \in I, \quad t \neq 8,$$

$$f(g(t)) = \frac{1}{1+(t+1)^2},$$

$$D_\beta(f \circ g)(t) = \frac{f(g(\beta(t))) - f(g(t))}{\beta(t) - t}$$

$$= \left(\frac{1}{1+(\beta(t)+1)^2} - \frac{1}{1+(t+1)^2} \right) \frac{1}{\beta(t) - t}$$

$$= -\frac{(\beta(t)+1)^2 - (t+1)^2}{\left(1 + \left(\frac{7}{8}t+2\right)^2\right)(1+(t+1)^2)} \cdot \frac{1}{\beta(t) - t}$$

$$= -\frac{\beta(t) + t + 2}{\left(1 + \frac{49}{64}t^2 + \frac{7}{2}t + 4\right)(1 + t^2 + 2t + 1)}$$

$$= -\frac{\frac{7}{8}t + t + 3}{\frac{1}{64}(49t^2 + 224t + 320)(t^2 + 2t + 2)}$$

$$= -\frac{\frac{15}{8}t + 3}{\frac{1}{64}(49t^2 + 224t + 320)(t^2 + 2t + 2)}$$

$$= -\frac{8(15t + 24)}{(49t^2 + 224t + 320)(t^2 + 2t + 2)}, \quad t \in I, \quad t \neq 8,$$

and

$$g(t) + h(\beta(t) - t)D_\beta g(t) = t + 1 + h\left(\frac{7}{8}t + 1 - t\right)$$

$$= t + 1 + h\left(1 - \frac{1}{8}t\right).$$

Therefore

$$f'(g(t) + h(\beta(t) - t)D_\beta g(t)) = -\frac{2\left(t + 1 + h\left(1 - \frac{1}{8}t\right)\right)}{\left(1 + \left(t + 1 + h\left(1 - \frac{1}{8}t\right)\right)^2\right)^2}, \quad t \in I, \quad h \in [0, 1],$$

and then

$$\left(\int_0^1 f'(g(t)) + h(\beta(t) - t)D_\beta g(t))dh\right) D_\beta g(t) = -\int_0^1 \frac{2\left(t + 1 + h\left(1 - \frac{1}{8}t\right)\right)}{\left(1 + \left(t + 1 + h\left(1 - \frac{1}{8}t\right)\right)^2\right)^2}dh$$

$$= -\frac{1}{1 - \frac{1}{8}t}\int_0^1 \frac{d\left(t + 1 + h\left(1 - \frac{1}{8}t\right)\right)^2}{\left(1 + \left(t + 1 + h\left(1 - \frac{1}{8}t\right)\right)^2\right)^2}$$

$$= \frac{1}{1 - \frac{1}{8}t}\frac{1}{1 + \left(t + 1 + h\left(1 - \frac{1}{8}t\right)\right)^2}\Bigg|_{h=0}^{h=1}$$

$$= \frac{1}{1 - \frac{1}{8}t}\left(\frac{1}{1 + \left(t + 1 + 1 - \frac{1}{8}t\right)^2} - \frac{1}{1 + (t+1)^2}\right)$$

$$= \frac{1}{1 - \frac{1}{8}t} \cdot \frac{1 + (t+1)^2 - 1 - \left(\frac{7}{8}t + 2\right)^2}{\left(1 + \left(2 + \frac{7}{8}t\right)^2\right)(1 + (t+1)^2)}$$

$$= \frac{1}{1 - \frac{1}{8}t} \cdot \frac{\left(t + 1 - \frac{7}{8}t - 2\right)\left(t + 1 + \frac{7}{8}t + 2\right)}{\left(1 + \left(2 + \frac{7}{8}t\right)^2\right)(1 + (t+1)^2)}$$

$$= -\frac{\frac{15}{8}t + 3}{\left(\frac{49}{64}t^2 + \frac{7}{2}t + 4 + 1\right)(t^2 + 2t + 1 + 1)}$$

$$= -\frac{\frac{1}{8}(15t + 24)}{\frac{1}{64}(49t^2 + 224t + 320)(t^2 + 2t + 2)}$$

$$= -\frac{8(15t + 24)}{(49t^2 + 224t + 320)(t^2 + 2t + 2)}$$

$$= D_\beta(f \circ g)(t), \quad t \in I, \quad t \neq 8.$$

Exercise 1.6.6. Let $I = \mathbb{R}$ and define

$$\beta(t) = \frac{1}{3}t + 1,$$
$$f(t) = t^3 + 2t^2 + 2, \quad \text{and}$$
$$g(t) = 2 + 3t, \quad t \in I.$$

Check (1.9).

Theorem 1.6.7. *Let $g: I \to I$ be β-differentiable, let $\gamma: I \to I$ be a general quantum operator such that*

$$g(\beta(t)) = \gamma(g(t)), \quad t \in I,$$

and let $f : \mathbb{R} \to \mathbb{R}$ be γ-differentiable on I. Then

$$D_\beta(f \circ g)(t) = D_\gamma f(g(t))D_\beta g(t), \quad t \in I. \tag{1.10}$$

Proof. Suppose $t = s_0$. Let $s_1 \in I$ be the unique fixed point of γ. Then

$$g(s_0) = \gamma(g(s_0)).$$

Hence

$$g(s_0) = s_1,$$

and (1.10) holds. Let $t \in I$ be such that $t \neq s_0$. Then

$$\begin{aligned}
D_\beta(f \circ g)(t) &= \frac{f(g(\beta(t))) - f(g(t))}{\beta(t) - t} \\
&= \frac{f(\gamma(g(t))) - f(g(t))}{\beta(t) - t} \\
&= \frac{f(\gamma(g(t))) - f(g(t))}{\gamma(g(t)) - g(t)} \cdot \frac{\gamma(g(t)) - g(t)}{\beta(t) - t} \\
&= \frac{f(\gamma(g(t))) - f(g(t))}{\gamma(g(t)) - g(t)} \cdot \frac{g(\beta(t)) - g(t)}{\beta(t) - t} \\
&= D_\gamma f(g(t))D_\beta g(t).
\end{aligned}$$

Hence, (1.10) holds. This completes the proof. $\qquad\square$

Example 1.6.8. Let $I = \mathbb{R}$ and define

$$\begin{aligned}
\beta(t) &= \frac{1}{2}t + 2, \\
\gamma(t) &= \frac{1}{2}t + 4, \\
f(t) &= t^2, \quad \text{and} \\
g(t) &= t + 4, \quad t \in I.
\end{aligned}$$

Then

$$\begin{aligned}
f(g(t)) &= (g(t))^2 \\
&= (t + 4)^2 \\
&= t^2 + 8t + 16, \quad t \in I.
\end{aligned}$$

We see that $s_0 = 4$ is the unique fixed point of β and $s_1 = 8$ is the unique fixed point of γ. Then

$$\begin{aligned}
D_\beta f(g(t)) &= \frac{(\beta(t))^2 + 8\beta(t) + 16 - t^2 - 8t - 16}{\beta(t) - t} \\
&= \frac{(\beta(t) - t)(\beta(t) + t) + 8(\beta(t) - t)}{\beta(t) - t}
\end{aligned}$$

$$= \beta(t) + t + 8$$
$$= \frac{1}{2}t + 2 + t + 8$$
$$= \frac{3}{2}t + 10, \quad t \in I, \quad t \neq 4,$$

and

$$D_\beta f(g(s_0)) = 2s_0 + 8$$
$$= 2 \cdot 4 + 8$$
$$= 8 + 8$$
$$= 16.$$

Next,

$$g(\beta(t)) = \beta(t) + 4$$
$$= \frac{1}{2}t + 2 + 4$$
$$= \frac{1}{2}t + 6,$$

and

$$\gamma(g(t)) = \frac{1}{2}g(t) + 4$$
$$= \frac{1}{2}(t + 4) + 4$$
$$= \frac{1}{2}t + 2 + 4$$
$$= \frac{1}{2}t + 6, \quad t \in I.$$

Thus,

$$g(\beta(t)) = \gamma(g(t)), \quad t \in I.$$

Moreover,

$$D_\gamma f(t) = \frac{f(\gamma(t)) - f(t)}{\gamma(t) - t}$$
$$= \frac{(\gamma(t))^2 - t^2}{\gamma(t) - t}$$
$$= \frac{(\gamma(t) - t)(\gamma(t) + t)}{\gamma(t) - t}$$
$$= \gamma(t) + t$$
$$= \frac{1}{2}t + 4 + t$$
$$D_\gamma f(t) = \frac{3}{2}t + 4.$$

Using this, we obtain

$$
\begin{aligned}
D_\gamma f(g(t)) &= \frac{3}{2}g(t) + 4 \\
&= \frac{3}{2}(t+4) + 4 \\
&= \frac{3}{2}t + 6 + 4 \\
&= \frac{3}{2}t + 10, \quad t \in I, \quad t \neq 4.
\end{aligned}
$$

Note that

$$
\begin{aligned}
g(s_0) &= s_1 \\
D_\gamma f(g(s_0)) &= D_\gamma f(s_1) \\
&= \frac{3}{2}s_1 + 4 \\
&= \frac{3}{2} \cdot 8 + 4 \\
&= 12 + 4 \\
&= 16,
\end{aligned}
$$

and

$$
\begin{aligned}
D_\beta g(t) &= \frac{g(\beta(t)) - g(t)}{\beta(t) - t} \\
&= \frac{\beta(t) + 4 - t - 4}{\beta(t) - t} \\
&= 1, \\
D_\beta g(s_0) &= 1.
\end{aligned}
$$

Hence

$$
\begin{aligned}
D_\gamma f(g(t)) D_\beta g(t) &= \left(\frac{3}{2}t + 10\right) \cdot 1 \\
&= \frac{3}{2}t + 10 \\
&= D_\beta (f \circ g)(t), \quad t \in I, \quad t \neq s_0,
\end{aligned}
$$

and

$$
\begin{aligned}
D_\gamma f(g(s_0)) D_\beta g(s_0) &= 16 \cdot 1 \\
&= 16 \\
&= D_\beta (f \circ g)(s_0).
\end{aligned}
$$

Thus, (1.10) holds.

Exercise 1.6.9. Let $I = \mathbb{R}$ and define

$$\beta(t) = \frac{1}{4}t + 1,$$

$$\gamma(t) = \frac{1}{4}t + 5,$$

$$g(t) = 2t + 6, \quad \text{and}$$

$$f(t) = t^3, \quad t \in I.$$

Check (1.10).

1.7 Higher-order β-derivatives

Let $n \in \mathbb{N}$. We denote by $S_k^{(n)}$ the set of all possible strings of length n containing k times β and $(n - k)$ times D_β. We also denote

$$f^{D_\beta \beta}(t) = (D_\beta f)(\beta(t)), \quad t \in I,$$

$$f^{\beta D_\beta}(t) = D_\beta(f \circ \beta)(t), \quad t \in I,$$

and accordingly define f^Λ for $\Lambda \in S_k^{(n)}$.

Definition 1.7.1. If f is β-differentiable n-times over I, then the higher-order derivatives of f are defined by

$$D_\beta^n f = D_\beta(D_\beta^{n-1} f), \quad n \in \mathbb{N},$$

where

$$D_\beta^0 f = f.$$

By $\mathscr{C}_\beta^k(I)$, $k \in \mathbb{N}_0$, we denote the set of continuous functions $f : I \to \mathbb{R}$ such that $D_\beta^l f$ exist and are continuous for each $l \in \{0, \ldots, k\}$.

Example 1.7.2. Let $I = \mathbb{R}$ and define

$$\beta(t) = \frac{2}{3}t + 1, \quad \text{and}$$

$$f(t) = t^2 + t, \quad t \in I.$$

Then $s_0 = 3$. For $t \neq 3$, we have

$$\begin{aligned}
D_\beta f(t) &= \frac{f(\beta(t)) - f(t)}{\beta(t) - t} \\
&= \frac{(\beta(t))^2 + \beta(t) - t^2 - t}{\beta(t) - t} \\
&= \frac{(\beta(t) - t)(\beta(t) + t) + (\beta(t) - t)}{\beta(t) - t}
\end{aligned}$$

$$= \beta(t) + t + 1$$
$$= \frac{2}{3}t + 1 + t + 1$$
$$= \frac{5}{3}t + 2,$$

and hence

$$D_\beta^2 f(t) = \frac{5}{3}.$$

Exercise 1.7.3. Let $I = \mathbb{R}$ and define

$$\beta(t) = \frac{3}{4}t + 1, \quad \text{and}$$
$$f(t) = t^2 - t + 2, \quad t \in I.$$

Find $D_\beta^2 f(t), t \in I, t \neq 4$.

Now we present the Leibniz formula for higher-order β-differentiation.

Theorem 1.7.4 (β-Leibniz formula). *If f^Λ exists for all $\Lambda \in S_k^{(n)}$, then*

$$D_\beta^n (fg) = \sum_{k=0}^{n} \left(\sum_{\Lambda \in S_k^{(n)}} f^\Lambda \right) D_\beta^k g. \tag{1.11}$$

Proof. We will use the principle of mathematical induction.
For $n = 1$, we have

$$S_0^{(1)} = 0, \quad S_1^{(1)} = \beta.$$

Hence

$$\sum_{k=0}^{1} \left(\sum_{\Lambda \in S_k^{(1)}} f^\Lambda \right) D_\beta^k g = \sum_{\Lambda \in S_0^{(1)}} f^\Lambda g + \sum_{\Lambda \in S_1^{(1)}} f^\Lambda D_\beta g$$
$$= D_\beta f g + f^\beta D_\beta g$$
$$= D_\beta(fg),$$

that is, the statement holds for $n = 1$. Assume that the statement is valid for some $n \in \mathbb{N}$. We will prove that

$$D_\beta^{n+1}(fg) = \left(\sum_{\Lambda \in S_k^{(n+1)}} f^\Lambda \right) D_\beta^k g.$$

Using (1.11), we have

$$D_\beta^{n+1}(fg) = D_\beta\left(D_\beta^n(fg)\right)$$

$$= D_\beta\left(\sum_{k=0}^{n}\left(\sum_{\Lambda\in S_k^{(n)}} f^\Lambda\right) D_\beta^k g\right)$$

$$= \sum_{k=0}^{n} D_\beta\left(\left(\sum_{\Lambda\in S_k^{(n)}} f^\Lambda\right) D_\beta^k g\right)$$

$$= \sum_{k=0}^{n}\left(\left(\sum_{\Lambda\in S_k^{(n)}} f^{\Lambda\beta}\right) D_\beta^{k+1} g + \left(\sum_{\Lambda\in S_k^{(n)}} D_\beta f^\Lambda\right) D_\beta^k g\right)$$

$$= \sum_{k=0}^{n}\left(\sum_{\Lambda\in S_k^{(n)}} f^{\Lambda\beta}\right) D_\beta^{k+1} g + \sum_{k=0}^{n}\left(\sum_{\Lambda\in S_k^{(n)}} f^{\Lambda D_\beta}\right) D_\beta^k g$$

$$= \sum_{k=1}^{n+1}\left(\sum_{\Lambda\in S_{k-1}^{(n)}} f^{\Lambda\beta}\right) D_\beta^k g + \sum_{k=0}^{n}\left(\sum_{\Lambda\in S_k^{(n)}} f^{\Lambda\beta}\right) D_\beta^k g$$

$$= \sum_{k=1}^{n}\left(\sum_{\Lambda\in S_{k-1}^{(n)}} f^{\Lambda\beta}\right) D_\beta^k g + \left(\sum_{\Lambda\in S_k^{(n)}} f^{\Lambda\beta}\right) D_\beta^{n+1} g + \sum_{k=1}^{n}\left(\sum_{\Lambda\in S_k^{(n)}} f^{\Lambda\beta}\right) D_\beta^k g + \left(\sum_{\Lambda\in S_0^{(n)}} f^{\Lambda\beta}\right) g$$

$$= \sum_{k=1}^{n}\left(\sum_{\Lambda\in S_{k-1}^{(n)}} f^{\Lambda\beta} + \sum_{\Lambda\in S_k^{(n)}} f^{\Lambda D_\beta}\right) D_\beta^k g + \left(\sum_{\Lambda\in S_n^{(n)}} f^{\Lambda D_\beta} + \sum_{\Lambda\in S_0^{(n)}} f^{\Lambda D_\beta}\right) g$$

$$= \sum_{k=1}^{n}\left(\sum_{\Lambda\in S_k^{(n+1)}} f^\Lambda\right) D_\beta^k g + \left(\sum_{\Lambda\in S_{n+1}^{(n+1)}} f^\Lambda\right) D_\beta^{n+1} g + \left(\sum_{\Lambda\in S_0^{(n+1)}} f^\Lambda\right) g$$

$$= \sum_{k=0}^{n+1}\left(\sum_{\Lambda\in S_k^{(n+1)}} f^\Lambda\right) D_\beta^k g.$$

Hence, the statement is true for all $n\in\mathbb{N}$. This completes the proof. $\qquad\square$

1.8 The β-Rolle theorem

In this section, we prove the Rolle theorem in the β-differential calculus. We start with the following useful result.

Theorem 1.8.1. *Let f be a β-differentiable function at $t \in I$. Then there exists a function g in a neighborhood U of t with*

$$\lim_{s \to t} g(s) = g(t) = 0$$

such that

$$f(\beta(t)) = f(s) + (D_\beta f(t) + g(s))(\beta(t) - s)$$

for every $s \in I$.

Proof. Define

$$g(s) = \begin{cases} \dfrac{f(\beta(t)) - f(s)}{\beta(t) - s} - D_\beta f(t) & \text{if} \quad s \neq \beta(t), \\[2ex] 0 & \text{if} \quad s = \beta(t). \end{cases}$$

For $s = \beta(t) \in U$, the statement is obvious. For $s \neq \beta(t)$, $s \in U$, we have

$$\frac{f(\beta(t)) - f(s)}{\beta(t) - s} = D_\beta f(t) + g(s),$$

whereupon we get the desired result. This completes the proof. $\qquad\qquad\square$

Definition 1.8.2. Let $y \colon I \to \mathbb{R}$ be $(k-1)$-times β-differentiable, $k \in \mathbb{N}$. We say that y has a general zero (GZ) of order greater than or equal to k at $t \in I$ provided

$$D_\beta^i y(t) = 0, \quad i \in \{0, \ldots, k-1\}, \tag{1.12}$$

or

$$D_\beta^i y(t) = 0 \quad \text{for} \quad i \in \{0, \ldots, k-2\} \tag{1.13}$$

and

$$D_\beta^{k-1} y(\beta^{-1}(t)) D_\beta^{k-1} y(t) < 0. \tag{1.14}$$

Remark 1.8.3. Note that in the case (1.13)–(1.14), we must have $\beta^{-1}(t) \neq t$. Otherwise, for $\beta^{-1}(t) = t$,

$$0 > D_\beta^{k-1} y(\beta^{-1}(t)) D_\beta^{k-1} y(t)$$
$$= \left(D_\beta^{k-1} y(t) \right)^2$$
$$\geq 0,$$

which is a contradiction.

Theorem 1.8.4. *Condition* (1.14) *holds if and only if*

$$D_\beta^j y(t) = 0, \quad j \in \{0, \ldots, k-2\}, \tag{1.15}$$

and

$$(-1)^{k-1}y(\beta^{-1}(t))D_\beta^{k-1}y(t) < 0. \tag{1.16}$$

Proof. Suppose (1.14) holds. Then $t < \beta(t)$, $\beta(\beta^{-1}(t)) = t$, and

$$
\begin{aligned}
D_\beta^{k-1}y(\beta^{-1}(t)) &= \frac{D_\beta^{k-2}y(\beta(\beta^{-1}(t))) - D_\beta^{k-2}y(\beta^{-1}(t))}{t - \beta^{-1}(t)} \\[2mm]
&= -\frac{D_\beta^{k-2}y(\beta^{-1}(t))}{t - \beta^{-1}(t)} \\[2mm]
&= \cdots \\[2mm]
&= (-1)^{k-1}\frac{y(\beta^{-1}(t))}{(t - \beta^{-1}(t))^{k-1}}.
\end{aligned} \tag{1.17}
$$

Hence, (1.14) yields

$$
\begin{aligned}
0 > D_\beta^{k-1}y(\beta^{-1}(t))D_\beta^{k-1}y(t) \\[2mm]
= (-1)^{k-1}\frac{y(\beta^{-1}(t))}{(t - \beta^{-1}(t))^{k-1}}D_\beta^{k-1}y(t).
\end{aligned}
$$

Thus, (1.16) holds. Conversely, assume that (1.16) holds. Let $\beta(t) = t = s_0$. Then $\beta^{-1}(t) = t$, and we have the following cases.

(i) Let $k = 1$. Then $(y(t))^2 < 0$, which is a contradiction.
(ii) Let $k \geq 2$. Then $y(t) = 0$, and

$$(-1)^{k-1}y(t)D_\beta^{k-1}y(t) < 0,$$

which is a contradiction.

Consequently, $\beta^{-1}(t) \neq t$. Hence, using (1.17), we obtain (1.14). This completes the proof. $\qquad\square$

Theorem 1.8.5. *Let $j \in \mathbb{N}_0$ and $t \in I$. Then*

$$D_\beta^i y(t) = 0, \quad 0 \leq i \leq j, \tag{1.18}$$

if and only if

$$D_\beta^i y(\beta^l(t)) = 0, \quad 0 \leq i \leq j - l, \quad 0 \leq l \leq j. \tag{1.19}$$

In this case,

$$D_\beta^{j+1-l}y(\beta^l(t)) = \prod_{s=0}^{l-1}(\beta^{s+1}(t)\beta^s(t))D_\beta^{j+1}y(t). \tag{1.20}$$

Proof. Suppose (1.18) holds. We consider two cases.
   Suppose that $j = 0$. Then $l = i = 0$, $y(t) = 0$, and

$$D_\beta^i y(\beta^l(t)) = y(t) = 0.$$

Next, suppose that $j > 0$. Then

$$y(t) = D_\beta y(t) = \cdots = D_\beta^j y(t) = 0.$$

We employ the principle of the mathematical induction on j.

Suppose that $j = 1$. Then $i, l \in \{0, 1\}$. If $l = 0$, then $i \in \{0, 1\}$, and (1.19) holds. If $l = 1$, then $i = 0$, and

$$y(\beta(t)) = y(t) + (\beta(t) - t)D_\beta y(t)$$
$$= 0.$$

Thus, (1.19) holds for $j = 1$.

Now, assume that (1.19) is true for some $j \in \mathbb{N}$, that is,

$$y(t) = D_\beta y(t) = \cdots = D_\beta^j y(t) = 0,$$
$$y(\beta(t)) = D_\beta y(\beta(t)) = \cdots = D_\beta^{j-1} y(\beta(t)) = 0,$$
$$y(\beta^2(t)) = D_\beta y(\beta^2(t)) = \cdots = D_\beta^{j-2} y(\beta^2(t)) = 0,$$
$$\vdots \tag{1.21}$$
$$y(\beta^l(t)) = D_\beta y(\beta^l(t)) = \cdots = D_\beta^{j-l} y(\beta^l(t)) = 0,$$
$$\vdots$$
$$y(\beta^j(t)) = 0.$$

We will prove (1.19) for $j + 1$, that is,

$$y(t) = D_\beta y(t) = \cdots = D_\beta^{j+1} y(t) = 0,$$
$$y(\beta(t)) = D_\beta y(\beta(t)) = \ldots = D_\beta^j y(\beta(t)) = 0,$$
$$y(\beta^2(t)) = D_\beta y(\beta^2(t)) = \cdots = D_\beta^{j-1} y(\beta^2(t)) = 0,$$
$$\vdots$$
$$y(\beta^l(t)) = D_\beta y(\beta^l(t)) = \cdots = D_\beta^{j-l+1} y(\beta^l(t)) = 0,$$
$$\vdots$$
$$y(\beta^{j+1}(t)) = 0.$$

Keeping in mind (1.21), it suffices to prove that

$$D_\beta^{j+1} y(t) = D_\beta^j y(\beta(t)) = \cdots = D_\beta^{j=l+1} y(\beta^l(t))$$
$$= \cdots = y(\beta^{j+1}(t)) = 0.$$

From (1.18), we have $D_\beta^{j+1} y(t) = 0$. Then

$$D_\beta^j y(\beta(t)) = D_\beta^j y(t) + (\beta(t) - t) D_\beta^{j+1} y(t)$$
$$= 0,$$
$$D_\beta^{j-1} y(\beta^2(t)) = D_\beta^{j-1} y(\beta(t)) + (\beta(t) - t) D_\beta^j y(\beta(t))$$
$$= 0,$$
$$\vdots$$
$$D_\beta y(\beta^j(t)) = 0,$$
$$y(\beta^{j+1}(t)) = y(\beta^j(t)) + (\beta(t) - t) D_\beta y(\beta^j(t))$$
$$= 0.$$

Hence, (1.19) holds for $j + 1$. Thus, by the principle of the mathematical induction, we conclude that (1.19) holds for all $j \in \mathbb{N}_0$.

Conversely, suppose (1.19) holds. We consider two cases.

Suppose $j = 0$. Then $i = l = 0$, and

$$D_\beta^i y(\beta^l(t)) = y(t) = 0.$$

Next, suppose that $j > 0$. We employ the principle of mathematical induction on j. Suppose $j = 1$. Then by (1.19) we have

$$y(t) = D_\beta y(t) = y(\beta(t)) = 0.$$

Thus, (1.18) holds for $j = 1$.

Now assume that (1.18) holds for some $j \in \mathbb{N}_0$. We will prove that (1.18) holds for $j + 1$. Since (1.19) holds for $j + 1$, by (1.21) we obtain

$$y(t) = D_\beta y(t) = \cdots = D_\beta^{j+1} y(t) = 0.$$

Hence, (1.18) holds for $j + 1$. Thus, by the principle of mathematical induction, it follows that (1.18) holds for all $j \in \mathbb{N}_0$.

Next, we will prove (1.20). Suppose (1.19). Then (1.21) holds, and we have

$$D_\beta^j y(\beta(t)) = D_\beta^j y(t) + (\beta(t) - t) D_\beta^{j+1} y(t)$$
$$= (\beta(t) - t) D_\beta^{j+1} y(t),$$
$$D_\beta^{j-1} y(\beta^2(t)) = D_\beta^{j-1} y(\beta(t)) + (\beta^2(t) - \beta(t)) D_\beta^j y(\beta(t))$$
$$= (\beta^2(t) - \beta(t)) D_\beta^j y(\beta(t))$$
$$= (\beta^2(t) - \beta(t))(\beta(t) - t) D_\beta^{j+1} y(t),$$
$$\vdots$$

$$D_\beta^{j+1-l} y(\beta^l(t)) = \prod_{s=0}^{l-1} (\beta^{s+1}(t) - \beta^s(t)) D_\beta^{j+1} y(t).$$

This completes the proof. $\square$

Before proving the Rolle theorem for β-differential calculus, we introduce the following definition.

Definition 1.8.6. Let $y \colon I \to \mathbb{R}$ be k times β-differentiable, $k \in \mathbb{N}$. We say that y has at least k GZs, counting multiplicities, provided y has a GZ of order greater than or equal to k at $t \in I$. By Theorem 1.8.5, it follows that if y has a GZ of order greater than or equal to k at t, then y has a GZ of order greater than or equal to $k - 1$ at $\beta(t)$. Thus we say that y has at least $k_1 + k_2$ GZs, counting multiplicities, provided y has a GZ of order greater than or equal to k_1 at $t_1 \in I$ and y has a GZ of order greater than or equal to k_2 at $t_2 \in I$ and $\beta^{k_1}(t_1) < t_2$.

Theorem 1.8.7 (β-Rolle theorem). *If y has at least k GZs on $[a, b]$, counting multiplicities, then $D_\beta y$ has at least $k - 1$ GZs on $[a, b]$, counting multiplicities.*

Proof. **1.** First, we prove that if y has a GZ of order less than or equal to m at t, then $D_\beta y$ has a GZ of order less than or equal to $m - 1$ at t.
Suppose y has a GZ of order less than or equal to m at t. Then

$$D_\beta^i y(t) = 0, \quad i \in \{0, \ldots, m - 1\},$$

or

$$D_\beta^i y(t) = 0, \quad i \in \{0, \ldots, m - 2\},$$

and

$$D_\beta^{m-1} y(\beta^{-1}(t)) D_\beta^{m-1} y(t) < 0.$$

If $m = 1$, then $y(t) = 0$ or $y(\beta^{-1}(t))y(t) < 0$. If $D_\beta y(t) = 0$ or $D_\beta y(\beta^{-1}(t))D_\beta y(t) < 0$, then $m > 1$, a contradiction. Therefore, $D_\beta y$ has no GZ at t. Let $m \geq 2$. Then

$$D_\beta^{i-1}(D_\beta y)(t) = 0, \quad i \in \{0, \ldots, m - 2\},$$

or

$$D_\beta^i(D_\beta y)(t) = 0, \quad i \in \{0, \ldots, m - 3\},$$

and

$$D_\beta^{m-2}(D_\beta y)(\beta^{-1}(t)) D_\beta^{m-2}(D_\beta y)(t) < 0.$$

Thus, $D_\beta y$ has a GZ of order less than or equal to $m - 1$ at t.
2. Now, we will prove that if y has a GZ of order less than or equal to m at t and y has a GZ of order less than or equal to 1 at s with $\beta^{m-1}(t) < s$, then $D_\beta y$ has at least m GZs in $[t, s)$ which is equivalent to the following:

(i)   If $y(r) = 0$ and $D_\beta y$ has no GZ in $[r, s)$, where $r < \beta^{-1}(s)$, then y has no GZ at s.

(ii)   If $y(\beta^{-1}(r))y(r) < 0$ and $D_\beta y$ has no GZ in $[r, s)$, where $r < \beta^{-1}(s)$, then y has no GZ at s.

If the assumptions of (i) hold, then $D_\beta y(\tau) > 0$, $\tau \in [r, s)$, or $D_\beta y(\tau) < 0$, $\tau \in [r, s)$. Thus

$$y(\beta^{-1}(s))y(s) = \left(\int_r^{\beta^{-1}(s)} D_\beta y(\tau)d_\beta\tau \right)\left(\int_r^s D_\beta y(\tau)d_\beta\tau \right)$$

$$> 0.$$

Hence, y has no GZ at s. If the assumptions of (ii) hold, then $\beta^{-1}(r) < r$, and

$$y(\beta^{-1}(r))D_\beta y(\beta^{-1}(r)) = y(\beta^{-1}(r))\frac{y(r) - y(\beta^{-1}(r))}{r - \beta^{-1}(r)}$$

$$< 0.$$

Since $D_\beta y$ has a constant sign on $[\beta^{-1}(r), s)$, we have

$$y(\beta^{-1}(r))D_\beta y(\tau) < 0, \quad \tau \in [\beta^{-1}(r), s).$$

Then

$$y(\beta^{-1}(r))y(t) = y(\beta^{-1}(r))\left(y(r) + \int_r^t D_\beta y(\tau)d_\beta\tau \right)$$

$$< 0, \quad t \in \{\beta^{-1}(s), s\}.$$

Hence

$$y(\beta^{-1}(r))y(\beta^{-1}(s)) < 0, \quad y(\beta^{-1}(r))y(s) < 0,$$

and

$$(y(\beta^{-1}(r)))^2 y(\beta^{-1}(s))y(s) > 0.$$

Therefore $y(\beta^{-1}(s))y(s) > 0$, and y has no GZ at s. This completes the proof. $\square$

1.9 Advanced practical problems

Problem 1.9.1. Let $I = \mathbb{R}$. Define

$$\beta(t) = 3t, \quad t \in I.$$

Prove that β is a first-kind general quantum operator.

Problem 1.9.2. Let $I = [-1, 1]$. Define

$$\beta(t) = \frac{1}{4}t + \frac{1}{4}, \quad t \in I.$$

Prove that β is a first-kind general quantum operator.

Problem 1.9.3. Let I and β be as in Problem 1.9.1. Find

1. $\beta^{-1}(t), t \in I$;
2. $\beta^k(t), t \in I, k \in \mathbb{N}_0$;
3. $\beta^{-k}(t), t \in I, k \in \mathbb{N}$.

Problem 1.9.4. Let I and β be as in Problem 1.9.2. Find

1. $\beta^{-1}(t), t \in I$;
2. $\beta^k(t), t \in I, k \in \mathbb{N}_0$;
3. $\beta^{-k}(t), t \in I, k \in \mathbb{N}$.

Problem 1.9.5. Let $I = [1, \infty)$. Define

$$\beta(t) = \log t + 1, \quad t \in I.$$

1. Prove that β is a first-kind general quantum operator.
2. Prove that

$$\beta^{-1}(t) = e^{t-1}, \quad t \in I.$$

3. Find

$$\lim_{k \to \infty} \beta^k(t), \quad t \in I.$$

4. Find

$$\lim_{k \to \infty} \beta^{-k}(t), \quad t \in I.$$

Problem 1.9.6. Let $I = \mathbb{R}$. Define

$$\beta(t) = \frac{1}{2}t \quad \text{and}$$
$$f(t) = t^2 + 4, \quad t \in I.$$

1. Prove that β is a first-kind general quantum operator.
2. Find $D_\beta f(t), t \in I$.

Problem 1.9.7. Let $I = \mathbb{R}$. Define

$$\beta(t) = \frac{3}{4}t + 1 \quad \text{and}$$
$$f(t) = \frac{t-1}{t^2+1}, \quad t \in I.$$

Find $D_\beta f(t), t \in I$.

Problem 1.9.8. Let $I = \mathbb{R}$. Define

$$\beta(t) = \frac{2}{3}t + \frac{1}{3}, \quad t \in I.$$

Find

1. $[0, 2]_\beta$;
2. $[3]_\beta$.

Problem 1.9.9. Let $I = \mathbb{R}$. Define

$$\beta(t) = \frac{2}{3}t + 1,$$
$$f(t) = t^2, \quad \text{and}$$
$$g(t) = t + 2, \quad t \in I.$$

Find $c \in \mathbb{R}$ such that

$$D_\beta(f \circ g)(0) = f'(g(c)) D_\beta g(0).$$

Problem 1.9.10. Let $I = \mathbb{R}$. Define

$$\beta(t) = \frac{1}{2}t + 1,$$
$$f(t) = \frac{t+1}{t^2+1}, \quad \text{and}$$
$$g(t) = 1 + t + t^2, \quad t \in I.$$

Check (1.9).

Problem 1.9.11. Let $I = \mathbb{R}$. Define

$$\beta(t) = \frac{1}{8}t + 1,$$
$$\gamma(t) = \frac{1}{8}t + 6,$$
$$g(t) = 5t + 2, \quad \text{and}$$
$$f(t) = t^3 + t, \quad t \in I.$$

Check (1.10).

Problem 1.9.12. Let $I = \mathbb{R}$. Define

$$\beta(t) = \frac{7}{8}t + 1 \quad \text{and}$$
$$f(t) = 2t^2 - 2t + 7, \quad t \in I.$$

Find $D_\beta^2 f(t)$ for $t \in I$ such that $t \neq 4$.

1.10 **Notes and references**

In this chapter, we defined general quantum operators of first and second kinds. We have introduced the concept of β-differentiation. The main properties and main rules of β-differentiation are deduced. Further, some chain rules and the Rolle theorem for β-differential calculus are discussed. The Leibniz formula for higher-order β-differentiation is proved.

General quantum integration*

In this chapter, we define the β-integral and deduce some of its basic properties and the rules for β-integration. We deduce mean value theorem and also present the β-Taylor formula. The improper integrals of the first and the second kind are introduced. Throughout this chapter, we suppose that $I \subseteq \mathbb{R}$ and $\beta : I \to \mathbb{R}$ is a first-kind general quantum operator. Also, let X be a Banach space with norm $\| \cdot \|$.

2.1 β-antiderivatives

Definition 2.1.1. A function $F : I \to X$ is said to be a β-antiderivative of a function $f : I \to X$ provided F is β-differentiable on I and

$$D_\beta F(t) = f(t), \quad t \in I.$$

Example 2.1.2. Let $I = (-\sqrt{2}, \sqrt{2})$. Define

$$\beta(t) = \frac{1}{2}t^3,$$

$$f(t) = \frac{1}{4}t^6 + \frac{1}{2}t^4 - \frac{1}{2}t^3 + t^2 - t + 1, \quad \text{and}$$

$$F(t) = t^3 - t^2 + t, \quad t \in I.$$

Note that here $s_0 = 0$. For $t \neq 0$, we have

$$D_\beta F(t) = \frac{F(\beta(t)) - F(t)}{\beta(t) - t}$$

$$= \frac{(\beta(t))^3 - (\beta(t))^2 + \beta(t) - t^3 + t^2 - t}{\beta(t) - t}$$

$$= \frac{(\beta(t) - t)((\beta(t))^2 + t\beta(t) + t^2) - (\beta(t) - t)(\beta(t) + t) + \beta(t) - t}{\beta(t) - t}$$

$$= (\beta(t))^2 + t\beta(t) + t^2 - \beta(t) - t + 1$$

$$= \left(\frac{1}{2}t^3\right)^2 + t\left(\frac{1}{2}t^3\right) + t^2 - \frac{1}{2}t^3 - t + 1$$

---

* This book has a companion website hosting complementary materials. Visit this URL to access it: https://www.elsevier.com/books-and-journals/book-companion/9780443328046.

Generalized Quantum Calculus with Applications. https://doi.org/10.1016/B978-0-44-332804-6.00007-8

$$= \frac{1}{4}t^6 + \frac{1}{2}t^4 - \frac{1}{2}t^3 + t^2 - t + 1$$
$$= f(t).$$

Next,

$$F'(0) = (3t^2 - 2t + 1)\Big|_{t=0}$$
$$= 1$$
$$= f(0).$$

Thus, F is a β-antiderivative of f.

Exercise 2.1.3. Let $I = \mathbb{R}$. Define

$$\beta(t) = \frac{1}{2}t + 2,$$
$$f(t) = \frac{3}{2}t + 1, \quad \text{and}$$
$$F(t) = t^2 - t + 1, \quad t \in I.$$

Prove that F is a β-antiderivative of f.

Definition 2.1.4. The vector space of all functions $g : I \to X$ that are continuous at s_0 with $g(s_0) = 0$ is denoted by Ω. Define the operator $T_\beta : \Omega \to \Omega$ as

$$T_\beta(g)(t) = g(\beta(t)), \quad t \in I, \quad g \in \Omega.$$

Denote

$$Y = \text{range}(\mathscr{I} - T_\beta),$$

where $\mathscr{I} : \Omega \to \Omega$ is the identity operator.

Theorem 2.1.5. *For each $g \in Y$, the series*

$$\sum_{k=0}^{\infty} g(\beta^k(t)), \quad t \in I, \tag{2.1}$$

is uniformly convergent on I.

Proof. Let $\varepsilon > 0$. By Corollary 1.1.15 and Definition 2.1.4, we get

$$\lim_{k \to \infty} g(\beta^k(t)) = g(s_0)$$
$$= 0.$$

Therefore, there is a $K \in \mathbb{N}$ such that

$$\|g(\beta^k(t))\| < \frac{\varepsilon}{(k+1)^2}, \quad t \in I,$$

for all $k \geq K$. Note that the series

$$\sum_{k=0}^{\infty} \frac{1}{(k+1)^2}$$

is convergent. Now, applying the Weierstrass criteria, we conclude that series (2.1) is uniformly convergent on I. This completes the proof. $\square$

Theorem 2.1.6. *The operator* $\mathscr{I} - T_\beta : \Omega \to \Omega$ *is one-to-one.*

Proof. Let $f, g \in \Omega$ be such that

$$(\mathscr{I} - T_\beta)(f)(t) = (\mathscr{I} - T_\beta)(g)(t), \quad t \in I,$$

that is,

$$f(t) - f(\beta(t)) = g(t) - g(\beta(t)), \quad t \in I.$$

We consider two cases. If $t = s_0$, then clearly

$$f(s_0) = g(s_0).$$

For $t \neq s_0$, we have

$$\frac{f(\beta(t)) - f(t)}{\beta(t) - t} = \frac{g(\beta(t)) - g(t)}{\beta(t) - t},$$

that is,

$$D_\beta f(t) = D_\beta g(t),$$

and thus

$$D_\beta (f - g)(t) = 0.$$

Applying Corollary 1.5.2, we get

$$f(t) - g(t) = f(s_0) - g(s_0)$$
$$= 0.$$

Thus, $\mathscr{I} - T_\beta : \Omega \to \Omega$ is injective. Now, let $h \in \Omega$ and set

$$l(t) = \sum_{k=0}^{\infty} h(\beta^k(t)), \quad t \in I.$$

By Theorem 2.1.5, it follows that $l \in \Omega$. Moreover, we have

$$(\mathscr{I} - T_\beta)(l)(t) = l(t) - T_\beta l(t)$$
$$= \sum_{k=0}^{\infty} h(\beta^k(t)) - T_\beta \sum_{k=0}^{\infty} h(\beta^k(t))$$

$$= \sum_{k=0}^{\infty} h(\beta^k(t)) - \sum_{k=0}^{\infty} h(\beta^{k+1}(t))$$

$$= \sum_{k=0}^{\infty} h(\beta^k(t)) - \sum_{k=1}^{\infty} h(\beta^k(t))$$

$$= h(\beta^0(t))$$

$$= h(t), \quad t \in I.$$

Therefore, $\mathscr{I} - T_\beta : \Omega \to \Omega$ is a surjective and hence one-to-one. This completes the proof. $\qquad\square$

Corollary 2.1.7. *The operator* $A \colon Y \to \Omega$ *defined by*

$$A(g)(t) = \sum_{k=0}^{\infty} g(\beta^k(t)), \quad t \in I,$$

is the inverse operator of the operator $\mathscr{I} - T_\beta$.

Theorem 2.1.8. *Let* $f \colon I \to X$ *be continuous at* s_0. *Then the function* $F \colon I \to X$ *defined by*

$$F(t) = \sum_{k=0}^{\infty} (\beta^k(t) - \beta^{k+1}(t)) f(\beta^k(t)), \quad t \in I, \tag{2.2}$$

is a β-antiderivative of f with $F(s_0) = 0$. Conversely, if F is a β-antiderivative of f that vanishes at s_0, then F is given by (2.2).

Proof. Suppose F is given by (2.2). Then

$$F(s_0) = \sum_{k=0}^{\infty} (\beta^k(s_0) - \beta^{k+1}(s_0)) f(\beta^k(t))$$

$$= 0,$$

and

$$\frac{F(\beta(t)) - F(t)}{\beta(t) - t} = \frac{1}{\beta(t) - t} \left(\sum_{k=0}^{\infty} (\beta^{k+1}(t) - \beta^{k+2}(t)) f(\beta^k(t)) - \sum_{k=0}^{\infty} (\beta^k(t) - \beta^{k+1}(t)) f(\beta^k(t)) \right)$$

$$= \frac{1}{\beta(t) - t} \left(\sum_{k=1}^{\infty} (\beta^k(t) - \beta^{k+1}(t)) f(\beta^k(t)) - \sum_{k=0}^{\infty} (\beta^k(t) - \beta^{k+1}(t)) f(\beta^k(t)) \right)$$

$$= \frac{1}{\beta(t) - t} (\beta(t) - t) f(t)$$

$$= f(t), \quad t \neq s_0, \quad t \in I.$$

Let $\varepsilon > 0$. Since f is continuous at s_0, there exists $\delta > 0$ such that for $0 < h < \delta$,

$$\| f(\beta^k(s_0 + h)) - f(s_0) \| < \varepsilon, \quad k \geq 0.$$

Hence, keeping in mind that

$$\frac{1}{h}\sum_{k=0}^{\infty}(\beta^k(s_0+h)-\beta^{k+1}(s_0+h))=1$$

and

$$f(s_0)=\frac{1}{h}\sum_{k=0}^{\infty}(\beta^k(s_0+h)-\beta^{k+1}(s_0+h))f(s_0),$$

for $0<h<\delta$ and $k\geq 0$, we have

$$\left\|\frac{1}{h}[F(s_0+h)-F(s_0)]-f(s_0)\right\|=\left\|\frac{1}{h}\sum_{k=0}^{\infty}(\beta^k(s_0+h)-\beta^{k+1}(s_0+h))f(\beta^k(s_0+h))-f(s_0)\right\|$$

$$=\left\|\frac{1}{h}\sum_{k=0}^{\infty}(\beta^k(s_0+h)-\beta^{k+1}(s_0+h))[f(\beta^k(s_0+h))-f(s_0)]\right\|$$

$$\leq\frac{1}{h}\sum_{k=0}^{\infty}(\beta^k(s_0+h)-\beta^{k+1}(s_0+h))|f(\beta^k(s_0+h))-f(s_0)|$$

$$<\varepsilon\frac{1}{h}\sum_{k=0}^{\infty}(\beta^k(s_0+h)-\beta^{k+1}(s_0+h))$$

$$=\varepsilon.$$

Thus

$$F'(s_0)=f(s_0),$$

and F is a β-antiderivative of f. Now, let F be a β-antiderivative of f with $F(s_0)=0$. Then

$$f(t)=D_\beta F(t)$$
$$=\frac{F(\beta(t))-F(t)}{\beta(t)-t}$$
$$=\frac{1}{t-\beta(t)}(\mathscr{I}-T_\beta)F(t),\quad t\neq s_0,\quad t\in I,$$

whereupon

$$(\mathscr{I}-T_\beta)F(t)=f(t)(t-\beta(t)),\quad t\in I.$$

Hence,

$$F(t)=(\mathscr{I}-T_\beta)^{-1}(t-\beta(t))f(t)$$
$$=\sum_{k=0}^{\infty}(\beta^k(t)-\beta^{k+1}(t))f(\beta^k(t)),\quad t\in I.$$

This completes the proof. $\square$

2.2 Definition of β-integral and examples

Definition 2.2.1. Let $f : I \to X$ and $a, b \in I$. Define the indefinite β-integral of f as

$$\int_{s_0}^{a} f(t)d_\beta t = \sum_{k=0}^{\infty} (\beta^k(a) - \beta^{k+1}(a)) f(\beta^k(a))$$

and the definite β-integral of f from a to b as

$$\int_{a}^{b} f(t)d_\beta t = \int_{s_0}^{b} f(t)d_\beta t - \int_{s_0}^{a} f(t)d_\beta t. \tag{2.3}$$

We can rewrite (2.3) in the form

$$\int_{a}^{b} f(t)d_\beta t = \sum_{k=0}^{\infty} (\beta^k(b) - \beta^{k+1}(b)) f(\beta^k(b)) - \sum_{k=0}^{\infty} (\beta^k(a) - \beta^{k+1}(a)) f(\beta^k(a)).$$

Example 2.2.2. Let $I = (-\sqrt{2}, \sqrt{2})$, and let

$$\beta(t) = \frac{1}{2}t^3, \quad t \in I.$$

By Example 1.1.11, we have

$$\beta^k(t) = \left(\frac{1}{2}\right)^{[k]_3} t^{3^k}$$

$$= \left(\frac{1}{2}\right)^{\frac{3^k-1}{2}} t^{3^k}$$

$$= \sqrt{2}\left(\frac{1}{2}\right)^{\frac{3^k}{2}} t^{3^k}$$

$$= \sqrt{2}\left(\frac{1}{\sqrt{2}}\right)^{3^k} t^{3^k}$$

$$= \sqrt{2}\left(\frac{t}{\sqrt{2}}\right)^{3^k}, \quad t \in I.$$

Hence,

$$\int_{0}^{x} t\,d_\beta t = \sum_{k=0}^{\infty} (\beta^k(x) - \beta^{k+1}(x)) \beta^k(x)$$

$$= \sqrt{2} \sum_{k=0}^{\infty} \left(\left(\frac{x}{\sqrt{2}}\right)^{3^k} - \left(\frac{x}{\sqrt{2}}\right)^{3^{k+1}}\right) \left(\frac{x}{\sqrt{2}}\right)^{3^k}$$

$$= \sqrt{2} \sum_{k=0}^{\infty} \left(\left(\frac{x}{\sqrt{2}}\right)^{2 \cdot 3^k} - \left(\frac{x}{\sqrt{2}}\right)^{4 \cdot 3^k}\right)$$

$$= \sqrt{2} \sum_{k=0}^{\infty} \left(\left(\frac{x^2}{2} \right)^{3^k} - \left(\frac{x^4}{4} \right)^{3^k} \right), \quad x \in I.$$

Example 2.2.3. Let $I = \mathbb{R}$, and let

$$\beta(t) = \frac{1}{2}t + 1, \quad t \in I.$$

Here $s_0 = 2$. We will compute

$$I_1 = \int_2^3 t \, d_\beta t,$$

$$I_2 = \int_2^4 t \, d_\beta t, \quad \text{and}$$

$$I_3 = \int_3^4 t \, d_\beta t.$$

By Example 1.1.11, we have

$$\beta^k(t) = \left(\frac{1}{2} \right)^k t + [k]_{\frac{1}{2}}, \quad t \in I, \quad k \in \mathbb{N}.$$

Moreover,

$$[k]_{\frac{1}{2}} = \frac{1 - \left(\frac{1}{2} \right)^k}{1 - \frac{1}{2}}$$

$$= 2 \left(1 - \left(\frac{1}{2} \right)^k \right),$$

and

$$[k]_{\frac{1}{2}} - [k+1]_{\frac{1}{2}} = \frac{1 - \left(\frac{1}{2} \right)^k}{1 - \frac{1}{2}} - \frac{1 - \left(\frac{1}{2} \right)^{k+1}}{1 - \frac{1}{2}}$$

$$= 2 \left(1 - \left(\frac{1}{2} \right)^k - 1 + \left(\frac{1}{2} \right)^{k+1} \right)$$

$$= 2 \left(- \left(\frac{1}{2} \right)^k + \left(\frac{1}{2} \right)^{k+1} \right), \quad k \in \mathbb{N}.$$

Hence,

$$I_1 = \int_2^3 t \, d_\beta t$$

$$= \sum_{k=0}^{\infty} \left(\beta^k(3) - \beta^{k+1}(3) \right) \beta^k(3)$$

$$= \sum_{k=0}^{\infty} \left(3\left(\frac{1}{2}\right)^k + [k]_{\frac{1}{2}} - 3\left(\frac{1}{2}\right)^{k+1} - [k+1]_{\frac{1}{2}} \right) \times \left(3\left(\frac{1}{2}\right)^k + [k]_{\frac{1}{2}} \right)$$

$$= \sum_{k=0}^{\infty} \left(3\left(\frac{1}{2}\right)^k - 3\left(\frac{1}{2}\right)^{k+1} - 2\left(\frac{1}{2}\right)^k + 2\left(\frac{1}{2}\right)^{k+1} \right) \times \left(3\left(\frac{1}{2}\right)^k + [k]_{\frac{1}{2}} \right)$$

$$= \sum_{k=0}^{\infty} \left(\left(\frac{1}{2}\right)^k - \left(\frac{1}{2}\right)^{k+1} \right) \left(3\left(\frac{1}{2}\right)^k + [k]_{\frac{1}{2}} \right)$$

$$= \sum_{k=0}^{\infty} \left(1 - \frac{1}{2} \right) \left(\frac{1}{2}\right)^k \left(3 \cdot \left(\frac{1}{2}\right)^k + [k]_{\frac{1}{2}} \right)$$

$$= \frac{1}{2} \sum_{k=0}^{\infty} \left(3\left(\frac{1}{4}\right)^k + [k]_{\frac{1}{2}} \left(\frac{1}{2}\right)^k \right)$$

$$= \frac{3}{2} \sum_{k=0}^{\infty} \left(\frac{1}{4}\right)^k + \frac{1}{2} \sum_{k=0}^{\infty} \left(\frac{1 - \left(\frac{1}{2}\right)^k}{1 - \frac{1}{2}} \right) \left(\frac{1}{2}\right)^k$$

$$= \frac{3}{2} \cdot \frac{1}{1 - \frac{1}{4}} + \sum_{k=0}^{\infty} \left(\frac{1}{2}\right)^k - \sum_{k=0}^{\infty} \left(\frac{1}{4}\right)^k$$

$$= 2 + \frac{1}{1 - \frac{1}{2}} - \frac{1}{1 - \frac{1}{4}}$$

$$= 2 + 2 - \frac{4}{3}$$

$$= 4 - \frac{4}{3}$$

$$= \frac{8}{3}.$$

Next,

$$I_2 = \int_2^4 t \, d_\beta t$$

$$= \sum_{k=0}^{\infty} \left(\beta^k(4) - \beta^{k+1}(4) \right) \beta^k(4)$$

$$= \sum_{k=0}^{\infty} \left(4\left(\frac{1}{2}\right)^k + [k]_{\frac{1}{2}} - 4\left(\frac{1}{2}\right)^{k+1} - [k+1] - \frac{1}{2} \right) \times \left(4\left(\frac{1}{2}\right)^k + [k]_{\frac{1}{2}} \right)$$

$$= \sum_{k=0}^{\infty} \left(4\left(\frac{1}{2}\right)^k - 4\left(\frac{1}{2}\right)^{k+1} - 2\left(\frac{1}{2}\right)^k + 2\left(\frac{1}{2}\right)^{k+1} \right) \times \left(4\left(\frac{1}{2}\right)^k + [k]_{\frac{1}{2}} \right)$$

$$= \sum_{k=0}^{\infty} \left(2\left(\frac{1}{2}\right)^k - 2\left(\frac{1}{2}\right)^{k+1} \right) \left(4\left(\frac{1}{2}\right)^k + \frac{1 - \left(\frac{1}{2}\right)^k}{1 - \frac{1}{2}} \right)$$

$$= 2\sum_{k=0}^{\infty} \left(\frac{1}{2}\right)^k \left(1 - \frac{1}{2}\right)\left(4\left(\frac{1}{2}\right)^k + 2 - 2\left(\frac{1}{2}\right)^k\right)$$

$$= \sum_{k=0}^{\infty} \left(\frac{1}{2}\right)^k \left(2 + 2\left(\frac{1}{2}\right)^k\right)$$

$$= 2\sum_{k=0}^{\infty} \left(\frac{1}{2}\right)^k + 2\sum_{k=0}^{\infty} \left(\frac{1}{4}\right)^k$$

$$= 2\frac{1}{1 - \frac{1}{2}} + 2\frac{1}{1 - \frac{1}{4}}$$

$$= 1 + \frac{8}{3}$$

$$= \frac{11}{3},$$

and

$$I_3 = \int_3^4 t\,d_\beta t$$

$$= \int_2^4 t\,d_\beta t - \int_2^3 t\,d_\beta t$$

$$= I_2 - I_1$$

$$= \frac{11}{3} - \frac{8}{3}$$

$$= 1.$$

Remark 2.2.4. Note that in the classical case,

$$\int_3^4 t\,dt = \frac{t^2}{2}\bigg|_{t=3}^{t=4}$$

$$= \frac{16}{2} - \frac{9}{2}$$

$$= \frac{7}{2}.$$

Example 2.2.5. Let $I = \mathbb{R}$, and let

$$\beta(t) = \frac{3}{4}t - 1, \quad t \in I.$$

Here $s_0 = -4$. We will find

$$I = \int_1^3 (t^2 - t + 1)\,d_\beta t.$$

For this, we will compute

$$I_1 = \int -4^1(t^2 - t + 1)\,d_\beta t \quad \text{and}$$

$$I_2 = \int_{-4}^{3} (t^2 - t + 1)d_\beta t.$$

By Example 1.1.11, we have

$$\beta^k(t) = \left(\frac{3}{4}\right)^k t - [k]_{\frac{3}{4}}$$

$$= \left(\frac{3}{4}\right)^k - \frac{1 - \left(\frac{3}{4}\right)^k}{1 - \frac{3}{4}}$$

$$= \left(\frac{3}{4}\right)^k - 4\left(1 - \left(\frac{3}{4}\right)^k\right), \quad t \in I.$$

Then

$$I_1 = \int_{-4}^{1} (t^2 - t + 1)d_\beta t$$

$$= \sum_{k=0}^{\infty} \left(\beta^k(1) - \beta^{k+1}(1)\right)\left((\beta^k(1))^2 - \beta^k(1) + 1\right)$$

$$= \sum_{k=0}^{\infty} \left(\left(\left(\frac{3}{4}\right)^k - 4 + 4\left(\frac{3}{4}\right)^k - \left(\frac{3}{4}\right)^{k+1} + 4 - 4\left(\frac{3}{4}\right)^{k+1}\right)\right.$$

$$\left. \times \left(\left(\left(\frac{3}{4}\right)^k - 4 + 4\left(\frac{3}{4}\right)^k\right)^2 - \left(\frac{3}{4}\right)^k + 4 - 4\left(\frac{3}{4}\right)^k + 1\right)\right)$$

$$= \sum_{k=0}^{\infty} \left(\left(5\left(\frac{3}{4}\right)^k - \frac{3}{4}\left(\frac{3}{4}\right)^k - 3\left(\frac{3}{4}\right)^k\right)\right.$$

$$\left. \times \left(\left(5\left(\frac{3}{4}\right)^k - 4\right)^2 - 5\left(\frac{3}{4}\right)^k + 5\right)\right)$$

$$= \sum_{k=0}^{\infty} \left(\frac{5}{4}\left(\frac{3}{4}\right)^k \left(25\left(\frac{9}{16}\right)^k - 40\left(\frac{3}{4}\right)^k + 16 - 5\left(\frac{3}{4}\right)^k + 5\right)\right)$$

$$= \frac{5}{4}\sum_{k=0}^{\infty} \left(\frac{3}{4}\right)^k \left(25\left(\frac{9}{16}\right)^k - 45\left(\frac{3}{4}\right)^k + 21\right)$$

$$= \frac{125}{4}\sum_{k=0}^{\infty} \left(\frac{27}{64}\right)^k - \frac{225}{4}\sum_{k=0}^{\infty} \left(\frac{9}{16}\right)^k + \frac{105}{4}\sum_{k=0}^{\infty} \left(\frac{3}{4}\right)^k$$

$$= \frac{125}{4}\left(\frac{1}{1 - \frac{27}{64}}\right) - \frac{225}{4}\left(\frac{1}{1 - \frac{9}{16}}\right) + \frac{105}{4}\left(\frac{1}{1 - \frac{3}{4}}\right)$$

$$= \frac{125 \times 16}{37} - \frac{225 \times 4}{7} + 105$$

$$= \frac{2000}{37} - \frac{900}{7} + 105$$

$$= \frac{(2000 \times 7) - (900 \times 37) + (105 \times 37 \times 7)}{259}$$

$$= \frac{14000 - 33300 + 27195}{259}$$

$$= \frac{7895}{259}.$$

Next,

$$I_2 = \int_{-4}^{3} (t^2 - t + 1)\, d_\beta t$$

$$= \sum_{k=0}^{\infty} \left(\beta^k(3) - \beta^{k+1}(3) \right) \left((\beta^k(3))^2 - \beta^k(3) + 1 \right)$$

$$= \sum_{k=0}^{\infty} \left(\left(3\left(\frac{3}{4}\right)^k - 4 + 4\left(\frac{3}{4}\right)^k - 3\left(\frac{3}{4}\right)^{k+1} + 4 - 4\left(\frac{3}{4}\right)^{k+1} \right) \right.$$

$$\left. \times \left(\left(3\left(\frac{3}{4}\right)^k - 4 + 4\left(\frac{3}{4}\right)^k \right)^2 - 3\left(\frac{3}{4}\right)^k + 4 - 4\left(\frac{3}{4}\right)^k + 1 \right) \right)$$

$$= \sum_{k=0}^{\infty} \left(\left(7\left(\frac{3}{4}\right)^k - \frac{21}{4}\left(\frac{3}{4}\right)^k \right) \right.$$

$$\left. \times \left(\left(7\left(\frac{3}{4}\right)^k - 4 \right)^2 - 7\left(\frac{3}{4}\right)^k + 5 \right) \right)$$

$$= \sum_{k=0}^{\infty} \left(\frac{7}{4}\left(\frac{3}{4}\right)^k \left(49\left(\frac{9}{16}\right)^k - 56\left(\frac{3}{4}\right)^k + 16 - 7\left(\frac{3}{4}\right)^k + 5 \right) \right)$$

$$= \frac{7}{4} \sum_{k=0}^{\infty} \left(\frac{3}{4}\right)^k \left(49\left(\frac{9}{16}\right)^k - 63\left(\frac{3}{4}\right)^k + 21 \right)$$

$$= \frac{343}{4} \sum_{k=0}^{\infty} \left(\frac{27}{64}\right)^k - \frac{441}{4} \sum_{k=0}^{\infty} \left(\frac{9}{16}\right)^k + \frac{147}{4} \sum_{k=0}^{\infty} \left(\frac{3}{4}\right)^k$$

$$= \frac{343}{4} \left(\frac{1}{1 - \frac{27}{64}} \right) - \frac{441}{4} \left(\frac{1}{1 - \frac{9}{16}} \right) + \frac{147}{4} \left(\frac{1}{1 - \frac{3}{4}} \right)$$

$$= \frac{343 \times 16}{37} - \frac{441 \times 4}{7} + 147$$

$$= \frac{5488}{37} - \frac{1764}{7} + 147$$

$$= \frac{(5488 \times 7) - (1764 \times 37) + (147 \times 37 \times 7)}{259}$$

$$= \frac{38416 - 65268 + 38073}{259}$$

$$= \frac{11221}{259}.$$

Therefore,

$$I = \int_1^3 (t^2 - t + 1)d_\beta t$$

$$= \int_{-4}^3 (t^2 - t + 1)d_\beta t - \int_{-4}^1 (t^2 - t + 1)d_\beta t$$

$$= I_2 - I_1$$

$$= \frac{11221}{259} - \frac{7895}{259}$$

$$= \frac{3326}{259}.$$

Exercise 2.2.6. Let $I = \mathbb{R}$, and let

$$\beta(t) = \frac{1}{2}t - 1, \quad t \in I.$$

Compute

1. $\displaystyle\int_{-2}^{-1} (t - 1)d_\beta t$;

2. $\displaystyle\int_{-2}^{1} (t - 1)d_\beta t$;

3. $\displaystyle\int_{-1}^{1} (t - 1)d_\beta t$.

2.3 Properties of β-integrals

In this section, we prove some important properties of β-integrals. Throughout this section, suppose that $[a, b] \subseteq I$.

Theorem 2.3.1. *The β-integral is a linear operator.*

Proof. Suppose $f, g : I \to X$ are β-integrable and $\alpha \in \mathbb{C}$ is arbitrarily chosen. Then we have

$$\int_a^b (f(t) + g(t))d_\beta t = \int_a^b (f + g)(t)d_\beta t$$

$$= \sum_{k=0}^\infty (\beta^k(b) - \beta^{k+1}(b))(f + g)(\beta^k(b)) - \sum_{k=0}^\infty (\beta^k(a) - \beta^{k+1}(a))(f + g)(\beta^k(a))$$

$$= \sum_{k=0}^\infty (\beta^k(b) - \beta^{k+1}(b))f(\beta^k(b)) - \sum_{k=0}^\infty (\beta^k(a) - \beta^{k+1}(a))f(\beta^k(a))$$

$$+ \sum_{k=0}^\infty (\beta^k(b) - \beta^{k+1}(b))g(\beta^k(b)) - \sum_{k=0}^\infty (\beta^k(a) - \beta^{k+1}(a))g(\beta^k(a))$$

$$= \int_a^b f(t)d_\beta t + \int_a^b g(t)d_\beta t$$

and

$$\int_a^b \alpha f(t)d_\beta t = \int_a^b (\alpha f)(t)d_\beta t$$

$$= \sum_{k=0}^{\infty}(\beta^k(b) - \beta^{k+1}(b))(\alpha f)(\beta^k(b)) - \sum_{k=0}^{\infty}(\beta^k(a) - \beta^{k+1}(a))(\alpha f)(\beta^k(a))$$

$$= \alpha \sum_{k=0}^{\infty}(\beta^k(b) - \beta^{k+1}(b))f(\beta^k(b)) - \alpha \sum_{k=0}^{\infty}(\beta^k(a) - \beta^{k+1}(a))f(\beta^k(a))$$

$$= \alpha\left(\sum_{k=0}^{\infty}(\beta^k(b) - \beta^{k+1}(b))f(\beta^k(b)) - \sum_{k=0}^{\infty}(\beta^k(a) - \beta^{k+1}(a))f(\beta^k(a)) \right)$$

$$= \alpha \int_a^b f(t)d_\beta t.$$

This completes the proof. $\qquad\square$

Theorem 2.3.2. *Let* $f : I \to X$ *be* β-*integrable over* $[a, b]$. *Then*

$$\int_a^a f(t)d_\beta t = 0.$$

Proof. We have

$$\int_a^a f(t)d_\beta t = \sum_{k=0}^{\infty}(\beta^k(a) - \beta^{k+1}(a))f(\beta^k(a)) - \sum_{k=0}^{\infty}(\beta^k(a) - \beta^{k+1}(a))f(\beta^k(a))$$

$$= 0. \qquad\square$$

Theorem 2.3.3. *Let* $f : I \to X$ *be* β-*integrable over* $[a, b]$. *Then*

$$\int_a^b f(t)d_\beta t = -\int_b^a f(t)d_\beta t.$$

Proof. We have

$$\int_a^b f(t)d_\beta t = \sum_{k=0}^{\infty}(\beta^k(b) - \beta^{k+1}(b))f(\beta^k(b)) - \sum_{k=0}^{\infty}(\beta^k(a) - \beta^{k+1}(a))f(\beta^k(a))$$

$$= -\left(-\sum_{k=0}^{\infty}(\beta^k(b) - \beta^{k+1}(b))f(\beta^k(b)) + \sum_{k=0}^{\infty}(\beta^k(a) - \beta^{k+1}(a))f(\beta^k(a)) \right)$$

$$= -\int_b^a f(t)d_\beta t. \qquad\square$$

Theorem 2.3.4. *Let* $f : I \to X$ *be* β-*integrable over* $[a, b]$. *Then, for each* $c \in [a, b]$, *we have*

$$\int_a^b f(t)d_\beta t = \int_a^c f(t)d_\beta t + \int_c^b f(t)d_\beta t.$$

Proof. We have

$$\int_a^b f(t)d_\beta t = \sum_{k=0}^{\infty}(\beta^k(b) - \beta^{k+1}(b))f(\beta^k(b)) - \sum_{k=0}^{\infty}(\beta^k(a) - \beta^{k+1}(a))f(\beta^k(a))$$

$$= \sum_{k=0}^{\infty}(\beta^k(c) - \beta^{k+1}(c))f(\beta^k(c)) - \sum_{k=0}^{\infty}(\beta^k(a) - \beta^{k+1}(a))f(\beta^k(a))$$

$$+ \sum_{k=0}^{\infty}(\beta^k(b) - \beta^{k+1}(b))f(\beta^k(b)) - \sum_{k=0}^{\infty}(\beta^k(c) - \beta^{k+1}(c))f(\beta^k(c))$$

$$= \int_a^c f(t)d_\beta t + \int_c^b f(t)d_\beta t. \qquad \square$$

Theorem 2.3.5 (Fundamental theorem of β-differential calculus). *Let $f : I \to X$ be continuous at s_0. Then*

$$F(x) = \int_{s_0}^x f(t)d_\beta t, \quad x \in I,$$

is continuous at s_0, and $D_\beta F(x)$ exists for all $x \in I$. Moreover,

$$D_\beta F(x) = f(x), \quad x \in I.$$

Proof. We have

$$F(x) = \sum_{k=0}^{\infty}(\beta^k(x) - \beta^{k+1}(x))f(\beta^k(x)), \quad x \in I.$$

Hence, by Theorem 2.1.8, it follows that F is a β-antiderivative of f. This completes the proof. $\qquad \square$

Corollary 2.3.6. *Let $f : I \to X$ be continuous at x_0. Then*

$$\int_t^{\beta(t)} f(x)d_\beta x = (\beta(t) - t)f(t), \quad t \in I.$$

Proof. Set

$$F(t) = \int_{s_0}^t f(x)d_\beta x, \quad t \in I.$$

Then applying Theorem 2.3.5, we get

$$\int_t^{\beta(t)} f(x)d_\beta x = \int_{s_0}^{\beta(t)} f(x)d_\beta x - \int_{s_0}^t f(x)d_\beta x$$

$$= F(\beta(t)) - F(t)$$

$$= (\beta(t) - t)D_\beta F(t)$$

$$= (\beta(t) - t)f(t), \quad t \in I.$$

This completes the proof. $\qquad \square$

Corollary 2.3.7. *If $f : I \to X$ is continuous at s_0, and let $\Phi : I \to X$ be a β-antiderivative of f on I. Then*

$$\int_a^b f(t)d_\beta t = \Phi(b) - \Phi(a).$$

Proof. We have

$$\Phi(x) = \int_{s_0}^x f(t)d_\beta t, \quad x \in I.$$

Hence,

$$\int_a^b f(t)d_\beta t = \int_{s_0}^b f(t)d_\beta t - \int_{s_0}^a f(t)d_\beta t$$
$$= \Phi(b) - \Phi(a).$$

This completes the proof. $\qquad\qquad\square$

The β-analogue of the integration-by-parts formula is presented by the following:

Theorem 2.3.8. *Let $f, g : I \to X$ be β-differentiable functions on I such that $D_\beta f$ and $D_\beta g$ are continuous at s_0. Then*

$$\int_a^b f(t)D_\beta g(t)d_\beta t = (f(t)g(t))\Big|_{t=a}^{t=b} - \int_a^b D_\beta f(t)g(\beta(t))d_\beta t.$$

Proof. Since f and g are β-differentiable on I, it follows that fg is β-differentiable on I. Also, we have that fg is a β-antiderivative of $f D_\beta g + g^\beta D_\beta f$. Hence the result follows by Corollary 2.3.7. This completes the proof. $\qquad\square$

2.4 Inequalities and β-integrals

In this section, we start our investigation with the following useful lemma. Suppose that $[c, d] \subseteq I$. Denote by D the set of all real-valued functions on $[c, d]_\beta$.

Lemma 2.4.1. *Let $f \in D$. Then*

$$\int_c^d f(t)h(\beta(t))d_\beta t = 0 \qquad\qquad (2.4)$$

for each $h \in D$ with $h(c) = h(d) = 0$ if and only if

$$f(t) = 0, \quad t \in [c, d]_\beta. \qquad\qquad (2.5)$$

Proof. Suppose (2.5) holds. Then

$$f(\beta^k(t)) = 0, \quad t \in [c, d], \quad k \in \mathbb{N}.$$

Hence,

$$\int_c^d f(t)h(\beta(t))d_\beta t = \sum_{k=0}^{\infty} (\beta^k(d) - \beta^{k+1}(d)) f(\beta^k(d))h(\beta^{k+1}(d))$$

$$- \sum_{k=0}^{\infty} (\beta^k(c) - \beta^{k+1}(c)) f(\beta^k(c))h(\beta^{k+1}(c))$$

$$= 0,$$

that is, (2.4) holds.

Conversely, suppose that (2.4) holds. We assume that there is $l \in [c, d]_\beta$ such that $f(l) \neq 0$. Suppose $l \neq s_0$. Then either $l = \beta^k(c)$, or $l = \beta^k(d)$ for some $k \in \mathbb{N}_0$. If $l = \beta^k(c)$ for some $k \in \mathbb{N}_0$, then we define

$$h(t) = \begin{cases} f(l) & \text{if } t = \beta(l), \\ 0 & \text{otherwise.} \end{cases}$$

This yields that $h \in D$ and

$$h(c) = h(d) = 0.$$

Hence

$$0 = \int_c^d f(t)h(\beta(t))d_\beta t$$

$$= \int_c^{\beta(c)} f(t)h(\beta(t))d_\beta t$$

$$= (\beta(c) - c) f(c)h(\beta(c))$$

$$= (\beta(c) - c)(f(c))^2$$

$$\neq 0,$$

which is a contradiction. Thus, $l \neq \beta^k(c)$ for all $k \in \mathbb{N}_0$. The case $l = \beta^k(d)$ for some $k \in \mathbb{N}_0$ is treated similarly.

Next, let $c = s_0$. Without loss of generality, assume that $f(s_0) > 0$. Since f is continuous at s_0, we have

$$\lim_{k \to \infty} f(\beta^k(d)) = \lim_{k \to \infty} f(\beta^k(c))$$

$$= f(s_0).$$

Thus, there exists $k_0 \in \mathbb{N}_0$ such that

$$f(\beta^k(d)) > 0,$$

$$f(\beta^k(c)) > 0, \quad k > k_0.$$

For $s_0 \notin [c,d]$, define

$$h(t) = \begin{cases} f(\beta^k(c)) & \text{if} \quad t = \beta^{k+1}(c), \quad k > k_0, \\ f(\beta^k(d)) & \text{if} \quad t = \beta^{k+1}(d), \quad k > k_0, \\ 0 & \text{otherwise.} \end{cases}$$

Hence $h \in D$ and

$$\int_c^d f(t)h(\beta(t))d_\beta t$$

$$= \sum_{k=k_0}^{\infty} (\beta^k(d) - \beta^{k+1}(d)) f(\beta^k(d))h(\beta^{k+1}(d)) - \sum_{k=k_0}^{\infty} (\beta^k(c) - \beta^{k+1}(c)) f(\beta^k(c))h(\beta^{k+1}(c))$$

$$= \sum_{k=k_0}^{\infty} (\beta^k(d) - \beta^{k+1}(d))(f(\beta^k(d)))^2 h - \sum_{k=k_0}^{\infty} (\beta^k(c) - \beta^{k+1}(c))(f(\beta^k(c)))^2$$

$$\neq 0,$$

which is a contradiction. Now, let $s_0 = c$. Then define

$$h(t) = \begin{cases} f(\beta^k(d)) & \text{if} \quad t = \beta^{k+1}(d), \quad k > k_0, \\ 0 & \text{otherwise.} \end{cases}$$

Therefore

$$\int_c^d f(t)h(\beta(t))d_\beta t = \sum_{k=k_0}^{\infty} (\beta^k(d) - \beta^{k+1}(d)) f(\beta^k(d))h(\beta^{k+1}(d))$$

$$= \sum_{k=k_0}^{\infty} (\beta^k(d) - \beta^{k+1}(d))(f(\beta^k(d)))^2$$

$$\neq 0,$$

which is a contradiction. Finally, the case $s_0 = d$ is treated similarly. This completes the proof. $\quad\square$

Definition 2.4.2. Let $g \colon [r]_\beta \times (-\theta, \theta) \to \mathbb{R}$, and let $\theta_0 \in (-\theta, \theta)$. We say that $g(t, \cdot)$ is continuous at θ_0 uniformly in $t \in [r]_\beta$ provided for every $\varepsilon > 0$, there exists $\delta = \delta(\varepsilon) > 0$ such that whenever $|\theta - \theta_0| < \delta$, we have

$$|g(t, \theta) - g(t, \theta_0)| < \varepsilon, \quad t \in [r]_\beta.$$

Definition 2.4.3. Let $g \colon [r]_\beta \times (-\theta, \theta) \to \mathbb{R}$, and let $\theta_0 \in (-\theta, \theta)$. We say that $g(t, \cdot)$ is differentiable at θ_0 uniformly in $t \in [r]_\beta$ with derivative $g_\theta(t, \cdot)$ provided for every $\varepsilon > 0$, there exists $\delta = \delta(\varepsilon) > 0$ such that whenever $|\theta - \theta_0| < \delta$, we have

$$\left| \frac{g(t, \theta) - g(t, \theta_0)}{\theta - \theta_0} - g_\theta(t, \theta_0) \right| < \varepsilon, \quad t \in [r]_\beta.$$

Lemma 2.4.4. *Assume that $g(t, \cdot)$ is differentiable at θ_0 uniformly in $t \in [r]_\beta$. Also, let*

$$G(\theta) = \int_{s_0}^{r} g(t, \theta) d_\beta t$$

and

$$\int_{s_0}^{r} g_\theta(t, \theta_0) d_\beta t$$

exist for $\theta \in \mathbb{R}$. Then G is differentiable at θ_0, and

$$G'(\theta_0) = \int_{s_0}^{r} g_\theta(t, \theta_0) d_\beta t.$$

Proof. Let $\varepsilon > 0$. Since $g(t, \cdot)$ is differentiable at θ_0 uniformly in $t \in [r]_\beta$, there exists $\delta = \delta(\varepsilon) > 0$ such that

$$\left| \frac{g(t, \theta) - g(t, \theta_0)}{\theta - \theta_0} - g_\theta(t, \theta_0) \right| < \frac{\varepsilon}{r - s_0}, \quad t \in [r]_\beta.$$

Hence,

$$\left| \frac{G(\theta) - G(\theta_0)}{\theta - \theta_0} - \int_{s_0}^{r} g_\theta(t, \theta_0) d_\beta t \right| = \left| \int_{s_0}^{r} \left(\frac{g(t, \theta) - g(t, \theta_0)}{\theta - \theta_0} - g_\theta(t, \theta_0) \right) d_\beta t \right|$$

$$\leq \int_{s_0}^{r} \left| \frac{g(t, \theta) - g(t, \theta_0)}{\theta - \theta_0} - g_\theta(t, \theta_0) \right| d_\beta t$$

$$< \frac{\varepsilon}{r - s_0} (r - s_0)$$

$$= \varepsilon.$$

This completes the proof. $\qquad\square$

Lemma 2.4.5. *Let $f : I \to X$ and $g : I \to \mathbb{R}$ be β-integrable on I. If*

$$\| f(t) \| \leq g(t), \quad t \in [a, b]_\beta, \quad a, b \in I, \quad a < b,$$

then for $x, y \in [a, b]$, $x < s_0 < y$, we have

$$\left\| \int_{s_0}^{y} f(t) d_\beta t \right\| \leq \int_{s_0}^{y} g(t) d_\beta t,$$

$$\left\| \int_{s_0}^{x} f(t) d_\beta t \right\| \leq - \int_{s_0}^{x} g(t) d_\beta t, \quad \text{and}$$

$$\left\| \int_{x}^{y} f(t) d_\beta t \right\| \leq \int_{x}^{y} g(t) d_\beta t.$$

Proof. Since $y > s_0$, we have

$$\beta^{k+1}(y) < \beta^{k}(y), \quad k \in \mathbb{N}.$$

Then

$$\left\| \int_{s_0}^{y} f(t)\,d_\beta t \right\| = \left\| \sum_{k=0}^{\infty} (\beta^k(y) - \beta^{k+1}(y)) f(\beta^k(y)) \right\|$$

$$\leq \sum_{k=0}^{\infty} (\beta^k(y) - \beta^{k+1}(y)) \left\| f(\beta^k(y)) \right\|$$

$$\leq \sum_{k=0}^{\infty} (\beta^k(y) - \beta^{k+1}(y)) g(\beta^k(y))$$

$$= \int_{s_0}^{y} g(t)\,d_\beta t.$$

Also, since $x < s_0$, we have

$$\beta^k(x) < \beta^{k+1}(x), \quad k \in \mathbb{N}.$$

Now,

$$\left\| \int_{s_0}^{x} f(t)\,d_\beta t \right\| = \left\| \sum_{k=0}^{\infty} (\beta^k(x) - \beta^{k+1}(x)) f(\beta^k(x)) \right\|$$

$$\leq \sum_{k=0}^{\infty} (\beta^{k+1}(x) - \beta^k(x)) \left\| f(\beta^k(x)) \right\|$$

$$\leq - \sum_{k=0}^{\infty} (\beta^k(x) - \beta^{k+1}(x)) g(\beta^k(x))$$

$$= - \int_{s_0}^{x} g(t)\,d_\beta t.$$

Thus,

$$\left\| \int_{x}^{y} f(t)\,d_\beta t \right\| = \left\| \int_{s_0}^{y} f(t)\,d_\beta t - \int_{s_0}^{x} f(t)\,d_\beta t \right\|$$

$$\leq \left\| \int_{s_0}^{y} f(t)\,d_\beta t \right\| + \left\| \int_{s_0}^{x} f(t)\,d_\beta t \right\|$$

$$\leq \int_{s_0}^{y} g(t)\,d_\beta t - \int_{s_0}^{x} g(t)\,d_\beta t$$

$$- \int_{x}^{y} g(t)\,d_\rho t.$$

This completes the proof. $\qquad\qquad\square$

Lemma 2.4.6. *Let $f : I \to X$ and $g : I \to \mathbb{R}$ be β-differentiable on I. Suppose that*

$$\| D_\beta f(t) \| \leq D_\beta g(t), \quad t \in [a, b]_\beta, \quad a, b \in I, \quad a < b.$$

Then

$$\|f(y) - f(x)\| \le g(y) - g(x)$$

for $x, y \in [a, b]_\beta$, $x < s_0 < y$.

Proof. Applying Lemma 2.4.5, we get

$$\begin{aligned}
\|f(y) - f(x)\| &= \left\| \int_x^y D_\beta f(t) d_\beta t \right\| \\
&\le \int_x^y \|D_\beta f(t)\| d_\beta t \\
&\le \int_x^y D_\beta g(t) d_\beta t \\
&= g(y) - g(x).
\end{aligned}$$

This completes the proof. $\qquad\square$

Theorem 2.4.7 (Mean value theorem). *Let $f : I \to X$ be β-differentiable on I. Then*

$$\|f(y) - f(x)\| \le \sup_{t \in I} \|D_\beta f(t)\|(y - x)$$

for $x, y \in [a, b]_\beta$ such that $x < s_0 < y$.

Proof. We have

$$\begin{aligned}
\|f(y) - f(x)\| &= \left\| \int_x^y D_\beta f(t) d_\beta t \right\| \\
&\le \int_x^y \|D_\beta f(t)\| d_\beta t \\
&\le \sup_{t \in I} \|D_\beta f(t)\|(y - x), \quad x, y \in [a, b]_\beta, \ x < s_0 < y.
\end{aligned}$$

This completes the proof. $\qquad\square$

Theorem 2.4.8. *Let $k : I \times I \to \mathbb{R}$ be continuous and β-differentiable with respect to its first argument. Also, let*

$$g(x) = \int_{s_0}^x k(x, s) d_\beta s, \quad x \in I.$$

Then

$$D_\beta g(x) = k(\beta(x), x) + \int_{s_0}^x D_{\beta x} k(x, s) d_\beta s, \quad x \in I,$$

where $D_{\beta x}$ denotes the β-derivative with respect to x.

Proof. Suppose $x \neq s_0$. Then

$$g(\beta(x)) - g(x) = \int_{s_0}^{\beta(x)} k(\beta(x), s) d_\beta s - \int_{s_0}^x k(x, s) d_\beta s$$

$$= \int_{s_0}^{x} (k(\beta(x), s) - k(x, s)) d_\beta s + \int_{x}^{\beta(x)} k(\beta(x), s) d_\beta s$$

$$= \int_{s_0}^{x} (k(\beta(x), s) - k(x, s)) d_\beta s + (\beta(x) - x) k(\beta(x), x),$$

whereupon

$$D_\beta g(x) = \frac{g(\beta(x)) - g(x)}{\beta(x) - x}$$

$$= \int_{x_0}^{x} \frac{k(\beta(x), s) - k(x, s)}{\beta(x) - x} d_\beta s + k(\beta(x), x)$$

$$= k(\beta(x), x) + \int_{s_0}^{x} D_{\beta x} k(x, s) d_\beta s.$$

Now, suppose $x = s_0$. For $h \neq 0$ such that $s_0 + h \in I$, we have

$$g(s_0 + h) = \sum_{k=0}^{\infty} (\beta^k(s_0 + h) - \beta^{k+1}(s_0 + h)) k(s_0 + h, \beta^k(s_0 + h))$$

$$= \sum_{k=0}^{\infty} (\beta^k(s_0 + h) - \beta^{k+1}(s_0 + h))(k(s_0 + h, \beta^k(s_0 + h)) - k(s_0, s_0))$$

$$+ \sum_{k=0}^{\infty} (\beta^k(s_0 + h) - \beta^{k+1}(s_0 + h)) k(s_0, s_0).$$

Therefore,

$$\frac{g(s_0 + h)}{h} = \sum_{k=0}^{\infty} \frac{\beta^k(s_0 + h) - \beta^{k+1}(s_0 + h)}{h} \times (k(s_0 + h, \beta^k(s_0 + h)) - k(s_0, s_0))$$

$$+ \sum_{k=0}^{\infty} \frac{\beta^k(s_0 + h) - \beta^{k+1}(s_0 + h)}{h} k(s_0, s_0).$$

Hence,

$$\lim_{h \to 0} \frac{g(s_0 + h)}{h} = \lim_{h \to 0} \sum_{k=0}^{\infty} \frac{\beta^k(s_0 + h) - \beta^{k+1}(s_0 + h)}{h} \times (k(s_0 + h, \beta^k(s_0 + h)) - k(s_0, s_0))$$

$$+ \lim_{h \to 0} \sum_{k=0}^{\infty} \frac{\beta^k(s_0 + h) - \beta^{k+1}(s_0 + h)}{h} k(s_0, s_0)$$

$$= 0 + k(s_0, s_0)$$

$$= k(s_0, s_0).$$

This completes the proof. $\qquad\Box$

2.5 General quantum monomials

We introduce the general quantum monomials $h_k : I \times I \to \mathbb{R}$, $k \in \mathbb{N}_0$, by

$$h_0(x, y) = 1,$$

$$h_k(x, y) = \int_y^x h_{k-1}(z, y)d_\beta z, \quad k \in \mathbb{N}, \quad x, y \in I.$$

Note that

$$D_{\beta x} h_k(x, y) = h_{k-1}(x, y), \quad k \in \mathbb{N}, \quad x, y \in I.$$

Example 2.5.1. Let $I = \mathbb{R}$, and let

$$\beta(t) = \frac{1}{2}t + 1, \quad t \in I.$$

Then, for $x, y \in I$, we have

$$h_0(x, y) = 1,$$

$$h_1(x, y) = \int_y^x h_0(z, y)d_\beta z$$

$$= \int_y^x d_\beta z,$$

$$= x - y, \quad \text{and}$$

$$h_2(x, y) = \int_y^x h_1(z, y)d_\beta z$$

$$= \int_y^x (z - y)d_\beta z$$

$$= \int_y^x z d_\beta z - y \int_y^x d_\beta z$$

$$= \int_2^x z d_\beta z - \int_2^y z d_\beta z - y(x - y)$$

$$= \left(\left(1 - \frac{1}{2}\right)x - 1\right)\sum_{k=0}^{\infty}\left(\frac{1}{2}\right)^k\left(\left(\frac{1}{2}\right)^k x + \frac{1 - \left(\frac{1}{2}\right)^k}{1 - \frac{1}{2}}\right)$$

$$- \left(\left(1 - \frac{1}{2}\right)y - 1\right)\sum_{k=0}^{\infty}\left(\frac{1}{2}\right)^k\left(\left(\frac{1}{2}\right)^k y + \frac{1 - \left(\frac{1}{2}\right)^k}{1 - \frac{1}{2}}\right) - y(x - y)$$

$$= \left(\frac{x}{2} - 1\right)\sum_{k=0}^{\infty}\left(\frac{1}{2}\right)^k\left(\left(\frac{1}{2}\right)^k x + 2 - 2\left(\frac{1}{2}\right)^k\right)$$

$$- \left(\frac{y}{2} - 1\right)\sum_{k=0}^{\infty}\left(\frac{1}{2}\right)^k\left(\left(\frac{1}{2}\right)^k y + 2 - 2\left(\frac{1}{2}\right)^k\right) - y(x - y)$$

$$= \left(\frac{x}{2} - 1\right)\left(\sum_{k=0}^{\infty}\left(\frac{1}{4}\right)^k x + 2\sum_{k=0}^{\infty}\left(\frac{1}{2}\right)^k - 2\sum_{k=0}^{\infty}\left(\frac{1}{4}\right)^k\right)$$

$$- \left(\frac{y}{2} - 1\right)\left(\sum_{k=0}^{\infty}\left(\frac{1}{4}\right)^k y + 2\sum_{k=0}^{\infty}\left(\frac{1}{2}\right)^k - 2\sum_{k=0}^{\infty}\left(\frac{1}{4}\right)^k\right) - y(x - y)$$

$$= \left(\frac{x}{2} - 1\right)\left(\frac{1}{1 - \frac{1}{4}}x + \frac{2}{1 - \frac{1}{2}} - \frac{2}{1 - \frac{1}{4}}\right)$$

$$- \left(\frac{y}{2} - 1\right)\left(\frac{1}{1 - \frac{1}{4}}y + \frac{2}{1 - \frac{1}{2}} - \frac{2}{1 - \frac{1}{4}}\right) - y(x - y)$$

$$= \left(\frac{x}{2} - 1\right)\left(\frac{4}{3}x + 4 - \frac{8}{3}\right) - \left(\frac{y}{2} - 1\right)\left(\frac{4}{3}y + 4 - \frac{8}{3}\right) - y(x - y)$$

$$= \left(\frac{x}{2} - 1\right)\left(\frac{4}{3}x + \frac{4}{3}\right) - \left(\frac{y}{2} - 1\right)\left(\frac{4}{3}y + \frac{4}{3}\right) - y(x - y)$$

$$= \frac{2}{3}((x - 2)(x + 1) - (y - 2)(y + 1)) - y(x - y)$$

$$= \frac{2}{3}(x^2 + x - 2x - 2 - y^2 - y + 2y + 2) - y(x - y)$$

$$= \frac{2}{3}(x^2 - y^2 - x + y) - y(x - y)$$

$$= \frac{2}{3}(x - y)(x + y - 1) - y(x - y)$$

$$= \frac{2}{3}(x - y)\left(x + y - 1 - \frac{3}{2}y\right)$$

$$= \frac{2}{3}(x - y)\left(x - \frac{y}{2} - 1\right), \quad x, y \in I.$$

Theorem 2.5.2. *We have*

$$h_{k+m+1}(x, s_0) = \int_{s_0}^{x} h_k(x, \beta(y))h_m(y, s_0)d_\beta y, \quad x \in I.$$

Proof. Let

$$g(x) = \int_{s_0}^{x} h_k(x, \beta(y))h_m(y, s_0)d_\beta y, \quad x \in I.$$

Then

$$D_\beta g(x) = h_k(\beta(x), \beta(x))h_k(x, s_0) + \int_{s_0}^{x} D_{\beta x}h_k(x, \beta(y))h_m(y, s_0)d_\beta y$$

$$= \int_{s_0}^{x} h_{k-1}(x, \beta(y))h_m(y, s_0)d_\beta y$$

$$\vdots$$

$$D_\beta^k g(x) = \int_{s_0}^{x} h_m(y, s_0)d_\beta y$$

$$= h_{m+1}(x, s_0), \quad x \in I.$$

Hence,

$$D_\beta^{k-1} g(x) - D_\beta^{k-1} g(s_0) = \int_{s_0}^x h_{m+1}(y, s_0) d_\beta y$$

$$= h_{m+2}(x, s_0), \quad x \in I,$$

that is,

$$D_\beta^{k-1} g(x) = h_{m+2}(x, s_0), \quad x \in I.$$

Continuing, we get

$$D_\beta g(x) = h_{k+m}(x, s_0), \quad x \in I,$$

whereupon

$$g(x) - g(s_0) = \int_{s_0}^x h_{k+m}(y, s_0) d_\beta y$$

$$= h_{k+m+1}(x, s_0), \quad x \in I,$$

that is,

$$g(x) = h_{k+m+1}(x, s_0), \quad x \in I.$$

This completes the proof. $\square$

Theorem 2.5.3. *For $k \in \mathbb{N}_0$, we have*

$$h_k(x, s_0) \geq \frac{(x - s_0)^k}{k!}, \quad k \in \mathbb{N}, \quad x \geq s_0, \quad x \in I. \tag{2.6}$$

Proof. Let

$$g(x) = (x - s_0)^{k+1}, \quad x \in I, \quad x \geq s_0.$$

We have

$$D_\beta g(x) = \sum_{l=0}^k (\beta(x) - s_0)^{k-l}(x - s_0)^l, \quad x \in I, \quad x \geq s_0.$$

Note that (2.6) holds for $k = 0$ and $k = 1$. Assume that (2.6) holds for some $k \in \mathbb{N}$. We will prove the inequality for $k + 1$. We have

$$h_{k+1}(x, s_0) = \int_{s_0}^x h_k(z, s_0) d_\beta z$$

$$\geq \frac{1}{k!} \int_{s_0}^x (z - s_0)^k d_\beta z$$

$$= \frac{1}{(k+1)!} \int_{s_0}^x \sum_{j=0}^k (z - s_0)^k d_\beta z$$

$$= \frac{1}{(k+1)!} \int_{s_0}^{x} \sum_{j=0}^{k} (z-s_0)^{k-j}(z-s_0)^j \, d_\beta z$$

$$\geq \frac{1}{(k+1)!} \int_{s_0}^{x} \sum_{j=0}^{k} (\beta(z)-s_0)^{k-j}(z-s_0)^j \, d_\beta z$$

$$= \frac{1}{(k+1)!} \int_{s_0}^{x} D_\beta g(z) d_\beta z$$

$$= \frac{1}{(k+1)!} g(z) \Big|_{z=s_0}^{z=x}$$

$$= \frac{(x-s_0)^{k+1}}{(k+1)!}, \qquad x \geq s_0, \quad x \in I.$$

By the principle of mathematical induction, it follows that (2.6) holds for all $k \in \mathbb{N}$. This completes the proof. $\qquad\square$

Theorem 2.5.4. *The functions $h_n(\cdot, \cdot)$, $n \in \mathbb{N}$, satisfy the relationship*

$$h_n(\beta^k(t), t) = 0$$

for all $t \in I$, $n \in \mathbb{N}$, and $0 \leq k \leq n-1$.

Proof. Let $n \in \mathbb{N}$. Then

$$h_n(\beta^0(t), t) = h_n(t, t)$$

$$= \int_t^t h_{n-1}(z, t) d_\beta z$$

$$= 0.$$

Assume that

$$h_{n-1}(\beta^k(t), t) = 0 \quad \text{and} \quad h_n(\beta^k(t), t) = 0$$

for some $k \in \{0, \ldots, n-2\}$. We will prove that

$$h_n(\beta^{k+1}(t), t) = 0.$$

In view of Theorem 1.3.2, we have

$$h_n(\beta^{k+1}(t), t) = h_n(\beta^k(t), t) + (\beta(t) - t) D_\beta h_n(\beta^k(t), t).$$

Then

$$h_n(\beta^{k+1}(t), t) = h_n(\beta^k(t), t) + (\beta(t) - t) h_{n-1}(\beta^k(t), t)$$

$$= 0.$$

This completes the proof. $\qquad\square$

Remark 2.5.5. For $t, s \in I$, define the general quantum monomials

$$g_0(t, s) = 1,$$

$$g_n(t, s) = \int_s^t g_{n-1}(\beta(z), s)\, d_\beta z, \quad n \in \mathbb{N}.$$

Then we have

$$D_\beta g_n(t, s) = g_{n-1}(\beta(t), s), \quad t, s \in I, \quad n \in \mathbb{N}.$$

2.6 β-Taylor formula

Theorem 2.6.1. *Let $m \in \mathbb{N}$ and suppose that f is m-times β-differentiable at $\beta^{m-1}(t)$. Then*

$$f(t) = \sum_{k=0}^{m-1} (-1)^k D_\beta^k f(\beta^{m-1-k}(t)) h_k(\beta^{m-1}(t), t), \quad t \in I.$$

Proof. For $m = 1$, we have

$$\sum_{k=0}^{0} (-1)^k D_\beta^k f(\beta^{0-k}(t)) h_k(\beta^0(t), t) = (-1)^0 f(t) h_0(t, t)$$

$$= f(t), \quad t \in I.$$

Therefore the result holds for $m = 1$. Assume that

$$f(t) = \sum_{k=0}^{m-1} (-1)^k D_\beta^k f(\beta^{m-1-k}(t)) h_k(\beta^{m-1}(t), t), \quad t \in I,$$

for some $m \in \mathbb{N}$. We will prove the result for $m + 1$, that is,

$$f(t) = \sum_{k=0}^{m} (-1)^k D_\beta^k f(\beta^{m-k}(t)) h_k(\beta^m(t), t), \quad t \in I.$$

Using the fact that

$$h_k(\beta^m(t), t) = h_k(\beta^{m-1}(t), t) + (\beta(t) - t) h_{k-1}(\beta^{m-1}(t), t), \quad t \in I,$$

and

$$h_m(\beta^{m-1}(t), t) = 0, \quad t \in I,$$

we obtain

$$\sum_{k=0}^{m} (-1)^k D_\beta^k f(\beta^{m-k}(t)) h_k(\beta^m(t), t)$$

$$= f(\beta^m(t)) + \sum_{k=1}^{m} (-1)^k D_\beta^k f(\beta^{m-k}(t)) h_k(\beta^m(t), t)$$

$$= f(\beta^m(t)) + \sum_{k=1}^{m}(-1)^k D_\beta^k f(\beta^{m-k}(t))h_k(\beta^{m-1}(t),t)$$

$$+ (\beta(t) - t)\sum_{k=1}^{m}(-1)^k D_\beta^k f(\beta^{m-k}(t))h_{k-1}(\beta^{m-1}(t),t)$$

$$= f(\beta^m(t)) + \sum_{k=1}^{m-1}(-1)^k D_\beta^k f(\beta^{m-k}(t))h_k(\beta^{m-1}(t),t)$$

$$+ (-1)^m(\beta(t) - t)D_\beta^m f(t)h_m(\beta^{m-1}(t),t)$$

$$+ (\beta(t) - t)\sum_{k=0}^{m-1}(-1)^{k+1} D_\beta^{k+1} f(\beta^{m-k-1}(t))h_k(\beta^{m-1}(t),t)$$

$$= \sum_{k=0}^{m-1}(-1)^k D_\beta^k f(\beta^{m-k}(t))h_k(\beta^{m-1}(t),t)$$

$$- (\beta(t) - t)\sum_{k=0}^{m-1}(-1)^k D_\beta^{k+1} f(\beta^{m-k-1}(t))h_k(\beta^{m-1}(t),t)$$

$$= \sum_{k=0}^{m-1}(-1)^k \left(D_\beta^k f(\beta^{m-k}(t)) - (\beta(t) - t)D_\beta^{k+1} f(\beta^{m-k-1}(t)) \right) \times h_k(\beta^{m-1}(t),t)$$

$$= \sum_{k=0}^{m-1}(-1)^k D_\beta^k f\left(\beta^{m-1-k}(t) \right) h_k(\beta^{m-1}(t),t)$$

$$= f(t), \quad t \in I.$$

Hence, by the principle of mathematical induction, the result holds for all $m \in \mathbb{N}$. This completes the proof. $\qquad\square$

Theorem 2.6.2. *Let $n \in \mathbb{N}$. If f is n-times β-differentiable and p_k, $0 \leq k \leq n - 1$, are β-differentiable at some $t \in I$ with*

$$D_\beta p_{k+1}(t) = p_k(t), \quad 0 \leq k \leq n - 2, \quad n \in \mathbb{N} \setminus \{1\},$$

then

$$D_\beta\left(\sum_{k=0}^{n-1}(-1)^k D_\beta^k f\left(\beta^{-k}(\cdot) \right) p_k(\cdot) \right)(t) = (-1)^{n-1} D_\beta^n f\left(\beta^{-(n-1)}(t) \right) p_{n-1}(t) + f(\beta(t))D_\beta p_0(t), \quad t \in I.$$

Proof. We have

$$D_\beta\left(\sum_{k=0}^{n-1}(-1)^k D_\beta^k f\left(\beta^{-k}(\cdot) \right) p_k(\cdot) \right)(t)$$

$$= \sum_{k=0}^{n-1}(-1)^k D_\beta\left(D_\beta^k f\left(\beta^{-k}(\cdot) \right) p_k(\cdot) \right)(t)$$

$$= \sum_{k=0}^{n-1}(-1)^k \left(D_\beta^{k+1} f\left(\beta^{-k}(t)\right) p_k(t) + D_\beta^k f\left(\beta^{1-k}(t)\right) D_\beta p_k(t) \right)$$

$$= \sum_{k=0}^{n-1}(-1)^k D_\beta^{k+1} f\left(\beta^{-k}(t)\right) p_k(t) + \sum_{k=0}^{n-1}(-1)^k D_\beta^k f\left(\beta^{1-k}(t)\right) D_\beta p_k(t)$$

$$= \sum_{k=0}^{n-2}(-1)^k D_\beta^{k+1} f\left(\beta^{-k}(t)\right) p_k(t) + (-1)^{n-1} D_\beta^n f\left(\beta^{-(n-1)}(t)\right) p_{n-1}(t)$$

$$+ \sum_{k=1}^{n-1}(-1)^k D_\beta^k f\left(\beta^{1-k}(t)\right) D_\beta p_k(t) + f(\beta(t)) D_\beta p_0(t)$$

$$= \sum_{k=0}^{n-2}(-1)^k D_\beta^{k+1} f\left(\beta^{-k}(t)\right) p_k(t) - \sum_{k=0}^{n-2}(-1)^k D_\beta^{k+1} f\left(\beta^{-k}(t)\right) p_{k+1}(t)$$

$$+ (-1)^{n-1} D_\beta^n f\left(\beta^{-(n-1)}(t)\right) p_{n-1}(t) + f(\beta(t)) D_\beta p_0(t)$$

$$= (-1)^{n-1} D_\beta^n f\left(\beta^{-(n-1)}(t)\right) p_{n-1}(t) + f(\beta(t)) D_\beta p_0(t), \quad t \in I. \qquad \square$$

Theorem 2.6.3 (β-Taylor formula I). *Let $n \in \mathbb{N}$. Suppose that f is n-times differentiable on I, and let $a, t \in I$. Then*

$$f(t) = \sum_{k=0}^{n-1}(-1)^k D_\beta^k f\left(\beta^{-k}(a)\right) h_k(a, t) + (-1)^{n-1} \int_a^{\beta^{n-1}(t)} D_\beta^n f\left(\beta^{-(n-1)}(\tau)\right) h_{n-1}(\tau, t) d_\beta \tau, \quad t \in I.$$

Proof. Applying Theorem 2.6.2, we obtain

$$D_\beta \left(\sum_{k=0}^{n-1}(-1)^k D_\beta^k f\left(\beta^{-k}(\cdot)\right) h_k(\cdot, t) \right)(\tau) = (-1)^{n-1} D_\beta^n f\left(\beta^{-(n-1)}(\tau)\right) h_{n-1}(\tau, t) + f(\beta(\tau)) D_\beta h_0(\tau, t)$$

$$= (-1)^{n-1} D_\beta^n f\left(\beta^{-(n-1)}(\tau)\right) h_{n-1}(\tau, t), \quad \tau, t \in I.$$

Now, integrating the last equation from a to $\beta^{n-1}(t)$, we get

$$\int_a^{\beta^{n-1}(t)} D_\beta \left(\sum_{k=0}^{n-1}(-1)^k D_\beta^k f\left(\beta^{-k}(\cdot)\right) h_k(\cdot, t) \right)(\tau) d_\beta \tau$$

$$= (-1)^{n-1} \int_a^{\beta^{n-1}(t)} D_\beta^n f\left(\beta^{-(n-1)}(\tau)\right) h_{n-1}(\tau, t) d_\beta \tau, \quad t \in I,$$

that is,

$$\sum_{k=0}^{n-1}(-1)^k D_\beta^k f\left(\beta^{n-1-k}(t)\right) h_k(\beta^{n-1}(t), t) - \sum_{k=0}^{n-1}(-1)^k D_\beta^k f\left(\beta^{-k}(a)\right) h_k(a, t)$$

$$= (-1)^{n-1} \int_a^{\beta^{n-1}(t)} D_\beta^n f\left(\beta^{-(n-1)}(\tau)\right) h_{n-1}(\tau, t) d_\beta \tau, \quad t \in I.$$

In view of Theorem 2.6.1, we get

$$f(t) - \sum_{k=0}^{n-1}(-1)^k D_\beta^k f\left(\beta^{-k}(a)\right) h_k(a,t) = (-1)^{n-1}\int_a^{\beta^{n-1}(t)} D_\beta^n f\left(\beta^{-(n-1)}(\tau)\right) h_{n-1}(\tau,t)d_\beta\tau, \quad t\in I.$$

This completes the proof. $\qquad\square$

Theorem 2.6.4. *The functions g_n and h_n satisfy the relationship*

$$h_n(t,s) = (-1)^n g_n(s,t), \quad s,t\in I.$$

Proof. Let $n\in\mathbb{N}$, and let

$$f(\tau) = g_n(\tau,s), \quad \tau,s\in I.$$

Then we have

$$D_\beta f(\tau) = g_{n-1}(\beta(\tau),s),$$
$$D_\beta^2 f(\tau) = g_{n-2}(\beta^2(\tau),s),$$

$$\vdots$$

$$D_\beta^k f(\tau) = g_{n-k}(\beta^k(\tau),s), \quad 0\le k\le n-1,$$
$$D_\beta^n f(\tau) = g_0(\beta^n(\tau),s)$$
$$= 1, \quad \tau,s\in I.$$

Hence,

$$D_\beta^k f(\beta^{-k}(s)) = g_{n-k}(\beta^k(\beta^{-k}(s)),s)$$
$$= g_{n-k}(s,s)$$
$$= 0, \quad 0\le k\le n-1.$$

Now, applying Theorem 2.6.3 for $a=s$, we get

$$f(t) = g_n(t,s)$$
$$= \sum_{k=0}^{n}(-1)^k D_\beta^k f(\beta^{-k}(s))h_k(s,t)$$
$$+ (-1)^n\int_s^{\beta^n(t)} D_\beta^{n+1} f(\beta^{-n}(\tau))h_{n+1}(\tau,t)d_\beta\tau$$
$$= \sum_{k=0}^{n-1}(-1)^k D_\beta^k f(\beta^{-k}(s))h_k(s,t) + (-1)^n D_\beta f(\beta^{-n}(s))h_n(s,t)$$
$$= (-1)^n h_n(s,t).$$

This completes the proof. $\qquad\square$

By Theorem 2.6.4 we get the following β-Taylor formula.

Theorem 2.6.5 (β-Taylor formula II). *Let $n \in \mathbb{N}$. Suppose that f is n-times differentiable on I, and let $a, t \in I$. Then*

$$f(t) = \sum_{k=0}^{n-1} D_\beta^k f\left(\beta^{-k}(a)\right) g_k(t, a) + \int_a^{\beta^{n-1}(t)} D_\beta^n f\left(\beta^{-(n-1)}(\tau)\right) g_{n-1}(t, \tau) d_\beta \tau, \quad t \in I,$$

where g_n, $n \in \mathbb{N}$, are defined in Remark 2.5.5.

2.7 Improper β-integrals of the first-kind

Throughout this section, suppose that $I = [a, \infty)$, where $a \in \mathbb{R}$, and $s_0 \in I$.

Definition 2.7.1. Suppose that a function $f : I \to \mathbb{R}$ is β-integrable from a to every point $A \in I$, $A \geq a$. Assume that the β-integral

$$F(A) = \int_a^A f(t) d_\beta t$$

has a finite limit as A tends to ∞. Then this limit is called the improper β-integral of f of the first-kind from a to ∞, and we write

$$\int_a^\infty f(t) d_\beta t = \lim_{A \to \infty} \int_a^A f(t) d_\beta t. \tag{2.7}$$

In this case, we say that the improper β-integral

$$\int_a^\infty f(t) d_\beta t \tag{2.8}$$

exists or that it is convergent. Otherwise, we say that the improper β-integral (2.8) is divergent.

Example 2.7.2. Let $I = [1, \infty)$, and let

$$\beta(t) = \frac{1}{2} t + 1, \quad t \in I.$$

Here $s_0 = 2$. Consider the improper β-integral

$$J = \int_1^\infty \frac{3t + 2}{(1 + t^2)(t^2 + 4t + 8)} d_\beta t.$$

We will investigate the improper β-integral J for its convergence. Let

$$f(t) = \frac{1}{1 + t^2}, \quad t \in I.$$

For $t \neq 2$, we have

$$D_\beta f(t) = \frac{f(\beta(t)) - f(t)}{\beta(t) - t}$$

$$= \frac{\frac{1}{(\beta(t))^2+1} - \frac{1}{1+t^2}}{\beta(t) - t}$$

$$= \frac{\frac{t^2+1-(\beta(t))^2-1}{(t^2+1)((\beta(t))^2+1)}}{\beta(t) - t}$$

$$= -\frac{(\beta(t))^2 - t^2}{(\beta(t) - t)(t^2 + 1)((\beta(t))^2 + 1)}$$

$$= -\frac{(\beta(t) - t)(\beta(t) + t)}{(\beta(t) - t)(t^2 + 1)((\beta(t))^2 + 1)}$$

$$= -\frac{\beta(t) + t}{(t^2 + 1)((\beta(t))^2 + 1)}$$

$$= -\frac{\frac{1}{2}t + 1 + t}{(t^2 + 1)\left(\left(\frac{1}{2}t + 1\right)^2 + 1\right)}$$

$$= -\frac{\frac{3}{2}t + 1}{(t^2 + 1)\left(\frac{1}{4}t^2 + t + 1 + 1\right)}$$

$$= -\frac{\frac{1}{2}(3t + 2)}{\frac{1}{4}(t^2 + 1)(t^2 + 4t + 8)}$$

$$= -\frac{2(3t + 2)}{(t^2 + 1)(t^2 + 4t + 8)}.$$

Next, since

$$f'(t) = -\frac{2t}{(1 + t^2)^2}, \quad t \in I,$$

we have

$$f'(2) = -\frac{4}{25}.$$

Therefore

$$D_\beta f(t) = -\frac{2(3t + 2)}{(t^2 + 1)(t^2 + 4t + 8)}, \quad t \in I,$$

which yields

$$\frac{3t + 2}{(1 + t^2)(t^2 + 4t + 8)} = -\frac{1}{2} D_\beta f(t), \quad t \in I.$$

Consequently,

$$J = \int_1^\infty \frac{3t + 2}{(1 + t^2)(t^2 + 4t + 8)} d_\beta t$$

$$= \lim_{A \to \infty} \int_1^A \frac{3t + 2}{(1 + t^2)(t^2 + 4t + 8)} d_\beta t$$

$$= -\frac{1}{2} \lim_{A \to \infty} \int_1^A D_\beta f(t) \, d_\beta t$$

$$= -\frac{1}{2} \lim_{A \to \infty} \left(f(t) \Big|_{t=1}^{t=A} \right)$$

$$= -\frac{1}{2} \left(\frac{1}{1+t^2} \Big|_{t=1}^{t=A} \right)$$

$$= -\frac{1}{2} \lim_{A \to \infty} \left(\frac{1}{1+A^2} - \frac{1}{2} \right)$$

$$= -\frac{1}{2} \left(-\frac{1}{2} \right)$$

$$= \frac{1}{4}.$$

Thus, the improper β-integral J is convergent.

Example 2.7.3. Let $I = [1, \infty)$, and let

$$\beta(t) = \frac{2}{3}t + 1, \quad t \in I.$$

Consider the improper β-integral

$$J = \int_1^\infty \frac{5t+3}{t^2(2t+3)^2} \, d_\beta t.$$

Here $s_0 = 3$. We will investigate the improper β-integral J for its convergence. Let

$$f(t) = \frac{1}{t^2}, \quad t \in I.$$

For $t \neq 3$, we have

$$D_\beta f(t) = \frac{f(\beta(t)) - f(t)}{\beta(t) - t}$$

$$= \frac{\frac{1}{(\beta(t))^2} - \frac{1}{t^2}}{\beta(t) - t}$$

$$= \frac{t^2 - (\beta(t))^2}{t^2(\beta(t))^2(\beta(t) - t)}$$

$$= -\frac{(\beta(t) - t)(\beta(t) + t)}{t^2(\beta(t))^2(\beta(t) - t)}$$

$$= -\frac{\beta(t) + t}{t^2(\beta(t))^2}$$

$$= -\frac{\frac{2}{3}t + 1 + t}{t^2 \left(\frac{2}{3}t + 1 \right)^2}$$

$$= -\frac{\frac{1}{3}(5t+3)}{\frac{t^2}{9}(2t+3)^2}$$

$$= -\frac{3(5t+3)}{t^2(2t+3)^2}.$$

Next, since

$$f'(t) = -\frac{2}{t^3}, \quad t \in I,$$

we have

$$f'(3) = -\frac{2}{27}.$$

Therefore

$$D_\beta f(t) = -\frac{3(5t+3)}{t^2(2t+3)^2}, \quad t \in I,$$

which yields

$$\frac{5t+3}{t^2(2t+3)^2} = -\frac{1}{3}D_\beta f(t), \quad t \in I.$$

Consequently,

$$J = \int_1^\infty \frac{5t+3}{t^2(2t+3)^2}d_\beta t$$

$$= \lim_{A \to \infty} \int_1^A \frac{5t+3}{t^2(2t+3)^2}d_\beta t$$

$$= -\frac{1}{3} \lim_{A \to \infty} \int_1^A D_\beta f(t)d_\beta t$$

$$= -\frac{1}{3} \lim_{A \to \infty} \left(f(t)\Big|_{t=1}^{t=A} \right)$$

$$= -\frac{1}{3} \lim_{A \to \infty} \left(\frac{1}{t^2}\Big|_{t=1}^{t=A} \right)$$

$$= -\frac{1}{3} \lim_{A \to \infty} \left(\frac{1}{A^2} - 1 \right)$$

$$= \frac{1}{3}.$$

Thus, the improper β-integral J is convergent.

Example 2.7.4. Let $I = [1, \infty)$, and let

$$\beta(t) = \frac{1}{2}t + 1, \quad t \in I.$$

Here $s_0 = 2$. Consider the improper β-integral

$$J = \int_1^\infty \frac{\sqrt{2}}{\sqrt{2t} + \sqrt{t+2}}\, d_\beta t.$$

Let

$$f(t) = \sqrt{t}, \quad t \in I.$$

Then, for $t \neq 2$, we have

$$\begin{aligned}
D_\beta f(t) &= \frac{f(\beta(t)) - f(t)}{\beta(t) - t} \\[2mm]
&= \frac{\sqrt{\beta(t)} - \sqrt{t}}{(\sqrt{\beta(t)} - \sqrt{t})(\sqrt{\beta(t)} + \sqrt{t})} \\[2mm]
&= \frac{1}{\sqrt{\beta(t)} + \sqrt{t}} \\[2mm]
&= \frac{1}{\sqrt{\frac{t+2}{2}} + \sqrt{t}} \\[2mm]
&= \frac{\sqrt{2}}{\sqrt{t+2} + \sqrt{2t}}.
\end{aligned}$$

Next, since

$$f'(t) = \frac{1}{2\sqrt{t}}, \quad t \in I,$$

we have

$$\begin{aligned}
f'(2) &= \frac{1}{2\sqrt{2}} \\[2mm]
&= \frac{\sqrt{2}}{4}.
\end{aligned}$$

Therefore,

$$D_\beta f(t) = \frac{\sqrt{t}}{\sqrt{t+2} + \sqrt{2t}}, \quad t \in I.$$

Hence,

$$\begin{aligned}
J &= \int_1^\infty \frac{\sqrt{2}}{\sqrt{t+2} + \sqrt{2t}}\, d_\beta t \\[2mm]
&= \lim_{A \to \infty} \int_1^A \frac{\sqrt{2}}{\sqrt{t+2} + \sqrt{2t}}\, d_\beta t \\[2mm]
&= \lim_{A \to \infty} \int_1^A D_\beta f(t)\, d_\beta t
\end{aligned}$$

$$= \lim_{A \to \infty} f(t)\Big|_{t=1}^{t=A}$$

$$= \lim_{A \to \infty} \left(\sqrt{t}\,\Big|_{t=1}^{t=A} \right)$$

$$= \lim_{A \to \infty} (\sqrt{A} - 1)$$

$$= \infty.$$

Thus, the improper β-integral J is divergent.

Exercise 2.7.5. Let $I = [1, \infty)$, and let

$$\beta(t) = \frac{2}{3}t + 1, \quad t \in I.$$

Investigate the following improper β-integrals for their convergence and divergence:

1. $\displaystyle \int_1^\infty (t^4 + 2t^2 + t + 3)d_\beta t;$

2. $\displaystyle \int_1^\infty \frac{3}{(t+1)(2t+3)}d_\beta t;$

3. $\displaystyle \int_1^\infty \frac{3}{(2t+5)(4t+21)}d_\beta t.$

Theorem 2.7.6 (The Cauchy criterion). *The improper β-integral* (2.8) *is convergent if and only if for every $\varepsilon > 0$, there exists $A_0 > a$ such that*

$$\left| \int_{A_1}^{A_2} f(t)d_\beta t \right| < \varepsilon \tag{2.9}$$

for all $A_1, A_2 \in I$ satisfying $A_1 > A_0$ and $A_2 > A_0$.

Proof. The convergence of (2.8) is equivalent to the existence of the limit

$$\lim_{A \to \infty} F(A),$$

where $F(A) = \displaystyle \int_a^A f(t)d_\beta t$. Using the Cauchy criterion for the existence of limit of a function, it follows that the existence of (2.8) is equivalent to condition (2.9). This completes the proof. $\qquad \square$

Example 2.7.7. Let $I = [1, \infty)$, and let

$$\beta(t) = \frac{2}{3}t + 1, \quad t \in I.$$

Consider the improper β-integral

$$\int_1^\infty \frac{3}{2(t+5)(t+9)}d_\beta t.$$

Here $s_0 = 3$. Let

$$f(t) = \frac{1}{t+5}, \quad t \in [1, \infty).$$

Then, for $t \neq 3$, we have

$$
\begin{aligned}
D_\beta f(t) &= \frac{f(\beta(t)) - f(t)}{\beta(t) - t} \\
&= \frac{\frac{1}{\beta(t)+5} - \frac{1}{t+5}}{\beta(t) - t} \\
&= \frac{t + 5 - \beta(t) - 5}{(\beta(t) - t)(t+5)(\beta(t)+5)} \\
&= -\frac{\beta(t) - t}{(\beta(t) - t)(t+5)(\beta(t)+5)} \\
&= -\frac{1}{(t+5)\left(\frac{2}{3}t + 6\right)} \\
&= -\frac{3}{(t+5)(2t+18)} \\
&= -\frac{3}{2(t+5)(t+9)}.
\end{aligned}
$$

Next, since

$$f'(t) = -\frac{1}{(t+5)^2}, \quad t \in I,$$

we have

$$f'(3) = -\frac{1}{64}.$$

Therefore,

$$D_\beta f(t) = -\frac{3}{2(t+5)(t+9)}, \quad t \in I.$$

Let $\varepsilon > 0$. Take

$$A > \max\left\{\frac{2 - 5\varepsilon}{\varepsilon}, 1\right\}.$$

Then $A > 0$, and

$$A > \frac{2 - 5\varepsilon}{\varepsilon},$$

that is,

$$A\varepsilon > 2 - 5\varepsilon,$$

$$(A + 5)\varepsilon > 2,$$

and, finally,

$$\frac{2}{A+5} < \varepsilon.$$

Let $A_1, A_2 > A$. Then

$$\left| \int_{A_1}^{A_2} \frac{3}{2(t+5)(t+9)} d_\beta t \right| = \left| -\int_{A_1}^{A_2} D_\beta F(t) d_\beta t \right|$$

$$= \left| -F(t) \Big|_{t=A_1}^{t=A_2} \right|$$

$$= |F(A_1) - F(A_2)|$$

$$= \left| \frac{1}{A_1+5} - \frac{1}{A_2+5} \right|$$

$$< \frac{1}{A_1+5} + \frac{1}{A_2+5}$$

$$< \frac{1}{A+5} + \frac{1}{A+5}$$

$$= \frac{2}{A+5}$$

$$< \varepsilon.$$

Now applying the Cauchy criterion, it follows that the considered improper β-integral is convergent.

Exercise 2.7.8. Let $I = \mathbb{R}$, and let

$$\beta(t) = \frac{9}{10}t + 1, \quad t \in I.$$

Using the Cauchy criterion, prove that the following improper β-integrals are convergent:

1. $\displaystyle\int_1^\infty \frac{1}{(t+2)(9t+30)} d_\beta t;$

2. $\displaystyle\int_1^\infty \frac{1}{(2t+3)(9t+25)} d_\beta t;$

3. $\displaystyle\int_1^\infty \frac{1}{(t+7)(9t+80)} d_\beta t.$

Definition 2.7.9. The improper β-integral (2.8) is said to be absolutely convergent provided the improper β-integral

$$\int_a^\infty |f(t)| d_\beta t \tag{2.10}$$

is convergent.

Theorem 2.7.10. *If the improper β-integral (2.8) is absolutely convergent, then it is convergent.*

Proof. Suppose (2.8) is absolutely convergent. Then the improper β-integral (2.10) is convergent. Hence, applying the Cauchy criterion, it follows that for every $\varepsilon > 0$, there exists $A > a$ such that for all $A_1, A_2 > A$, we have

$$\left| \int_{A_1}^{A_2} |f(t)| d_\beta t \right| < \varepsilon.$$

Thus, for $A_1, A_2 > A$,

$$\left| \int_{A_1}^{A_2} f(t) d_\beta t \right| \leq \left| \int_{A_1}^{A_2} |f(t)| d_\beta t \right|$$

$$< \varepsilon.$$

This completes the proof. $\qquad\square$

Theorem 2.7.11. *The improper β-integral (2.8) with $f(t) \geq 0$ for $t \geq a$ is convergent if and only if there exists a constant $M > 0$ such that*

$$\int_a^A f(t) d_\beta t \leq M, \quad A \geq a.$$

Proof. Let

$$F(A) = \int_a^A f(t) d_\beta t.$$

Suppose there exists $M > 0$ such that

$$F(A) \leq M, \quad A \geq a.$$

Then

$$0 \leq \int_a^\infty f(t) d_\beta t$$

$$= \lim_{A \to \infty} F(A)$$

$$\leq M.$$

Hence, (2.8) is convergent. Conversely, suppose that (2.8) is convergent. Assume that $F(A)$, $A \geq a$, is unbounded. Then

$$\int_a^\infty f(t) d_\beta t = \lim_{A \to \infty} F(A)$$

$$= \infty,$$

which is contradiction. This completes the proof. $\qquad\square$

The β-analogue of the direct comparison test for the improper β-integral of the first-kind is presented by the following.

Theorem 2.7.12. *Assume that*

$$0 \le f(t) \le g(t), \quad t \in [a, \infty). \tag{2.11}$$

Then the convergence of the improper β-integral

$$\int_a^\infty g(t)d_\beta t \tag{2.12}$$

implies the convergence of the improper β-integral (2.8), whereas the divergence of the improper β-integral (2.8) implies the divergence of the improper β-integral (2.11).

Proof. From (2.11), it follows that

$$0 \le \int_a^A f(t)d_\beta t$$

$$\le \int_a^A g(t)d_\beta t, \quad A \in [a, \infty).$$

This completes the proof. $\qquad\qquad\qquad\qquad\qquad\qquad\qquad\qquad\square$

Example 2.7.13. Let $I = [1, \infty)$, and let

$$\beta(t) = \frac{1}{2}t + 1, \quad t \in I.$$

Consider the improper β-integral

$$K = \int_1^\infty \frac{3t+2}{(1+t)(1+t^2)^4(t^2+4t+8)^3}d_\beta t.$$

Here

$$f(t) = \frac{3t+2}{(1+t)(1+t^2)^4(t^2+4t+8)^3}, \quad t \in I.$$

Let

$$g(t) = \frac{3t+2}{(1+t^2)(t^2+4t+8)}, \quad t \in I.$$

Then, we have

$$0 \le f(t)$$

$$\le g(t), \quad t \in I.$$

From Example 2.7.2, we have that the improper β-integral

$$\int_1^\infty g(t)d_\beta t$$

is convergent. Now, applying Theorem 2.7.12, we conclude that K is convergent.

Example 2.7.14. Let $I = [1, \infty)$, and let

$$\beta(t) = \frac{2}{3}t + 1, \quad t \in I.$$

Consider the improper β-integral

$$K = \int_1^\infty \frac{5t}{(t+1)^2(2t+3)^4(t+7)^{10}} d_\beta t.$$

Here

$$f(t) = \frac{5t}{(t+1)^2(2t+3)^4(t+7)^{10}}, \quad t \in I.$$

Let

$$g(t) = \frac{5t+3}{t^2(2t+3)^2}, \quad t \in I.$$

Then, we have

$$0 \leq f(t)$$
$$\leq g(t), \quad t \in I.$$

From Example 2.7.3, it follows that the improper β-integral

$$\int_1^\infty g(t)d_\beta t$$

is convergent. Now, applying Theorem 2.7.12, we conclude that K is convergent.

Example 2.7.15. Let $I = [1, \infty)$, and let

$$\beta(t) = \frac{1}{2}t + 1, \quad t \in I.$$

Consider the improper β-integral

$$K = \int_1^\infty \frac{2}{\sqrt{t+2}} d_\beta t.$$

Here

$$f(t) = \frac{2}{\sqrt{t+2}}, \quad t \in I.$$

Let

$$g(t) = \frac{\sqrt{2}}{\sqrt{2t} + \sqrt{t+2}}, \quad t \in I.$$

Then, we have

$$0 \leq g(t)$$
$$\leq f(t), \quad t \in I.$$

From Example 2.7.4, we obtain that the improper β-integral

$$\int_1^\infty g(t)d_\beta t$$

is divergent. Now, applying Theorem 2.7.12, we conclude that the considered improper β-integral is divergent.

Example 2.7.16. Let $I = \mathbb{R}$, and let

$$\beta(t) = \frac{2}{3}t + 1, \quad t \in I.$$

Consider the improper β-integral

$$K = \int_1^\infty \frac{1}{(t+5)^{10}(t+9)^8(t^2+1)}d_\beta t.$$

Here

$$f(t) = \frac{1}{(t+5)^{10}(t+9)^8(t^2+1)}, \quad t \in I.$$

Let

$$g(t) = \frac{3}{2(t+5)(t+9)}, \quad t \in I.$$

Then, we have

$$0 \leq f(t)$$
$$\leq g(t), \quad t \in I.$$

From Example 2.7.7, we note that the improper β-integral

$$\int_1^\infty g(t)d_\beta t$$

is convergent. Now, applying Theorem 2.7.12, we conclude that K is convergent.

Exercise 2.7.17. Let $I = [1, \infty)$, and let

$$\beta(t) = \frac{2}{3}t + 1, \quad t \in I.$$

Investigate the following improper β-integrals for their convergence and divergence:

1. $\displaystyle\int_1^\infty (2t^8 + t^4 + 4t^2 + 2t + 6)d_\beta t;$

2. $\displaystyle\int_1^\infty \frac{1}{(t+1)^2(50t+3)^2}d_\beta t;$

3. $\displaystyle\int_{1}^{\infty} \frac{1}{(2t+5)^2(4t+23)^3(t^2+1)}\,d_\beta t.$

Hint. Use Exercise 2.7.5 and Theorem 2.7.12.

Answer. 1. Divergent.
2. Convergent.
3. Convergent.

Theorem 2.7.18. *Suppose*

$$|f(t)| \le g(t), \quad t \in I, \quad t \ge a.$$

Then the convergence of (2.12) *implies the convergence of* (2.8).

Proof. Since (2.12) is convergent, from Theorem 2.7.12 we get that (2.10) is convergent. Therefore (2.8) is absolutely convergent. Now, applying Theorem 2.7.10, we conclude that (2.8) is convergent. This completes the proof. $\qquad\square$

The β-analogue of the limit comparison test for the improper β-integral of the first-kind is presented in the following:

Theorem 2.7.19 (Comparison criterion). *Let*

$$\int_{a}^{\infty} f(t)d_\beta t \quad and \quad \int_{a}^{\infty} g(t)d_\beta t$$

be improper β-integrals of the first-kind with positive integrands. Suppose that the limit

$$\lim_{t\to\infty} \frac{f(t)}{g(t)} = L \tag{2.13}$$

exists (finite) and is nonzero. Then the improper β-integrals

$$\int_{a}^{\infty} f(t)d_\beta t \quad and \quad \int_{a}^{\infty} g(t)d_\beta t$$

are simultaneously convergent or divergent.

Proof. Let $\varepsilon \in (0, L)$. Keeping in mind (2.13), it follows that there exists $A_0 > a$ such that

$$L - \varepsilon \le \frac{f(t)}{g(t)} \le L + \varepsilon, \quad t \ge A_0.$$

This gives

$$(L - \varepsilon)g(t) \le f(t)$$
$$\le (L + \varepsilon)g(t), \quad t \ge A_0.$$

Therefore,

$$(L - \varepsilon) \int_a^\infty g(t)d_\beta t \leq \int_a^\infty f(t)d_\beta t$$

$$\leq (L + \varepsilon) \int_a^\infty g(t)d_\beta t. \qquad (2.14)$$

Now, we have the following cases.

1. Suppose

$$\int_a^\infty g(t)d_\beta t$$

  is convergent. Then

$$(L + \varepsilon) \int_a^\infty g(t)d_\beta t$$

  is convergent. Hence, using (2.14) and Theorem 2.7.12, we conclude that

$$\int_a^\infty f(t)d_\beta t$$

  is convergent.

2. Suppose

$$\int_a^\infty f(t)d_\beta t$$

  is convergent. Then

$$\frac{1}{L - \varepsilon} \int_a^\infty f(t)d_\beta t$$

  is convergent. Hence, using (2.14) and Theorem 2.7.12, we conclude that

$$\int_a^b g(t)d_\beta t$$

  is convergent.

3. Suppose

$$\int_a^\infty g(t)d_\beta t$$

  is divergent. Then

$$(L - \varepsilon) \int_a^\infty g(t)d_\beta t$$

  is divergent. Hence, using (2.14) and Theorem 2.7.12, we conclude that

$$\int_a^\infty f(t)d_\beta t$$

  is divergent.

4. Suppose

$$\int_a^\infty f(t)\,d_\beta t$$

is divergent. Then

$$\frac{1}{L+\varepsilon}\int_a^\infty f(t)\,d_\beta t$$

is divergent. Hence, using (2.14) and Theorem 2.7.12, we conclude that

$$\int_a^b g(t)\,d_\beta t$$

is divergent.

This completes the proof. $\qquad\square$

Example 2.7.20. Let $I = [1, \infty)$, and let

$$\beta(t) = \frac{1}{2}t + 1, \quad t \in I.$$

Consider the improper β-integral

$$K = \int_1^\infty \frac{t+1}{(2t^2+4)(3t^2+t+8)}\,d_\beta t.$$

Let

$$f(t) = \frac{t+1}{(2t^2+4)(3t^2+t+8)} \quad \text{and}$$

$$g(t) = \frac{3t+2}{(1+t^2)(t^2+4t+8)}, \quad t \in I.$$

Then

$$\frac{f(t)}{g(t)} = \frac{(t+1)(1+t^2)(t^2+4t+8)}{(3t+2)(2t^2+4)(3t^2+t+8)}, \quad t \in I,$$

and

$$\lim_{t\to\infty}\frac{f(t)}{g(t)} = \lim_{t\to\infty}\frac{(t+1)(1+t^2)(t^2+4t+8)}{(3t+2)(2t^2+4)(3t^2+t+8)}$$

$$= \frac{1}{18}$$

$$\neq 0.$$

Thus, in view of Theorem 2.7.19, K and the improper β-integral

$$J = \int_1^\infty \frac{3t+2}{(1+t^2)(t^2+4t+8)}\,d_\beta t$$

are simultaneously convergent or divergent. By Example 2.7.2, we have that J is convergent. Therefore, K is convergent.

Example 2.7.21. Let $I = [1, \infty)$, and let

$$\beta(t) = \frac{2}{3}t + 1, \quad t \in I.$$

Consider the improper β-integral

$$K = \int_1^\infty \frac{t+2}{(t+1)^2(t+4)^2} d_\beta t.$$

Let

$$f(t) = \frac{5t+3}{t^2(2t+3)^2} \quad \text{and}$$

$$g(t) = \frac{t+2}{(t+1)^2(t+4)^2}, \quad t \in I.$$

Then

$$\frac{f(t)}{g(t)} = \frac{(5t+3)(t+1)^2(t+4)^2}{t^2(t+2)(2t+3)^2}, \quad t \in I.$$

Hence,

$$\lim_{t \to \infty} \frac{f(t)}{g(t)} = \lim_{t \to \infty} \frac{f(t)}{g(t)} = \frac{(5t+3)(t+1)^2(t+4)^2}{t^2(t+2)(2t+3)^2}$$

$$= \frac{5}{4}$$

$$\neq 0.$$

Now, applying Theorem 2.7.19, we conclude that K and

$$J = \int_1^\infty \frac{5t+3}{t^2(2t+3)^2} d_\beta t$$

are simultaneously convergent or divergent. By Example 2.7.3, we have that J is convergent, and thus K is convergent.

Example 2.7.22. Let $I = [1, \infty)$, and let

$$\beta(t) = \frac{1}{2}t + 1, \quad t \in I.$$

Consider the improper β-integral

$$K = \int_1^\infty \frac{1}{\sqrt{t+2}} d_\beta t.$$

Let

$$f(t) = \frac{1}{\sqrt{t+2}} \quad \text{and}$$

$$g(t) = \frac{\sqrt{2}}{\sqrt{2t} + \sqrt{t+2}}, \quad t \in I.$$

Then

$$\frac{f(t)}{g(t)} = \frac{\sqrt{2t} + \sqrt{t+2}}{\sqrt{2t+4}}, \quad t \in I.$$

Hence,

$$\lim_{t \to \infty} \frac{f(t)}{g(t)} = \lim_{t \to \infty} \frac{\sqrt{2t} + \sqrt{t+2}}{\sqrt{2t+4}}$$

$$= \frac{\sqrt{2}+1}{\sqrt{2}}$$

$$\neq 0.$$

Now, applying Theorem 2.7.19, we conclude that K and

$$J = \int_1^\infty \frac{\sqrt{2}}{\sqrt{2t} + \sqrt{t+2}} d_\beta t$$

are simultaneously convergent or divergent. By Example 2.7.4, we have that J is divergent, and thus K is divergent.

Exercise 2.7.23. Let $I = [1, \infty)$, and let

$$\beta(t) = \frac{2}{3}t + 1, \quad t \in I.$$

Investigate the following improper β-integrals for their convergence and divergence:

1. $\displaystyle \int_1^\infty (2t^4 + 4t^2 + 2t + 1) d_\beta t;$

2. $\displaystyle \int_1^\infty \frac{1}{(t+1)^2} d_\beta t;$

3. $\displaystyle \int_1^\infty \frac{1}{(t+5)(t+23)} d_\beta t.$

Theorem 2.7.24. *Let f be β-integrable on $[a, A]$ for all $A \geq a$, $A, a \in I$. Suppose that (2.8) is absolutely convergent. Also, suppose that g is monotone on $[a, \infty)$ and*

$$\lim_{t \to \infty} g(t) = 0. \tag{2.15}$$

Then the improper β-integral of the first kind

$$\int_a^\infty f(t)g(t) d_\beta t \tag{2.16}$$

is convergent.

Proof. Let $\varepsilon > 0$. Since (2.8) is absolutely convergent, by the Cauchy criterion (Theorem 2.7.6), it follows that there exists $A_0 \geq 0$ such that

$$\left| \int_{A_1}^{A_2} |f(t)| d_\beta t \right| < \varepsilon$$

for all $A_1, A_2 \geq A_0$. From (2.15), it follows that there exists $A' \geq a$ such that

$$|g(t)| \leq 1, \quad t \geq A'.$$

Let

$$A'' = \max\{A', A_0\}.$$

Then, for all $A_1, A_2 \geq A''$, we have

$$\left| \int_{A_1}^{A_2} f(t)g(t) d_\beta t \right| \leq \left| \int_{A_1}^{A_2} |f(t)||g(t)| d_\beta t \right|$$

$$\leq \left| \int_{A_1}^{A_2} |f(t)| d_\beta t \right|$$

$$< \varepsilon.$$

Now, in view of the Cauchy criterion, we conclude that (2.16) is convergent. This completes the proof. $\square$

Example 2.7.25. Let $I = [1, \infty)$, and let

$$\beta(t) = \frac{1}{2}t + 1, \quad t \in I.$$

Consider the improper β-integral

$$K = \int_1^\infty \frac{3t + 2}{(1 + t^2)(t^2 + 4t + 8)(t^4 + 1)} d_\beta t.$$

Let

$$f(t) = \frac{3t + 2}{(1 + t^2)(t^2 + 4t + 8)} \quad \text{and}$$

$$g(t) = \frac{1}{1 + t^4}, \quad t \in [1, \infty).$$

Then, we have

$$g'(t) = -\frac{4t^3}{(1 + t^4)^2}$$

$$< 0, \quad t \in [1, \infty).$$

Thus, g is decreasing on $[1, \infty)$. Next,

$$\lim_{t \to \infty} g(t) = \lim_{t \to \infty} \frac{1}{1 + t^4}$$
$$= 0.$$

From Example 2.7.2, we have that

$$\int_1^\infty f(t) d_\beta t$$

is convergent. Hence, using Theorem 2.7.24, we conclude that K is convergent.

Example 2.7.26. Let $I = [1, \infty)$, and let

$$\beta(t) = \frac{2}{3}t + 1, \quad t \in I.$$

Consider the improper β-integral

$$K = \int_1^\infty \frac{5t + 3}{t^2(2t + 3)^2(t^4 + 1)} d_\beta t.$$

Let

$$f(t) = \frac{5t + 3}{t^2(2t + 3)^2}, \quad t \in I,$$

and let g be as in Example 2.7.25. From this example, it follows that g satisfies all conditions of Theorem 2.7.24. Therefore, by Example 2.7.3, it follows that

$$\int_1^\infty f(t) d_\beta t$$

is absolutely convergent. Now, applying Theorem 2.7.24, we conclude that K is convergent.

Exercise 2.7.27. Let $I = \mathbb{R}$, and let

$$\beta(t) = \frac{9}{10}t + 1, \quad t \in I.$$

Investigate the following improper β-integrals for their convergence and divergence:

1. $\displaystyle \int_1^\infty \frac{1}{(t + 2)(9t + 30)t^3} d_\beta t;$

2. $\displaystyle \int_1^\infty \frac{1}{(2t + 3)(9t + 25)(1 + t^8)} d_\beta t;$

3. $\displaystyle \int_1^\infty \frac{1}{(t + 7)(9t + 80)(1 + t^{10})} d_\beta t.$

2.8 Improper β-integrals of the second-kind

Suppose that $a, b \in I$ and $a < b$.

Definition 2.8.1. Let $f : [a, b) \to \mathbb{R}$ be β-integrable on every interval $[a, c]$ with $a < c < b$. Also, let f be unbounded on $[a, b)$. The formal expression

$$\int_a^b f(t)d_\beta t \tag{2.17}$$

is called the improper β-integral of the second-kind. In this case, we say that the β-integral (2.17) is improper at $t = b$. We also say that the point $t = b$ is singularity of f. If the left-sided limit

$$\lim_{c \to b-} \int_a^c f(t)d_\beta t \tag{2.18}$$

exists as a finite number, then the improper β-integral (2.17) is said to be convergent. In such a case, we call this limit the value of the improper β-integral (2.17), and we write

$$\int_a^b f(t)d_\beta t = \lim_{c \to b-} \int_a^c f(t)d_\beta t.$$

Otherwise, the improper β-integral (2.17) is said to be divergent.

Remark 2.8.2. As above, we can define integral (2.17) as the improper β-integral of the second-kind at $t = a$.

Remark 2.8.3. All theorems in the previous section have exact analogues for improper β-integrals of the second-kind.

Example 2.8.4. Let $I = [0, \infty)$, and let

$$\beta(t) = t^3, \quad t \in I.$$

Consider the improper β-integral

$$J = \int_0^{\frac{1}{2}} \frac{1}{t^{\frac{1}{2}}(1+t)} d_\beta t.$$

Here

$$f(t) = \frac{1}{t^{\frac{1}{2}}(t+1)}, \quad t \in I,$$

and $s_0 = 1$. We see that f is unbounded on $\left(0, \dfrac{1}{2}\right]$. Note that J is the improper β-integral of the second-kind. It is improper at 0. Let

$$g(t) = \sqrt{t}, \quad t \in I.$$

Then, for $t \in \left[0, \dfrac{1}{2}\right]$, we have

$$
\begin{aligned}
D_\beta g(t) &= \frac{g(\beta(t)) - g(t)}{\beta(t) - t} \\
&= \frac{\sqrt{\beta(t)} - \sqrt{t}}{\beta(t) - t} \\
&= \frac{\sqrt{\beta(t)} - \sqrt{t}}{(\sqrt{\beta(t)} - \sqrt{t})(\sqrt{\beta(t)} + \sqrt{t})} \\
&= \frac{1}{\sqrt{\beta(t)} + \sqrt{t}} \\
&= \frac{1}{\sqrt{t^3} + \sqrt{t}} \\
&= \frac{1}{\sqrt{t}(t + 1)} \\
&= f(t).
\end{aligned}
$$

Hence,

$$
\begin{aligned}
J &= \lim_{s \to 0-} \int_s^{\frac{1}{2}} f(t) d_\beta t \\
&= \lim_{s \to 0-} \left(g(t) \Big|_{t=s}^{\left| t=\frac{1}{2} \right.} \right) \\
&= \lim_{s \to 0-} \left(\sqrt{t} \Big|_{t=s}^{\left| t=\frac{1}{2} \right.} \right) \\
&= \lim_{s \to 0-} \left(\frac{1}{\sqrt{2}} - \sqrt{s} \right) \\
&= \frac{1}{\sqrt{2}}.
\end{aligned}
$$

Thus, the improper β-integral J is convergent.

Exercise 2.8.5. Let $I = \left[-\dfrac{1}{2}, \infty\right)$, and let

$$
\beta(t) = t^3, \quad t \in I.
$$

Prove that the improper β-integral

$$
\int_{-\frac{1}{2}}^{0} \frac{1}{t^2 + t\sqrt[3]{t} + t\sqrt[3]{t^2}} d_\beta t
$$

is convergent.

2.9 **Advanced practical problems**

Problem 2.9.1. Let $I = \mathbb{R}$, and let

$$\beta(t) = \frac{2}{3}t + 1,$$

$$f(t) = \frac{10}{3}t - 1, \quad \text{and}$$

$$F(t) = 2t^2 - 3t - 1, \quad t \in I.$$

Then prove that F is a β-antiderivative of f.

Problem 2.9.2. Let $I = \mathbb{R}$, and let

$$\beta(t) = \frac{2}{3}t - 1, \quad t \in I.$$

Compute

1. $\displaystyle\int_{-3}^{-1} (t-1)d_\beta t;$

2. $\displaystyle\int_{-3}^{1} (t-1)d_\beta t;$

3. $\displaystyle\int_{-1}^{1} (t-1)d_\beta t.$

Problem 2.9.3. Let $I = \mathbb{R}$, and let

$$\beta(t) = \frac{1}{8}t - 1 \quad \text{and}$$

$$f(t) = \frac{t+1}{t^2+1}, \quad t \in I.$$

Prove that

$$\int_{1}^{4} D_\beta f(t)d_\beta t = -\frac{12}{17}.$$

Problem 2.9.4. Let $I = \mathbb{R}$, and let

$$\beta(t) = \frac{1}{4}t + 1,$$

$$f(t) = t^2 - t + 2, \quad \text{and}$$

$$g(t) = \frac{2+t}{1+t^2}, \quad t \in I.$$

Prove that

1. $\displaystyle\int_{-1}^{3} f(t)D_\beta g(t)d_\beta t = 2 - \int_{-1}^{3} D_\beta f(t)g(\beta(t))d_\beta t;$

2. $\displaystyle\int_{-1}^{3} f(\beta(t))D_\beta g(t)d_\beta t = 2 - \int_{-1}^{3} D_\beta f(t)g(t)d_\beta t.$

Problem 2.9.5. Let $I = [1, \infty)$, and let

$$\beta(t) = \frac{3}{4}t + 1, \quad t \in I.$$

Investigate the following improper β-integrals for their convergence and divergence:

1. $\displaystyle\int_{1}^{\infty} (t^5 - 3t^2 + 1)d_\beta t;$

2. $\displaystyle\int_{1}^{\infty} \frac{4}{(t + 1)(3t + 8)}d_\beta t;$

3. $\displaystyle\int_{1}^{\infty} \frac{4}{(t + 3)(3t + 16)}d_\beta t.$

Problem 2.9.6. Let $I = \mathbb{R}$, and let

$$\beta(t) = \frac{1}{2}t + 3, \quad t \in I.$$

Using the Cauchy criterion, prove that the following improper β-integrals are convergent:

1. $\displaystyle\int_{1}^{\infty} \frac{1}{(t + 1)(t + 8)}d_\beta t;$

2. $\displaystyle\int_{1}^{\infty} \frac{1}{(2t + 1)(t + 7)}d_\beta t;$

3. $\displaystyle\int_{1}^{\infty} \frac{1}{(t + 3)(t + 12)}d_\beta t.$

Problem 2.9.7. Let $I = \mathbb{R}$, and let

$$\beta(t) = \frac{9}{10}t + 1, \quad t \in I.$$

Investigate the following improper β-integrals for their convergence and divergence:

1. $\displaystyle\int_{1}^{\infty} \frac{1}{10(t + 2)^4(9t + 30)^{50}}d_\beta t;$

2. $\displaystyle\int_{1}^{\infty} \frac{1}{(2t + 3)^{10}(9t + 25)^4}d_\beta t;$

3. $\displaystyle\int_{1}^{\infty} \frac{1}{(t^2 + 49)(9t^2 + 83)}d_\beta t.$

Problem 2.9.8. Let $I = \mathbb{R}$, and let

$$\beta(t) = \frac{1}{2}t + 3, \quad t \in I.$$

Investigate the following improper β-integrals for their convergence and divergence:

1. $\displaystyle\int_1^\infty \frac{1}{(t+1)^2(t+8)^4}\,d_\beta t.$

2. $\displaystyle\int_1^\infty \frac{1}{(2t+11)^5(t+7)^8}\,d_\beta t;$

3. $\displaystyle\int_1^\infty \frac{1}{(2t+9)^2(t+12)^4}\,d_\beta t.$

Problem 2.9.9. Let $I = \mathbb{R}$, and let

$$\beta(t) = \frac{9}{10}t + 1, \quad t \in I.$$

Investigate the following improper β-integrals for their convergence and divergence:

1. $\displaystyle\int_1^\infty \frac{1}{10(t+2)^2}\,d_\beta t;$

2. $\displaystyle\int_1^\infty \frac{1}{(3t+4)(t+25)}\,d_\beta t;$

3. $\displaystyle\int_1^\infty \frac{1}{(t+49)(t+83)}\,d_\beta t.$

Problem 2.9.10. Let $I = \mathbb{R}$, and let

$$\beta(t) = \frac{1}{2}t + 3, \quad t \in I.$$

Investigate the following improper β-integrals for their convergence and divergence:

1. $\displaystyle\int_1^\infty \frac{1}{(t+3)(2t+11)}\,d_\beta t;$

2. $\displaystyle\int_1^\infty \frac{1}{(2t+11)(t+9)}\,d_\beta t;$

3. $\displaystyle\int_1^\infty \frac{1}{(t+9)(t+11)}\,d_\beta t.$

Problem 2.9.11. Let $I = \mathbb{R}$, and let

$$\beta(t) = \frac{1}{2}t + 3, \quad t \in I.$$

Investigate the following improper β-integrals for their convergence and divergence:

1. $\displaystyle\int_1^\infty \frac{1}{(t+1)(t+8)(1+\frac{2}{t}+t^4)}\,d_\beta t;$

2. $\displaystyle\int_1^\infty \frac{1}{(2t+1)(t+7)(1+t^6)}\,d_\beta t;$

3. $\displaystyle\int_1^\infty \frac{1}{(t+3)(t+12)(1+t^7)}\,d_\beta t.$

Problem 2.9.12. Let $I = \left[-\dfrac{1}{2}, \dfrac{1}{2} \right]$, and let

$$\beta(t) = t^5, \quad t \in I.$$

Prove that the improper β-integral

$$\int_{-\frac{1}{2}}^{0} \frac{1}{t^{\frac{10}{3}} + t^2 + t^{\frac{2}{3}}} \, d_\beta t$$

is convergent.

2.10 Notes and references

In this chapter, we have defined the β-integral and proved some of its properties. The basic rules for β-integration are deduced. The β-analogue of the Taylor formula is derived. The improper integrals of the first- and second-kind are investigated for β-calculus. Some of the results in this chapter can be found in [10,11].

CHAPTER

β-elementary functions$^{\circledast}$

3

Let $I \subseteq \mathbb{R}$, and let $\beta : I \to \mathbb{R}$ be a first-kind general quantum operator.

3.1 β-regressive functions

Definition 3.1.1. A function $f : I \to \mathbb{R}$ is said to be β-regressive provided

$$1 + (\beta(t) - t)f(t) \neq 0, \quad t \in I.$$

The set of all β-regressive and continuous functions on I is denoted by $\mathscr{R}_\beta$ or $\mathscr{R}_\beta(I)$.

A function $f : I \to \mathbb{R}$ is said to be positively β-regressive provided

$$1 + (\beta(t) - t)f(t) > 0, \quad t \in I.$$

The set of all positively β-regressive and continuous functions on I is denoted by $\mathscr{R}_\beta^+$ or $\mathscr{R}_\beta^+(I)$.

Definition 3.1.2. In $\mathscr{R}_\beta$, define the β-circle plus $\oplus_\beta$ by

$$(f \oplus_\beta g)(t) = f(t) + g(t) + (\beta(t) - t)f(t)g(t), \quad t \in I.$$

Example 3.1.3. Let $I = \mathbb{R}$, and let

$$\beta(t) = \frac{2}{3}t + 1,$$
$$f(t) = 1 + t, \quad \text{and}$$
$$g(t) = t^2, \quad t \in I.$$

We will find

$$(f \oplus_\beta g)(t), \quad t \in I.$$

By definition, we have

$$(f \oplus_\beta g)(t) = f(t) + g(t) + (\beta(t) - t)f(t)g(t)$$

$^{\circledast}$ This book has a companion website hosting complementary materials. Visit this URL to access it: https://www.elsevier.com/books-and-journals/book-companion/9780443328046.

Generalized Quantum Calculus with Applications. https://doi.org/10.1016/B978-0-44-332804-6.00008-X

$$= 1 + t + t^2 + \left(\frac{2}{3}t + 1 - t\right)(1 + t)t^2$$

$$= 1 + t + t^2 + \left(1 - \frac{1}{3}t\right)(t^3 + t^2)$$

$$= 1 + t + t^2 - \frac{1}{3}t^3 + t^3 - \frac{1}{3}t^4 + t^2$$

$$= 1 + t + 2t^2 + \frac{2}{3}t^3 - \frac{1}{3}t^4, \quad t \in I.$$

Example 3.1.4. Let $I = (-\sqrt{2}, \sqrt{2})$, and let

$$\beta(t) = \frac{1}{2}t^3,$$
$$f(t) = t, \quad \text{and}$$
$$g(t) = t^2, \quad t \in I.$$

We will find

$$(f \oplus_\beta g)(t), \quad t \in I.$$

By definition, we have

$$(f \oplus_\beta g)(t) = f(t) + g(t) + (\beta(t) - t)f(t)g(t)$$

$$= t + t^2 + \left(\frac{1}{2}t^3 - t\right) \cdot t \cdot t^2$$

$$= t + t^2 + \left(\frac{1}{2}t^3 - t\right)t^3$$

$$= t + t^2 - t^4 + \frac{1}{2}t^6, \quad t \in I.$$

Exercise 3.1.5. Let $I = (-\sqrt[3]{3}, \sqrt[3]{3})$, and let

$$\beta(t) = \frac{1}{3}t^4,$$
$$f(t) = 2 - t, \quad \text{and}$$
$$g(t) = 1 + t, \quad t \in I.$$

Find

$$(f \oplus_\beta g)(t), \quad t \in I.$$

Theorem 3.1.6. *The set $\mathscr{R}_\beta$ is an abelian group under the operation $\oplus_\beta$.*

Proof. Let $f, g, h \in \mathscr{R}_\beta$. Then

$$1 + (\beta(t) - t)f(t) \neq 0,$$
$$1 + (\beta(t) - t)g(t) \neq 0, \quad \text{and}$$
$$1 + (\beta(t) - t)h(t) \neq 0, \quad t \in I.$$

Now, we have

$$1 + (\beta(t) - t)(f \oplus_\beta g)(t) = 1 + (\beta(t) - t)(f(t) + g(t) + (\beta(t) - t)f(t)g(t))$$
$$= 1 + (\beta(t) - t)f(t) + (\beta(t) - t)g(t) + (\beta(t) - t)^2 f(t)g(t)$$
$$= (1 + (\beta(t) - t)f(t)) + (\beta(t) - t)g(t)(1 + (\beta(t) - t)f(t))$$
$$= (1 + (\beta(t) - t)f(t))(1 + (\beta(t) - t)g(t))$$
$$\neq 0, \quad t \in I.$$

Thus, $f \oplus_\beta g \in \mathscr{R}_\beta$. Next,

$$(f \oplus_\beta g) \oplus_\beta h(t) = (f(t) + g(t) + (\beta(t) - t)f(t)g(t)) + h(t)$$
$$+ (\beta(t) - t)(f(t) + g(t) + (\beta(t) - t)f(t)g(t))h(t)$$
$$= f(t) + g(t) + (\beta(t) - t)f(t)g(t) + h(t) + (\beta(t) - t)f(t)h(t)$$
$$+ (\beta(t) - t)g(t)h(t) + (\beta(t) - t)^2 f(t)g(t)h(t)$$
$$= f(t) + (g(t) + h(t) + (\beta(t) - t)g(t)h(t))$$
$$+ (\beta(t) - t)(g(t) + h(t) + (\beta(t) - t)g(t)h(t))f(t)$$
$$= f(t) + (g \oplus_\beta h)(t) + (\beta(t) - t)(g \oplus_\beta h)(t)f(t)$$
$$= f(t) \oplus_\beta (g \oplus_\beta h)(t), \quad t \in I.$$

Therefore, in $\mathscr{R}_\beta$, the associativity law holds with respect to $\oplus_\beta$. Note that $0 \in \mathscr{R}_\beta$ and

$$(0 \oplus_\beta f)(t) = 0 + f(t) + (\beta(t) - t) \cdot 0 \cdot f(t)$$
$$= f(t), \quad t \in I.$$

Let

$$l(t) = -\frac{f(t)}{1 + (\beta(t) - t)f(t)}, \quad t \in I.$$

We will show that $l \in \mathscr{R}_\beta$. Indeed, we have

$$1 + (\beta(t) - t)l(t) = 1 - \frac{(\beta(t) - t)f(t)}{1 + (\beta(t) - t)f(t)}$$
$$= \frac{1}{1 + (\beta(t) - t)f(t)}$$
$$\neq 0, \quad t \in I.$$

Then

$$(f \oplus l)(t) = f(t) + l(t) + (\beta(t) - t)f(t)l(t)$$
$$= f(t) - \frac{f(t)}{1 + (\beta(t) - t)f(t)} - \frac{(\beta(t) - t)(f(t))^2}{1 + (\beta(t) - t)f(t)}$$
$$= \frac{f(t) + (\beta(t) - t)(f(t))^2 - f(t) - (\beta(t) - t)(f(t))^2}{1 + (\beta(t) - t)f(t)}$$
$$= 0, \quad t \in I.$$

Thus, every element of $\mathscr{R}_\beta$ has an inverse with respect to the β-circle addition $\oplus_\beta$. Also,

$$
\begin{aligned}
(f \oplus_\beta g)(t) &= f(t) + g(t) + (\beta(t) - t) f(t) g(t) \\
&= g(t) + f(t) + (\beta(t) - t) g(t) f(t) \\
&= (g \oplus_\beta f)(t), \quad t \in I,
\end{aligned}
$$

that is, the commutativity law holds with respect to $\oplus_\beta$. Therefore, $(\mathscr{R}_\beta, \oplus_\beta)$ is an abelian group. This completes the proof. $\qquad\square$

Definition 3.1.7. The group $(\mathscr{R}_\beta, \oplus_\beta)$ is said to be a β-regressive group.

Definition 3.1.8. For $f \in \mathscr{R}_\beta$, define the β-circle minus as

$$
(\ominus f)(t) = -\frac{f(t)}{1 + (\beta(t) - t) f(t)}, \quad t \in I.
$$

From the proof of Theorem 3.1.6, it follows that if $f \in \mathscr{R}_\beta$, then $\ominus f \in \mathscr{R}_\beta$.

Example 3.1.9. Let I, β, f be as in Example 3.1.3. Then

$$
\begin{aligned}
(\ominus_\beta f)(t) &= -\frac{f(t)}{1 + (\beta(t) - t) f(t)} \\
&= -\frac{1 + t}{1 + \left(\frac{2}{3}t + 1 - t\right)(1 + t)} \\
&= -\frac{1 + t}{1 + \left(1 - \frac{1}{3}t\right)(1 + t)} \\
&= -\frac{1 + t}{1 + 1 + t - \frac{1}{3}t - \frac{1}{3}t^2} \\
&= -\frac{1 + t}{2 + \frac{2}{3}t - \frac{1}{3}t^2} \\
&= -\frac{3(1 + t)}{6 + 2t - t^2} \\
&= \frac{3(1 + t)}{t^2 - 2t - 6}, \quad t \in I.
\end{aligned}
$$

Next,

$$
\begin{aligned}
(\ominus_\beta g)(t) &= -\frac{g(t)}{1 + (\beta(t) - t) g(t)} \\
&= -\frac{t^2}{1 + \left(\frac{2}{3}t + 1 - t\right)t^2} \\
&= -\frac{t^2}{1 + \left(1 - \frac{1}{3}t\right)t^2}
\end{aligned}
$$

$$= -\frac{3t^2}{3 + (3-t)t^2}$$

$$= -\frac{3t^2}{3 + 3t^2 - t^3}$$

$$= \frac{3t^2}{t^3 - 3t^2 - 3}, \quad t \in I.$$

Example 3.1.10. Let I, β, and f be as in Example 3.1.4. Then

$$(\ominus_\beta f)(t) = -\frac{f(t)}{1 + (\beta(t) - t)f(t)}$$

$$= -\frac{t}{1 + \left(\frac{1}{2}t^3 - t\right)t}$$

$$= -\frac{2t}{2 + (t^3 - 2t)t}$$

$$= -\frac{2t}{t^4 - 2t^2 + 2}, \quad t \in I,$$

and

$$(\ominus_\beta g)(t) = -\frac{g(t)}{1 + (\beta(t) - t)g(t)}$$

$$= -\frac{t^2}{1 + \left(\frac{1}{2}t^2 - t\right)t^2}$$

$$= -\frac{2t^2}{2 + (t^3 - 2t)t^2}$$

$$= -\frac{2t^2}{t^5 - 2t^3 + 2}, \quad t \in I.$$

Exercise 3.1.11. Let $I = \mathbb{R}$, and let

$$\beta(t) = \frac{1}{2}t \quad \text{and}$$

$$f(t) = t - t^2, \quad t \in I.$$

Find

$$(\ominus_\beta f)(t), \quad t \in I.$$

Definition 3.1.12. For $f, g \in \mathscr{R}_\beta$, define

$$f \ominus_\beta g = f \oplus_\beta (\ominus_\beta g).$$

By definition, we have

$$(f \ominus_\beta g)(t) = f(t) + (\ominus_\beta g)(t) + (\beta(t) - t)f(t)(\ominus_\beta g)(t)$$

$$= f(t) - \frac{g(t)}{1+(\beta(t)-t)g(t)} - \frac{(\beta(t)-t)f(t)g(t)}{1+(\beta(t)-t)g(t)}$$

$$= \frac{f(t) + (\beta(t)-t)f(t)g(t) - g(t) - (\beta(t)-t)f(t)g(t)}{1+(\beta(t)-t)g(t)}$$

$$= \frac{f(t) - g(t)}{1+(\beta(t)-t)g(t)}.$$

Thus,

$$(f \ominus_\beta g)(t) = \frac{f(t) - g(t)}{1+(\beta(t)-t)g(t)}, \quad t \in I.$$

Example 3.1.13. Let I, β, f, and g be as in Example 3.1.3. Then

$$(f \ominus_\beta g)(t) = \frac{f(t) - g(t)}{1+(\beta(t)-t)g(t)}$$

$$= \frac{1+t-t^2}{1+\left(\frac{2}{3}t + 1 - t\right)t^2}$$

$$= \frac{1+t-t^2}{1+\left(1 - \frac{1}{3}t\right)t^2}$$

$$= \frac{3(1+t-t^2)}{3+(3-t)t^2}$$

$$= \frac{3(1+t-t^2)}{3+3t^2-t^3}$$

$$= \frac{3(t^2-t-1)}{t^3-3t^2-3}, \quad t \in I,$$

and

$$(g \ominus_\beta f)(t) = \frac{g(t) - f(t)}{1+(\beta(t)-t)f(t)}$$

$$= \frac{t^2-t-1}{1+\left(\frac{2}{3}t + 1 - t\right)(1+t)}$$

$$= \frac{t^2-t-1}{1+\left(1 - \frac{1}{3}t\right)(t+1)}$$

$$= \frac{3(t^2-t-1)}{3+(3-t)(t+1)}$$

$$= \frac{3(t^2-t-1)}{3+3+3t-t^2-t}$$

$$= \frac{3(t^2-t-1)}{-t^2+2t+6}$$

$$= -\frac{3(t^2-t-1)}{t^2-2t-6}, \quad t \in I.$$

Example 3.1.14. Let I, β, f, and g be as in Example 3.1.4. Then

$$
\begin{aligned}
(f \ominus_\beta g)(t) &= \frac{f(t) - g(t)}{1 + (\beta(t) - t)g(t)} \\
&= \frac{t - t^2}{1 + \left(\frac{1}{2}t^3 - t\right)t^2} \\
&= \frac{2t(1 - t)}{2 + (t^3 - 2t)t^2} \\
&= \frac{2t(1 - t)}{t^5 - 2t^3 + 2}, \quad t \in I,
\end{aligned}
$$

and

$$
\begin{aligned}
(g \ominus_\beta f)(t) &= \frac{g(t) - f(t)}{1 + (\beta(t) - t)f(t)} \\
&= \frac{t^2 - t}{1 + \left(\frac{1}{2}t^3 - t\right)t} \\
&= \frac{2t(t - 1)}{2 + (t^3 - 2t)t} \\
&= \frac{2t(t - 1)}{t^4 - 2t^2 + 2}, \quad t \in I.
\end{aligned}
$$

Exercise 3.1.15. Let $I = \mathbb{R}$, and let

$$
\begin{aligned}
\beta(t) &= \frac{1}{4}t, \\
f(t) &= 2 - t^2, \quad \text{and} \\
g(t) &= 2 + t^2, \quad t \in I.
\end{aligned}
$$

Find

1. $(f \oplus_\beta g)(t), t \in I$;
2. $(\ominus_\beta f)(t), t \in I$;
3. $(\ominus_\beta g)(t), t \in I$;
4. $(f \ominus_\beta g)(t), t \in I$;
5. $(g \ominus_\beta f)(t), t \in I$.

Theorem 3.1.16. *Let* $f, g \in \mathcal{R}_\beta$. *Then*

1. $\ominus_\beta(\ominus_\beta f) = f$;
2. $f \ominus_\beta f = 0$;
3. $f \ominus_\beta g \in \mathcal{R}_\beta$;
4. $\ominus_\beta(f \ominus_\beta g) = g \ominus_\beta f$;
5. $(\ominus_\beta(f \oplus_\beta g)) = (\ominus_\beta f) \oplus_\beta (\ominus_\beta g)$;
6. $f \oplus_\beta (\ominus_\beta g) = f + g$.

Proof. **1.** We have

$$(\ominus_\beta(\ominus_\beta f))(t) = -\frac{(\ominus_\beta f)(t)}{1+(\beta(t)-t)(\ominus_\beta f)(t)}$$

$$= \frac{\dfrac{f(t)}{1+(\beta(t)-t)f(t)}}{1-\dfrac{(\beta(t)-t)f(t)}{1+(\beta(t)-t)f(t)}}$$

$$= \frac{\dfrac{f(t)}{1+(\beta(t)-t)f(t)}}{\dfrac{1}{1+(\beta(t)-t)f(t)}}$$

$$= f(t), \quad t \in I.$$

Thus,

$$(\ominus_\beta(\ominus_\beta f))(t) = f(t), \quad t \in I.$$

2. We have

$$(f \ominus_\beta f)(t) = (f \oplus_\beta (\ominus_\beta f))(t)$$

$$= f(t) + \left(-\frac{f(t)}{1+(\beta(t)-t)f(t)}\right) - (\beta(t)-t)\frac{(f(t))^2}{1+(\beta(t)-t)f(t)}$$

$$= \frac{f(t) + (\beta(t)-t)(f(t))^2 - f(t) - (\beta(t)-t)(f(t))^2}{1+(\beta(t)-t)f(t)}$$

$$= 0, \quad t \in I.$$

Thus,

$$(f \ominus_\beta f)(t) = 0, \quad t \in I.$$

3. Since $f, g \in \mathscr{R}_\beta$, we have

$$1 + (\beta(t)-t)f(t) \neq 0,$$
$$1 + (\beta(t)-t)g(t) \neq 0, \quad t \in I.$$

Then

$$1 + (\beta(t)-t)(f \ominus_\beta g)(t) = 1 + (\beta(t)-t)\frac{f(t)-g(t)}{1+(\beta(t)-t)g(t)}$$

$$= \frac{1+(\beta(t)-t)g(t)+(\beta(t)-t)f(t)-(\beta(t)-t)g(t)}{1+(\beta(t)-t)g(t)}$$

$$= \frac{1+(\beta(t)-t)f(t)}{1+(\beta(t)-t)g(t)}$$

$$\neq 0, \quad t \in I.$$

Thus, $f \ominus_\beta g \in \mathscr{R}_\beta$.

4. We have

$$(g \ominus_\beta f)(t) = \frac{g(t)-f(t)}{1+(\beta(t)-t)f(t)}, \quad t \in I.$$

Hence, using the computations in the previous item, we get

$$(\ominus_\beta(f \ominus_\beta g))(t) = -\frac{(f \ominus_\beta g)(t)}{1 + \beta(t)(f \ominus_\beta g)(t)}$$

$$= -\frac{\frac{f(t)-g(t)}{1+(\beta(t)-t)g(t)}}{1 + \frac{(\beta(t)-t)(f(t)-g(t))}{1+(\beta(t)-t)g(t)}}$$

$$= -\frac{f(t) - g(t)}{1 + (\beta(t) - t)g(t) + (\beta(t) - t)(f(t) - g(t))}$$

$$= \frac{g(t) - f(t)}{1 + (\beta(t) - t)f(t)}$$

$$= (g \ominus_\beta f)(t), \quad t \in I.$$

Thus,

$$(\ominus_\beta(f \ominus_\beta g))(t) = (g \ominus_\beta f)(t), \quad t \in I.$$

5. We have

$$(\ominus_\beta(f \oplus_\beta g))(t) = -\frac{(f \oplus_\beta g)(t)}{1 + (\beta(t) - t)(f \oplus_\beta g)(t)}$$

$$= -\frac{f(t) + g(t) + (\beta(t) - t)f(t)g(t)}{1 + (\beta(t) - t)(f(t) + g(t) + (\beta(t) - t)f(t)g(t))}$$

$$= -\frac{f(t) + g(t) + (\beta(t) - t)f(t)g(t)}{1 + (\beta(t) - t)f(t) + (\beta(t) - t)g(t)(1 + (\beta(t) - t)f(t))}$$

$$= -\frac{f(t) + g(t) + (\beta(t) - t)f(t)g(t)}{(1 + (\beta(t) - t)f(t))(1 + (\beta(t) - t)g(t))}, \quad t \in I.$$

Next,

$$((\ominus_\beta f) \oplus_\beta (\ominus_\beta g))(t) = (\ominus_\beta f)(t) + (\ominus_\beta g)(t) + (\beta(t) - t)(\ominus_\beta f)(t)(\ominus_\beta g)(t)$$

$$= -\frac{f(t)}{1 + (\beta(t) - t)f(t)} - \frac{g(t)}{1 + (\beta(t) - t)g(t)}$$

$$+ \frac{(\beta(t) - t)f(t)g(t)}{(1 + (\beta(t) - t)f(t))(1 + (\beta(t) - t)g(t))}$$

$$= -\frac{1}{(1 + (\beta(t) - t)f(t))(1 + (\beta(t) - t)g(t))}\Big(f(t) \times (1 + (\beta(t) - t)g(t))$$

$$+ g(t)(1 + (\beta(t) - t)f(t)) - (\beta(t) - t)f(t)g(t)\Big)$$

$$= -\frac{f(t) + g(t) + (\beta(t) - t)f(t)g(t)}{(1 + (\beta(t) - t)f(t))(1 + (\beta(t) - t)g(t))}$$

$$= (\ominus_\beta(f \oplus_\beta g))(t), \quad t \in I.$$

Thus,

$$(\ominus_\beta(f \oplus_\beta g))(t) = ((\ominus_\beta f) \oplus_\beta (\ominus_\beta g))(t), \quad t \in I.$$

6. We have

$$(f \oplus_\beta (\ominus_\beta g))(t) = f(t) \oplus_\beta \frac{g(t)}{1 + (\beta(t) - t)f(t)}$$

$$= f(t) + \frac{g(t)}{1 + (\beta(t) - t)f(t)} + (\beta(t) - t)\frac{f(t)g(t)}{1 + (\beta(t) - t)f(t)}$$

$$= f(t) + \frac{1 + (\beta(t) - t)f(t)}{1 + (\beta(t) - t)f(t)}g(t)$$

$$= f(t) + g(t), \quad t \in I.$$

Thus,

$$(f \oplus_\beta (\ominus_\beta g))(t) = (f + g)(t), \quad t \in I. \qquad \square$$

Definition 3.1.17. For $f \in \mathcal{R}_\beta$, define the β-circle square as

$$f^{\textcircled{2}\beta} = (-f)(\ominus_\beta f).$$

For $f \in \mathcal{R}_\beta$, we have

$$f^{\textcircled{2}\beta}(t) = (-f(t))(\ominus_\beta f)(t)$$

$$= -f(t)\frac{-f(t)}{1 + (\beta(t) - t)f(t)}$$

$$= \frac{(f(t))^2}{1 + (\beta(t) - t)f(t)}, \quad t \in I.$$

Hence,

$$\frac{(f(t))^2}{f^{\textcircled{2}\beta}(t)} = 1 + (\beta(t) - t)f(t), \quad t \in I.$$

Example 3.1.18. Let I, β, f, and g be as in Example 3.1.3. Then

$$f^{\textcircled{2}\beta}(t) = (-f(t))(\ominus_\beta f)(t)$$

$$= -(1 + t)\left(\frac{3(1 + t)}{t^2 - 2t - 6}\right)$$

$$= -\frac{3(1 + t)^2}{t^2 - 2t - 6}, \quad t \in I,$$

and

$$g^{\textcircled{2}\beta}(t) = (-g(t))(\ominus_\beta g)(t)$$

$$= -t^2\left(\frac{3t^2}{t^3 - 3t^2 - 3}\right)$$

$$= -\frac{3t^4}{t^3 - 3t^2 - 3}, \quad t \in I.$$

Example 3.1.19. Let I, β, f, and g be as in Example 3.1.4. Then

$$f^{\circled{2}\beta}(t) = (-f(t))(\ominus_\beta f)(t)$$
$$= -t\left(-\frac{2t}{t^4 - 2t^2 + 2}\right)$$
$$= \frac{2t^2}{t^4 - 2t^2 + 2}, \quad t \in I,$$

and

$$g^{\circled{2}\beta}(t) = (-g(t))(\ominus_\beta g)(t)$$
$$= -t^2\left(-\frac{2t^2}{t^5 - 2t^3 + 2}\right)$$
$$= \frac{2t^4}{t^5 - 2t^3 + 2}, \quad t \in I.$$

Exercise 3.1.20. Let $I = \mathbb{R}$, and let

$$\beta(t) = \frac{7}{8}t + 1 \quad \text{and}$$
$$f(t) = t^3, \quad t \in I.$$

Find

$$f^{\circled{2}\beta}(t), \quad t \in I.$$

Theorem 3.1.21. *Suppose that $f \in \mathscr{R}_\beta$. Then*

1. $(\ominus_\beta f)^{\circled{2}\beta} = f^{\circled{2}\beta}$;
2. $(f + (\ominus_\beta f))(t) = (\beta(t) - t)f^{\circled{2}\beta}(t), t \in I$;
3. $f \oplus_\beta f^{\circled{2}\beta} = f + f^2$.

Proof. **1.** We have

$$(\ominus_\beta f)(t) = -\frac{f(t)}{1 + (\beta(t) - t)f(t)}, \quad t \in I,$$

and

$$f^{\circled{2}\beta}(t) = \frac{(f(t))^2}{1 + (\beta(t) - t)f(t)}, \quad t \in I.$$

Hence,

$$(\ominus_\beta f)^{\circled{2}\beta}(t) = \frac{\left(-\frac{f(t)}{1+(\beta(t)-t)f(t)}\right)^2}{1 - \frac{(\beta(t)-t)f(t)}{1+(\beta(t)-t)f(t)}}$$
$$= \frac{\frac{(f(t))^2}{(1+(\beta(t)-t)f(t))^2}}{\frac{1}{1+(\beta(t)-t)f(t)}}$$

$$= \frac{(f(t))^2}{1 + (\beta(t) - t)f(t)}, \quad t \in I.$$

Thus,

$$(\ominus_\beta f)^{\oslash\beta}(t) = f^{\oslash\beta}(t), \quad t \in I.$$

2. We have

$$
\begin{aligned}
f(t) + (\ominus_\beta f)(t) &= f(t) - \frac{f(t)}{1 + (\beta(t) - t)f(t)} \\
&= f(t)\left(1 - \frac{1}{1 + (\beta(t) - t)f(t)}\right) \\
&= f(t)\left(\frac{1 + (\beta(t) - t)f(t) - 1}{1 + (\beta(t) - t)f(t)}\right) \\
&= (\beta(t) - t)\frac{(f(t))^2}{1 + (\beta(t) - t)f(t)} \\
&= (\beta(t) - t)f^{\oslash\beta}(t), \quad t \in I.
\end{aligned}
$$

Thus,

$$(f + (\ominus_\beta f))(t) = (\beta(t) - t)f^{\oslash\beta}(t), \quad t \in I.$$

3. We have

$$
\begin{aligned}
\left(f \oplus_\beta f^{\oslash\beta}\right)(t) &= f(t) + f^{\oslash\beta}(t) + (\beta(t) - t)f(t)f^{\oslash\beta}(t) \\
&= f(t) + \frac{(f(t))^2}{1 + (\beta(t) - t)f(t)} + \frac{(\beta(t) - t)(f(t))^3}{1 + (\beta(t) - t)f(t)} \\
&= f(t) + \frac{(1 + (\beta(t) - t)f(t))(f(t))^2}{1 + (\beta(t) - t)f(t)} \\
&= f(t) + (f(t))^2, \quad t \in I.
\end{aligned}
$$

Thus,

$$\left(f \oplus_\beta f^{\oslash\beta}\right)(t) = (f + f^2)(t), \quad t \in I. \qquad \square$$

3.2 β-exponential functions

Let $f : I \to \mathbb{C}$ be a continuous function at s_0. Then the series

$$\sum_{k=0}^{\infty} f(\beta^k(t))(\beta^k(t) - \beta^{k+1}(t))$$

is uniformly convergent on any compact interval $J \subseteq I$ containing s_0. Thus, both products

$$\prod_{k=0}^{\infty}\left(1 - f(\beta^k(t))(\beta^k(t) - \beta^{k+1}(t))\right)$$

and

$$\prod_{k=0}^{\infty}\left(1 + f\left(\beta^k(t)\right)(\beta^k(t) - \beta^{k+1}(t))\right)$$

are convergent on any compact interval $J \subseteq I$ containing s_0.

Definition 3.2.1. Let $f : I \to \mathbb{C}$ be a β-regressive and continuous function. Define the β-exponential functions $e_{f,\beta}$ and $E_{f,\beta}$ as

$$e_{f,\beta}(t) = \frac{1}{\prod\limits_{k=0}^{\infty}\left(1 - f(\beta^k(t))(\beta^k(t) - \beta^{k+1}(t))\right)}, \quad t \in I,$$

and

$$E_{f,\beta}(t) = \prod_{k=0}^{\infty}\left(1 + f(\beta^k(t))(\beta^k(t) - \beta^{k+1}(t))\right), \quad t \in I,$$

respectively. Clearly, $e_{f,\beta}(t) = \dfrac{1}{E_{-f,\beta}(t)}, t \in I.$

Example 3.2.2. Let $I = [1, 2]$, and let

$$\beta(t) = \frac{1}{2}t + \frac{1}{2} \quad \text{and}$$

$$f(t) = \frac{2}{t}, \quad t \in I.$$

Here $s_0 = 1$. Then

$$[k]_{\frac{1}{2}} = \frac{1 - \left(\frac{1}{2}\right)^k}{1 - \frac{1}{2}}$$

$$= 2\left(1 - \left(\frac{1}{2}\right)^k\right), \quad k \in \mathbb{N},$$

$$\beta^k(t) = \left(\frac{1}{2}\right)^k t + 2\left(1 - \left(\frac{1}{2}\right)^k\right)\frac{1}{2}$$

$$= \left(\frac{1}{2}\right)^k t + 1 - \left(\frac{1}{2}\right)^k$$

$$= \left(\frac{1}{2}\right)^k (t - 1) + 1, \quad k \in \mathbb{N}, \quad t \in I,$$

and

$$\beta^{k+1}(t) = \left(\frac{1}{2}\right)^{k+1}(t - 1) + 1, \quad k \in \mathbb{N}_0, \quad t \in I.$$

Hence,

$$\beta^k(t) - \beta^{k+1}(t) = \left(\frac{1}{2}\right)^k (t-1) + 1 - \left(\frac{1}{2}\right)^{k+1} (t-1) - 1$$

$$= \left(\frac{1}{2}\right)^{k+1} (t-1), \quad k \in \mathbb{N}, \quad t \in I,$$

and

$$f\left(\beta^k(t)\right) = \frac{2}{\beta^k(t)}$$

$$= \frac{2}{\left(\frac{1}{2}\right)^k t + 1 - \left(\frac{1}{2}\right)^k}$$

$$= \frac{2}{\left(\frac{1}{2}\right)^k (t-1) + 1}$$

$$= \frac{2^{k+1}}{t-1+2^k}, \quad k \in \mathbb{N}, \quad t \in I.$$

Therefore,

$$\prod_{k=0}^{\infty} \left(1 - f(\beta^k(t))(\beta^k(t) - \beta^{k+1}(t))\right) = \prod_{k=0}^{\infty} \left(1 - \frac{2^{k+1}}{t-1+2^k} \cdot \frac{t-1}{2^{k+1}}\right)$$

$$= \prod_{k=0}^{\infty} \left(1 - \frac{t-1}{t-1+2^k}\right), \quad t \in I,$$

and

$$\prod_{k=0}^{\infty} \left(1 + f(\beta^k(t))(\beta^k(t) - \beta^{k+1}(t))\right) = \prod_{k=0}^{\infty} \left(1 + \frac{2^{k+1}}{t-1+2^k} \cdot \frac{t-1}{2^{k+1}}\right)$$

$$= \prod_{k=0}^{\infty} \left(1 + \frac{t-1}{t-1+2^k}\right), \quad t \in I.$$

Consequently,

$$e_{\frac{2}{t}, \frac{1}{2}t+\frac{1}{2}}(t) = \frac{1}{\displaystyle\prod_{k=0}^{\infty} \left(1 - \frac{t-1}{t-1+2^k}\right)}, \quad t \in I,$$

and

$$E_{\frac{2}{t}, \frac{1}{2}t+\frac{1}{2}}(t) = \prod_{k=0}^{\infty} \left(1 + \frac{t-1}{t-1+2^k}\right), \quad t \in I.$$

Example 3.2.3. Let $I = [0, 2]$, and let

$$\beta(t) = \frac{1}{2}t \quad \text{and}$$
$$f(t) = (1 + i)t, \quad t \in I.$$

Then

$$\beta^k(t) = \left(\frac{1}{2}\right)^k t, \quad t \in I, \quad k \in \mathbb{N},$$

$$\beta^k(t) - \beta^{k+1}(t) = \left(\frac{1}{2}\right)^k t - \left(\frac{1}{2}\right)^{k+1} t$$
$$= \left(\frac{1}{2}\right)^{k+1} t, \quad t \in I, \quad k \in \mathbb{N},$$

and

$$f\left(\beta^k(t)\right)(\beta^k(t) - \beta^{k+1}(t)) = (1 + i)\left(\frac{1}{2}\right)^k t \left(\frac{1}{2}\right)^{k+1} t$$
$$= \frac{(1 + i)t^2}{2^{2k+1}}, \quad t \in I, \quad k \in \mathbb{N}.$$

Hence,

$$\prod_{k=0}^{\infty}\left(1 - f(\beta^k(t))(\beta^k(t) - \beta^{k+1}(t))\right) = \prod_{k=0}^{\infty}\left(1 - \frac{(1 + i)t^2}{2^{2k+1}}\right), \quad t \in I,$$

and

$$\prod_{k=0}^{\infty}\left(1 + f(\beta^k(t))(\beta^k(t) - \beta^{k+1}(t))\right) = \prod_{k=0}^{\infty}\left(1 + \frac{(1 + i)t^2}{2^{2k+1}}\right), \quad t \in I.$$

Consequently,

$$e_{(1+i)t, \frac{1}{2}t}(t) = \frac{1}{\prod\limits_{k=0}^{\infty}\left(1 - \frac{(1+i)t^2}{2^{2k+1}}\right)}, \quad t \in I,$$

and

$$E_{(1+i)t, \frac{1}{2}t}(t) = \prod_{k=0}^{\infty}\left(1 + \frac{(1 + i)t^2}{2^{2k+1}}\right), \quad t \in I.$$

Exercise 3.2.4. Let $I = [0, 3]$, and let

$$\beta(t) = \frac{1}{3}t \quad \text{and}$$
$$f(t) = 7t, \quad t \in I.$$

Find

1. $e_{7t,\frac{1}{3}t}$, $t \in I$;
2. $E_{7t,\frac{1}{3}t}(t)$, $t \in I$.

Now, we will prove some properties of the β-exponential functions.

Theorem 3.2.5. *Let $f : I \to \mathbb{C}$ be such that $f \in \mathscr{R}_\beta$, and let $\beta : I \to \mathbb{R}$ be a general quantum operator.* *Then*

1. $e_{f,\beta}(s_0) = E_{f,\beta}(s_0) = 1$;
2. $D_\beta e_{f,\beta}(t) = f(t) e_{f,\beta}(t)$, $t \in I$;
3. $D_\beta E_{f,\beta}(t) = f(t) E_{f,\beta}(\beta(t))$, $t \in I$;
4. $e_{f,\beta}(\beta(t)) = (1 + (\beta(t) - t) f(t)) e_{f,\beta}(t)$, $t \in I$;
5. $E_{f,\beta}(t) = (1 - (\beta(t) - t) f(t)) E_{f,\beta}(\beta(t))$, $t \in I$;
6. $\dfrac{1}{e_{f,\beta}(t)} = e_{\ominus_\beta f,\beta}(t)$, $t \in I$;
7. $e_{f,\beta}(t) e_{g,\beta}(t) = e_{f \oplus_\beta g}(t)$, $t \in I$;
8. $\dfrac{e_{f,\beta}(t)}{e_{g,\beta}(t)} = e_{f \ominus_\beta g,\beta}(t)$, $t \in I$.

Proof. **1.** We have

$$\beta^k(s_0) = s_0, \quad k \in \mathbb{N}.$$

Then

$$\prod_{k=0}^{\infty} \left(1 - f(\beta^k(s_0))(\beta^k(s_0) - \beta^{k+1}(s_0))\right) = \prod_{k=0}^{\infty} (1 - f(s_0)(s_0 - s_0))$$
$$= \prod_{k=0}^{\infty} 1$$
$$= 1,$$

and

$$\prod_{k=0}^{\infty} \left(1 + f(\beta^k(s_0))(\beta^k(s_0) - \beta^{k+1}(s_0))\right) = \prod_{k=0}^{\infty} (1 + f(s_0)(s_0 - s_0))$$
$$= \prod_{k=0}^{\infty} 1$$
$$= 1.$$

Hence,

$$e_{f,\beta}(s_0) = \frac{1}{\displaystyle\prod_{k=0}^{\infty} \left(1 - f(\beta^k(s_0))(\beta^k(s_0) - \beta^{k+1}(s_0))\right)}$$

$$= \frac{1}{1}$$
$$= 1,$$

and

$$E_{f,\beta}(s_0) = \prod_{k=0}^{\infty} \left(1 + f(\beta^k(s_0))(\beta^k(s_0) - \beta^{k+1}(s_0))\right)$$
$$= 1.$$

2. Suppose $t \neq s_0$. Then

$$D_\beta e_{f,\beta}(t) = \frac{1}{\beta(t) - t}(e_{f,\beta}(\beta(t)) - e_{f,\beta}(t))$$

$$= \frac{1}{\beta(t) - t}\left(\frac{1}{\displaystyle\prod_{k=0}^{\infty}\left(1 - f(\beta^{k+1}(t))(\beta^{k+1}(t) - \beta^{k+2}(t))\right)} - \frac{1}{\displaystyle\prod_{k=0}^{\infty}\left(1 - f(\beta^{k}(t))(\beta^{k}(t) - \beta^{k+1}(t))\right)}\right)$$

$$= \frac{1}{\beta(t) - t}\left(\frac{1}{\displaystyle\prod_{k=1}^{\infty}\left(1 - f(\beta^{k}(t))(\beta^{k}(t) - \beta^{k+1}(t))\right)} - \frac{1}{\displaystyle\prod_{k=0}^{\infty}\left(1 - f(\beta^{k}(t))(\beta^{k}(t) - \beta^{k+1}(t))\right)}\right)$$

$$= \frac{1}{\beta(t) - t}\left(\frac{1}{\displaystyle\prod_{k=1}^{\infty}\left(1 - f(\beta^{k}(t))(\beta^{k}(t) - \beta^{k+1}(t))\right)} - \frac{1}{(1 - f(t)(t - \beta(t)))\displaystyle\prod_{k=1}^{\infty}\left(1 - f(\beta^{k}(t))(\beta^{k}(t) - \beta^{k+1}(t))\right)}\right)$$

$$= \frac{1}{(\beta(t) - t)\displaystyle\prod_{k=1}^{\infty}\left(1 - f(\beta^{k}(t))(\beta^{k}(t) - \beta^{k+1}(t))\right)}$$
$$\times \left(1 - \frac{1}{(1 - f(t)(t - \beta(t)))}\right)$$

$$= \frac{1}{(\beta(t) - t) \prod_{k=1}^{\infty} \left(1 - f(\beta^k(t))(\beta^k(t) - \beta^{k+1}(t))\right)} \frac{1 - f(t)(t - \beta(t)) - 1}{(1 - f(t)(t - \beta(t)))}$$

$$= f(t) \frac{1}{\prod_{k=0}^{\infty} \left(1 - f(\beta^k(t))(\beta^k(t) - \beta^{k+1}(t))\right)}$$

$$= f(t) e_{f,\beta}(t), \quad t \in I.$$

Next, suppose $t = s_0$. Then

$$\lim_{h \to 0+} \frac{e_{f,\beta}(s_0 + h) - e_{f,\beta}(s_0)}{h} = \lim_{h \to 0+} \frac{e_{f,\beta}(s_0 + h) - 1}{h}$$

$$= \lim_{h \to 0+} \frac{1}{h} \left(\frac{1}{\prod_{k=0}^{\infty} \left(1 - f(\beta^k(s_0 + h))(\beta^k(s_0 + h) - \beta^{k+1}(s_0 + h))\right)} - 1 \right)$$

$$= \lim_{h \to 0+} \frac{1}{h} \left(\frac{1 - \prod_{k=0}^{\infty} \left(1 - f(\beta^k(s_0 + h))(\beta^k(s_0 + h) - \beta^{k+1}(s_0 + h))\right)}{\prod_{k=0}^{\infty} \left(1 - f(\beta^k(s_0 + h))(\beta^k(s_0 + h) - \beta^{k+1}(s_0 + h))\right)} \right)$$

$$= \lim_{h \to 0+} \left(\frac{1}{h} \cdot \frac{1 - (1 - f(s_0)h)}{1 - f(s_0)h} \right)$$

$$= \lim_{h \to 0+} \frac{f(s_0)h}{h(1 - f(s_0)h)}$$

$$= \lim_{h \to 0+} \frac{f(s_0)}{1 - f(s_0)h}$$

$$= f(s_0).$$

Similarly,

$$\lim_{h \to 0-} \frac{e_{f,\beta}(s_0 + h) - e_{f,\beta}(s_0)}{h} = f(s_0).$$

Thus,

$$D_\beta e_{f,\beta}(t) = f(t) e_{f,\beta}(t), \quad t \in I.$$

3. Suppose $t \neq s_0$. Then

$$D_\beta E_{f,\beta}(t) = \frac{E_{f,\beta}(\beta(t)) - E_{f,\beta}(t)}{\beta(t) - t}$$

$$= \frac{1}{\beta(t) - t} \left(\prod_{k=0}^{\infty} \left(1 + f(\beta^{k+1}(t))(\beta^{k+1}(t) - \beta^{k+2}(t))\right) \right.$$

$$\left. - \prod_{k=0}^{\infty} \left(1 + f(\beta^k(t))(\beta^k(t) - \beta^{k+1}(t))\right) \right)$$

$$= \frac{1}{\beta(t) - t} \left(\prod_{k=1}^{\infty} \left(1 + f(\beta^k(t))(\beta^k(t) - \beta^{k+1}(t)) \right) \right.$$

$$\left. - (1 + f(t)(t - \beta(t))) \prod_{k=1}^{\infty} \left(1 + f(\beta^k(t))(\beta^k(t) - \beta^{k+1}(t)) \right) \right)$$

$$= \frac{1}{\beta(t) - t} \prod_{k=1}^{\infty} \left(1 - f(\beta^k(t))(\beta^k(t) - \beta^{k+1}(t)) \right)$$

$$\times (1 - (1 + f(t)(t - \beta(t))))$$

$$= \frac{f(t)(\beta(t) - t)}{\beta(t) - t} \prod_{k=1}^{\infty} \left(1 + f(\beta^k(t))(\beta^k(t) - \beta^{k+1}(t)) \right)$$

$$= f(t) E_{f,\beta}(\beta(t)).$$

Next, suppose $t = s_0$. Then

$$\lim_{h \to 0+} \frac{E_{f,\beta}(s_0 + h) - E_{f,\beta}(s_0)}{h}$$

$$= \lim_{h \to 0+} \frac{1}{h} \left(\prod_{k=0}^{\infty} \left(1 - f(\beta^k(s_0 + h))(\beta^k(s_0 + h) - \beta^{k+1}(s_0 + h)) \right) - 1 \right)$$

$$= \lim_{h \to 0+} \frac{1}{h}(1 + f(s_0)(s_0 + h - s_0) - 1)$$

$$= \lim_{h \to 0+} \frac{f(s_0)h}{h}$$

$$= f(s_0).$$

Similarly,

$$\lim_{h \to 0-} \frac{E_{f,\beta}(s_0 + h) - E_{f,\beta}(s_0)}{h} = f(s_0).$$

Thus,

$$D_\beta E_{f,\beta}(t) = f(t) E_{f,\beta}(\beta(t)), \quad t \in I.$$

4. We have

$$e_{f,\beta}(\beta(t)) = e_{f,\beta}(t) + (\beta(t) - t) D_\beta e_{f,\beta}(t)$$

$$= e_{f,\beta}(t) + (\beta(t) - t) f(t) e_{f,\beta}(t)$$

$$= (1 + (\beta(t) - t) f(t)) e_{f,\beta}(t), \quad t \in I.$$

5. We have

$$E_{f,\beta}(\beta(t)) = E_{f,\beta}(t) + (\beta(t) - t) D_\beta E_{f,\beta}(t)$$

$$= E_{f,\beta}(t) + (\beta(t) - t) f(t) E_{f,\beta}(\beta(t)),$$

whereupon

$$E_{f,\beta}(t) = E_{f,\beta}(\beta(t)) - (\beta(t) - t)\,f(t)\,E_{f,\beta}(\beta(t))$$
$$= (1 - (\beta(t) - t)\,f(t))\,E_{f,\beta}(\beta(t)), \quad t \in I.$$

6. We have

$$e_{f,\beta}(t) = \frac{1}{\displaystyle\prod_{k=0}^{\infty}\left(1 - f(\beta^k(t))(\beta^k(t) - \beta^{k+1}(t))\right)}$$
$$= \prod_{k=0}^{\infty} \frac{1}{\left(1 - f(\beta^k(t))(\beta^k(t) - \beta^{k+1}(t))\right)}, \quad t \in I.$$

Hence,

$$\frac{1}{e_{f,\beta}(t)} = \prod_{k=0}^{\infty}\left(1 - f(\beta^k(t))(\beta^k(t) - \beta^{k+1}(t))\right), \quad t \in I.$$

Also,

$$e_{\ominus_\beta f,\beta}(t) = \prod_{k=0}^{\infty} \frac{1}{\left(1 - \ominus_\beta f(\beta^k(t))(\beta^k(t) - \beta^{k+1}(t))\right)}$$
$$= \prod_{k=0}^{\infty} \frac{1}{\left(1 + \dfrac{f(\beta^k(t))}{1 + (\beta^{k+1}(t) - \beta^k(t))f(\beta^k(t))}(\beta^k(t) - \beta^{k+1}(t))\right)}$$
$$= \prod_{k=0}^{\infty}\left(1 - f(\beta^k(t))(\beta^k(t) - \beta^{k+1}(t))\right)$$
$$= \frac{1}{e_{f,\beta}(t)}, \quad t \in I.$$

Thus,

$$e_{\ominus_\beta f,\beta}(t) = \frac{1}{e_{f,\beta}(t)}, \quad t \in I.$$

7. We have

$$e_{f,\beta}(t)e_{g,\beta}(t) = \left(\prod_{k=0}^{\infty} \frac{1}{\left(1 - f(\beta^k(t))(\beta^k(t) - \beta^{k+1}(t))\right)}\right)$$
$$\times \left(\prod_{k=0}^{\infty} \frac{1}{\left(1 - g(\beta^k(t))(\beta^k(t) - \beta^{k+1}(t))\right)}\right)$$
$$= \prod_{k=0}^{\infty} \frac{1}{\left(1 - \left[f(\beta^k(t)) + g(\beta^k(t)) + (\beta^k(t) - \beta^{k+1}(t))f(\beta^k(t))g(\beta^k(t))\right](\beta^k(t) - \beta^{k+1}(t))\right)}$$

$$= \prod_{k=0}^{\infty} \frac{1}{\left(1 - (f \oplus_{\beta} g)(\beta^k(t))(\beta^k(t) - \beta^{k+1}(t))\right)}$$

$$= e_{f \oplus_{\beta} g, \beta}(t), \quad t \in I.$$

Thus,

$$e_{f,\beta}(t)e_{g,\beta}(t) = e_{f \oplus_{\beta} g, \beta}(t), \quad t \in I.$$

8. We have

$$\frac{e_{f,\beta}(t)}{e_{g,\beta}(t)} = e_{f,\beta}(t)e_{\ominus_{\beta} g}(t)$$

$$= e_{f \oplus_{\beta}(\ominus_{\beta} g), \beta}(t)$$

$$= e_{f \ominus_{\beta} g, \beta}(t), \quad t \in I.$$

Thus,

$$\frac{e_{f,\beta}(t)}{e_{g,\beta}(t)} = e_{f \ominus_{\beta} g, \beta}(t), \quad t \in I. \qquad \square$$

3.3 β-trigonometric functions

In this section, we define the β-trigonometric functions and prove some of their properties. Let $I \subseteq \mathbb{R}$ and $f : I \to \mathbb{R}$ be a β-regressive and continuous function.

Definition 3.3.1. Define the β-trigonometric functions as follows:

$$\sin_{f,\beta}(t) = \frac{e_{if,\beta}(t) - e_{-if,\beta}(t)}{2i},$$

$$\cos_{f,\beta}(t) = \frac{e_{if,\beta}(t) + e_{-if,\beta}(t)}{2},$$

$$\operatorname{Sin}_{f,\beta}(t) = \frac{E_{if,\beta}(t) - E_{-if,\beta}(t)}{2i},$$

$$\operatorname{Cos}_{f,\beta}(t) = \frac{E_{if,\beta}(t) + E_{-if,\beta}(t)}{2}, \quad t \in I.$$

Some basic properties of the β-trigonometric functions are provided in the following:

Theorem 3.3.2. *Let $f : I \to \mathbb{C}$ be such that $f \in \mathcal{R}_{\beta}$, and let $\beta : I \to \mathbb{R}$ be a general quantum operator. Then, for $t \in I$, we have*

1. $D_{\beta} \sin_{f,\beta}(t) = f(t) \cos_{f,\beta}(t)$;
2. $D_{\beta} \cos_{f,\beta}(t) = -f(t) \sin_{f,\beta}(t)$;
3. $D_{\beta} \operatorname{Sin}_{f,\beta}(t) = f(t) \operatorname{Cos}_{f,\beta}(\beta(t))$;
4. $D_{\beta} \operatorname{Cos}_{f,\beta}(t) = -f(t) \operatorname{Sin}_{f,\beta}(\beta(t))$;
5. $(\cos_{f,\beta}(t))^2 + (\sin_{f,\beta}(t))^2 = e_{if,\beta}(t)e_{-if,\beta}(t)$;

6. $(\mathrm{Cos}_{f,\beta}(t))^2 + (\mathrm{Sin}_{f,\beta}(t))^2 = E_{if,\beta}(t)E_{-if,\beta}(t);$
7. $\cos_{f,\beta}(t) + i\sin_{f,\beta}(t) = e_{if,\beta}(t),\ t \in I;$
8. $\mathrm{Cos}_{f,\beta}(t) + i\,\mathrm{Sin}_{f,\beta}(t) = E_{if,\beta}(t).$

Proof. **1.** We have

$$
\begin{aligned}
D_\beta \sin_{f,\beta}(t) &= \frac{D_\beta e_{if,\beta}(t) - D_\beta e_{-if,\beta}(t)}{2i} \\
&= \frac{if(t)e_{if,\beta}(t) + if(t)e_{-if,\beta}(t)}{2i} \\
&= f(t)\frac{e_{if,\beta}(t) + e_{-if,\beta}(t)}{2} \\
&= f(t)\cos_{f,\beta}(t), \quad t \in I.
\end{aligned}
$$

Thus,

$$
D_\beta \sin_{f,\beta}(t) = f(t)\cos_{f,\beta}(t), \quad t \in I.
$$

2. We have

$$
\begin{aligned}
D_\beta \cos_{f,\beta}(t) &= \frac{D_\beta e_{if,\beta}(t) + D_\beta e_{-if,\beta}(t)}{2} \\
&= \frac{if(t)e_{if,\beta}(t) - if(t)e_{-if,\beta}(t)}{2} \\
&= -f(t)\frac{e_{if,\beta}(t) - e_{-if,\beta}(t)}{2i} \\
&= -f(t)\sin_{f,\beta}(t), \quad t \in I.
\end{aligned}
$$

Thus,

$$
D_\beta \cos_{f,\beta}(t) = -f(t)\sin_{f,\beta}(t), \quad t \in I.
$$

3. We have

$$
\begin{aligned}
D_\beta \mathrm{Sin}_{f,\beta}(t) &= \frac{D_\beta E_{if,\beta}(t) - D_\beta E_{-if,\beta}(t)}{2i} \\
&= \frac{if(t)E_{if,\beta}(\beta(t)) + if(t)E_{-if,\beta}(\beta(t))}{2i} \\
&= f(t)\frac{E_{if,\beta}(\beta(t)) + E_{-if,\beta}(\beta(t))}{2} \\
&= f(t)\mathrm{Cos}_{f,\beta}(\beta(t)), \quad t \in I.
\end{aligned}
$$

Thus,

$$
D_\beta \mathrm{Sin}_{f,\beta}(t) = f(t)\mathrm{Cos}_{f,\beta}(\beta(t)), \quad t \in I.
$$

4. We have

$$
D_\beta \mathrm{Cos}_{f,\beta}(t) = \frac{D_\beta E_{if,\beta}(t) + D_\beta E_{-if,\beta}(t)}{2}
$$

$$= \frac{if(t)E_{if,\beta}(\beta(t)) - if(t)E_{-if,\beta}(\beta(t))}{2}$$

$$= -f(t)\frac{E_{if,\beta}(\beta(t)) - E_{-if,\beta}(\beta(t))}{2i}$$

$$= -f(t)\mathrm{Sin}_{f,\beta}(\beta(t)), \quad t \in I.$$

Thus,

$$D_\beta \mathrm{Cos}_{f,\beta}(t) = -f(t)\mathrm{Sin}_{f,\beta}(\beta(t)), \quad t \in I.$$

5. We have

$$(\cos_{f,\beta}(t))^2 + (\sin_{f,\beta}(t))^2 = \left(\frac{e_{if,\beta}(t) + e_{-if,\beta}(t)}{2}\right)^2 + \left(\frac{e_{if,\beta}(t) - e_{-if,\beta}(t)}{2i}\right)^2$$

$$= \frac{(e_{if,\beta}(t))^2 + 2e_{if,\beta}(t)e_{-if,\beta}(t) + (e_{-if,\beta}(t))^2}{4}$$

$$- \frac{(e_{if,\beta}(t))^2 - 2e_{if,\beta}(t)e_{-if,\beta}(t) + (e_{-if,\beta}(t))^2}{4}$$

$$= e_{if,\beta}(t)e_{-if,\beta}(t), \quad t \in I.$$

Thus,

$$(\cos_{f,\beta}(t))^2 + (\sin_{f,\beta}(t))^2 = e_{if,\beta}(t)e_{-if,\beta}(t), \quad t \in I.$$

6. We have

$$(\mathrm{Cos}_{f,\beta}(t))^2 + (\mathrm{Sin}_{f,\beta}(t))^2 = \left(\frac{E_{if,\beta}(t) + E_{-if,\beta}(t)}{2}\right)^2 + \left(\frac{E_{if,\beta}(t) - E_{-if,\beta}(t)}{2i}\right)^2$$

$$= \frac{(E_{if,\beta}(t))^2 + 2E_{if,\beta}(t)E_{-if,\beta}(t) + (E_{-if,\beta}(t))^2}{4}$$

$$- \frac{(E_{if,\beta}(t))^2 - 2E_{if,\beta}(t)E_{-if,\beta}(t) + (E_{-if,\beta}(t))^2}{4}$$

$$= E_{if,\beta}(t)E_{-if,\beta}(t), \quad t \in I.$$

Thus,

$$(\mathrm{Cos}_{f,\beta}(t))^2 + (\mathrm{Sin}_{f,\beta}(t))^2 = E_{if,\beta}(t)E_{-if,\beta}(t), \quad t \in I.$$

7. We have

$$\cos_{f,\beta}(t) + i\sin_{f,\beta}(t) = \frac{e_{if,\beta}(t) + e_{-if,\beta}(t)}{2} + i\frac{e_{if,\beta}(t) - e_{-if,\beta}(t)}{2i}$$

$$= \frac{e_{if,\beta}(t) + e_{-if,\beta}(t) + e_{if,\beta}(t) - e_{-if,\beta}(t)}{2}$$

$$= e_{if,\beta}(t), \quad t \in I.$$

Thus,

$$\cos_{f,\beta}(t) + i\sin_{f,\beta}(t) = e_{if,\beta}(t), \quad t \in I.$$

8. We have

$$\mathrm{Cos}_{f,\beta}(t) + i\,\mathrm{Sin}_{f,\beta}(t) = \frac{E_{if,\beta}(t) + E_{-if,\beta}(t)}{2} + i\,\frac{E_{if,\beta}(t) - E_{-if,\beta}(t)}{2i}$$

$$= \frac{E_{if,\beta}(t) + E_{-if,\beta}(t) + E_{if,\beta}(t) - E_{-if,\beta}(t)}{2}$$

$$= E_{if,\beta}(t), \quad t \in I.$$

Thus,

$$\mathrm{Cos}_{f,\beta}(t) + i\,\mathrm{Sin}_{f,\beta}(t) = E_{if,\beta}(t), \quad t \in I. \qquad \square$$

Exercise 3.3.3. Prove that

$$\sin_{f,\beta}(t)\mathrm{Sin}_{f,\beta}(t) + \cos_{f,\beta}(t)\mathrm{Cos}_{f,\beta}(t) = 1, \quad t \in I.$$

3.4 β-hyperbolic functions

In this section, we define the β-hyperbolic functions and prove some of their properties. Let $I \subseteq \mathbb{R}$ and $f : I \to \mathbb{R}$ be a β-regressive and continuous function.

Definition 3.4.1. Define the β-hyperbolic functions as follows:

$$\sinh_{f,\beta}(t) = \frac{e_{f,\beta}(t) - e_{-f,\beta}(t)}{2},$$

$$\cosh_{f,\beta}(t) = \frac{e_{f,\beta}(t) + e_{-f,\beta}(t)}{2},$$

$$\mathrm{Sinh}_{f,\beta}(t) = \frac{E_{f,\beta}(t) - E_{-f,\beta}(t)}{2},$$

$$\mathrm{Cosh}_{f,\beta}(t) = \frac{E_{f,\beta}(t) + E_{-f,\beta}(t)}{2}, \quad t \in I.$$

We provide some basic properties of the β-hyperbolic functions in the following:

Theorem 3.4.2. *Let* $f : I \to \mathbb{C}$ *be such that* $f \in \mathscr{R}_\beta$, *and let* $\beta : I \to \mathbb{R}$ *be a general quantum operator. Then, for* $t \in I$, *we have*

1. $D_\beta \sinh_{f,\beta}(t) = f(t)\cosh_{f,\beta}(t)$;
2. $D_\beta \cosh_{f,\beta}(t) = f(t)\sinh_{f,\beta}(t)$;
3. $D_\beta \mathrm{Sinh}_{f,\beta}(t) = f(t)\mathrm{Cosh}_{f,\beta}(\beta(t))$;
4. $D_\beta \mathrm{Cosh}_{f,\beta}(t) = f(t)\mathrm{Sinh}_{f,\beta}(\beta(t))$;
5. $(\cosh_{f,\beta}(t))^2 - (\sinh_{f,\beta}(t))^2 = e_{f,\beta}(t)e_{-f,\beta}(t)$;
6. $(\mathrm{Cosh}_{f,\beta}(t))^2 - (\mathrm{Sinh}_{f,\beta}(t))^2 = E_{f,\beta}(t)E_{-f,\beta}(t)$;
7. $\cosh_{f,\beta}(t) + \sinh_{f,\beta}(t) = e_{f,\beta}(t), \ t \in I$;
8. $\mathrm{Cosh}_{f,\beta}(t) + \sinh_{f,\beta}(t) = E_{f,\beta}(t)$.

Proof. **1.** We have

$$D_\beta \sinh_{f,\beta}(t) = \frac{D_\beta e_{f,\beta}(t) - D_\beta e_{-f,\beta}(t)}{2}$$
$$= \frac{f(t)e_{f,\beta}(t) + f(t)e_{-f,\beta}(t)}{2}$$
$$= f(t)\frac{e_{f,\beta}(t) + e_{-f,\beta}(t)}{2}$$
$$= f(t)\cosh_{f,\beta}(t), \quad t \in I.$$

Thus,

$$D_\beta \sinh_{f,\beta}(t) = f(t)\cosh_{f,\beta}(t), \quad t \in I.$$

2. We have

$$D_\beta \cosh_{f,\beta}(t) = \frac{D_\beta e_{f,\beta}(t) + D_\beta e_{-f,\beta}(t)}{2}$$
$$= \frac{f(t)e_{f,\beta}(t) - f(t)e_{-f,\beta}(t)}{2}$$
$$= f(t)\frac{e_{f,\beta}(t) - e_{-f,\beta}(t)}{2}$$
$$= f(t)\sinh_{f,\beta}(t), \quad t \in I.$$

Thus,

$$D_\beta \cosh_{f,\beta}(t) = f(t)\sinh_{f,\beta}(t), \quad t \in I.$$

3. We have

$$D_\beta \mathrm{Sinh}_{f,\beta}(t) = \frac{D_\beta E_{f,\beta}(t) - D_\beta E_{-f,\beta}(t)}{2}$$
$$= \frac{f(t)E_{f,\beta}(\beta(t)) + f(t)E_{-f,\beta}(\beta(t))}{2}$$
$$= f(t)\frac{E_{f,\beta}(\beta(t)) + E_{-f,\beta}(\beta(t))}{2}$$
$$= f(t)\mathrm{Cosh}_{f,\beta}(\beta(t)), \quad t \in I.$$

Thus,

$$D_\beta \mathrm{Sinh}_{f,\beta}(t) = f(t)\mathrm{Cosh}_{f,\beta}(\beta(t)), \quad t \in I.$$

4. We have

$$D_\beta \mathrm{Cosh}_{f,\beta}(t) = \frac{D_\beta E_{f,\beta}(t) + D_\beta E_{-f,\beta}(t)}{2}$$
$$= \frac{f(t)E_{f,\beta}(\beta(t)) - f(t)E_{-f,\beta}(\beta(t))}{2}$$
$$= f(t)\frac{E_{f,\beta}(\beta(t)) - E_{-f,\beta}(\beta(t))}{2}$$

$$= f(t)\mathrm{Sinh}_{f,\beta}(\beta(t)), \quad t \in I.$$

Thus,

$$D_\beta \mathrm{Cosh}_{f,\beta}(t) = f(t)\mathrm{Sinh}_{f,\beta}(\beta(t)), \quad t \in I.$$

5. We have

$$(\cosh_{f,\beta}(t))^2 - (\sinh_{f,\beta}(t))^2 = \left(\frac{e_{f,\beta}(t) + e_{-f,\beta}(t)}{2}\right)^2 - \left(\frac{e_{f,\beta}(t) - e_{-f,\beta}(t)}{2}\right)^2$$

$$= \frac{(e_{f,\beta}(t))^2 + 2e_{f,\beta}(t)e_{-f,\beta}(t) + (e_{-f,\beta}(t))^2}{4}$$

$$- \frac{(e_{f,\beta}(t))^2 - 2e_{f,\beta}(t)e_{-f,\beta}(t) + (e_{-f,\beta}(t))^2}{4}$$

$$= e_{f,\beta}(t)e_{-f,\beta}(t), \quad t \in I.$$

Thus,

$$(\cosh_{f,\beta}(t))^2 - (\sinh_{f,\beta}(t))^2 = e_{f,\beta}(t)e_{-f,\beta}(t), \quad t \in I.$$

6. We have

$$(\mathrm{Cosh}_{f,\beta}(t))^2 - (\mathrm{Sinh}_{f,\beta}(t))^2 = \left(\frac{E_{f,\beta}(t) + E_{-f,\beta}(t)}{2}\right)^2 - \left(\frac{E_{f,\beta}(t) - E_{-f,\beta}(t)}{2}\right)^2$$

$$= \frac{(E_{f,\beta}(t))^2 + 2E_{f,\beta}(t)E_{-f,\beta}(t) + (E_{-f,\beta}(t))^2}{4}$$

$$- \frac{(E_{f,\beta}(t))^2 - 2E_{f,\beta}(t)E_{-f,\beta}(t) + (E_{-f,\beta}(t))^2}{4}$$

$$= E_{f,\beta}(t)E_{-f,\beta}(t), \quad t \in I.$$

Thus,

$$(\mathrm{Cosh}_{f,\beta}(t))^2 - (\mathrm{Sinh}_{f,\beta}(t))^2 = E_{f,\beta}(t)E_{-f,\beta}(t), \quad t \in I.$$

7. We have

$$\cosh_{f,\beta}(t) + \sinh_{f,\beta}(t) = \frac{e_{f,\beta}(t) + e_{-f,\beta}(t)}{2} + \frac{e_{f,\beta}(t) - e_{-f,\beta}(t)}{2}$$

$$= \frac{e_{f,\beta}(t) + e_{-f,\beta}(t) + e_{f,\beta}(t) - e_{-f,\beta}(t)}{2}$$

$$= e_{f,\beta}(t), \quad t \in I.$$

Thus,

$$\cosh_{f,\beta}(t) + \sinh_{f,\beta}(t) = e_{f,\beta}(t), \quad t \in I.$$

8. We have

$$\mathrm{Cosh}_{f,\beta}(t) + \mathrm{Sinh}_{f,\beta}(t) = \frac{E_{f,\beta}(t) + E_{-f,\beta}(t)}{2} + \frac{E_{f,\beta}(t) - E_{-f,\beta}(t)}{2}$$

$$= \frac{E_{f,\beta}(t) + E_{-f,\beta}(t) + E_{f,\beta}(t) - E_{-f,\beta}(t)}{2}$$

$$= E_{f,\beta}(t), \quad t \in I.$$

Thus,

$$\mathrm{Cosh}_{f,\beta}(t) + \mathrm{Sinh}_{f,\beta}(t) = E_{f,\beta}(t), \quad t \in I. \qquad \square$$

Exercise 3.4.3. Prove that

$$\cosh_{f,\beta}(t) - \sinh_{f,\beta}(t) = e_{-f,\beta}(t), \quad t \in I.$$

3.5 Advanced practical problems

Problem 3.5.1. Let $I = \mathbb{R}$, and let

$$\beta(t) = \frac{3}{4}t + 1,$$

$$f(t) = 1 - t, \quad \text{and}$$

$$g(t) = 1 + t, \quad t \in I.$$

Find

1. $(f \oplus_\beta g)(t), t \in I$;
2. $(\ominus_\beta f)(t), t \in I$;
3. $(\ominus_\beta g)(t), t \in I$;
4. $(f \ominus_\beta g)(t), t \in I$;
5. $(g \ominus_\beta f)(t), t \in I$;
6. $f^{\textcircled{2}\beta}(t), t \in I$;
7. $g^{\textcircled{2}\beta}(t), t \in I$.

Problem 3.5.2. Let $I = [0, 4]$, and let

$$\beta(t) = \frac{1}{4}t \quad \text{and}$$

$$f(t) = 9t, \quad t \in I.$$

Find

1. $e_{9t, \frac{1}{4}t}(t), t \in I$;
2. $E_{9t, \frac{1}{4}t}(t), t \in I$.

Problem 3.5.3. Prove that

$$\sin_{f,\beta}(t)\mathrm{Cos}_{f,\beta}(t) - \cos_{f,\beta}(t)\mathrm{Sin}_{f,\beta}(t) = 0, \quad t \in I.$$

Problem 3.5.4. Prove that

$$\mathrm{Cosh}_{f,\beta}(t) - \mathrm{Sinh}_{f,\beta}(t) = E_{-f,\beta}(t), \quad t \in I.$$

3.6 Notes and references

This chapter is devoted to introduce some β-elementary functions. The functions like β-exponential, β-trigonometric, and β-hyperbolic functions are defined. Some of their basic properties are provided. Some of the results in this chapter can be found in [14].

The β-Laplace transform⊛

Let $I \subseteq \mathbb{R}$ be such that $\sup I = \infty$ and $s_0 \in I$. We denote by $V([s_0, \infty))$ the set of all functions that are β-integrable over each compact subinterval of $[s_0, \infty)$.

4.1 Functions of exponential orders

Definition 4.1.1. A function $f \in V([s_0, \infty))$ is said to be of exponential order $\lambda \in \mathbb{R}$ provided there exists a constant $M > 0$ such that

$$|f(t)| \leq Me_{\lambda,\beta}(t), \quad t \in [s_0, \infty).$$

Theorem 4.1.2. *For $z, x \in \mathscr{R}_\beta$ with $z = x + iy$, $x, y \in \mathbb{R}$, we have*

$$|e_{\ominus_\beta z,\beta}(t)| \leq e_{\ominus_\beta x,\beta}(t), \quad t \in [s_0, \infty).$$

Proof. Note that

$$x \oplus_\beta \frac{iy}{1 + x(\beta(t) - t)} = x + \frac{iy}{1 + x(\beta(t) - t)} + (\beta(t) - t)\frac{ixy}{1 + x(\beta(t) - t)}$$
$$= x + \frac{iy(1 + x(\beta(t) - t))}{1 + x(\beta(t) - t)}$$
$$= x + iy, \quad t \in [s_0, \infty).$$

Then

$$e_{z,\beta}(t) = e_{x+iy,\beta}(t)$$
$$= e_{x \oplus_\beta \frac{iy}{1+x(\beta(t)-t)}}(t)$$
$$= e_{x,\beta}(t)e_{\frac{iy}{1+x(\beta(t)-t)}}(t), \quad t \in [s_0, \infty),$$

and

$$e_{\frac{iy}{1+x(\beta(t)-t)},\beta}(t) = \prod_{k=0}^{\infty} \frac{1}{1 - \frac{iy}{1+x(\beta^{k+1}(t)-\beta^k(t))}(\beta^k(t) - \beta^{k+1}(t))}$$

⊛ This book has a companion website hosting complementary materials. Visit this URL to access it: https://www.elsevier.com/books-and-journals/book-companion/9780443328046.

Generalized Quantum Calculus with Applications. https://doi.org/10.1016/B978-0-44-332804-6.00009-1

$$= \prod_{k=0}^{\infty} \frac{1 + x(\beta^k(t) - \beta^{k+1}(t))}{1 + (x + iy)(\beta^k(t) - \beta^{k+1}(t))}, \quad t \in [s_0, \infty).$$

Hence,

$$\left| e_{\frac{iy}{1+x(\beta(t)-t)},\beta}(t) \right| = \left| \prod_{k=0}^{\infty} \frac{1 + x(\beta^k(t) - \beta^{k+1}(t))}{1 + (x + iy)(\beta^k(t) - \beta^{k+1}(t))} \right|$$

$$= \prod_{k=0}^{\infty} \frac{|1 + x(\beta^k(t) - \beta^{k+1}(t))|}{|1 + (x + iy)(\beta^k(t) - \beta^{k+1}(t))|}$$

$$\geq \prod_{k=0}^{\infty} \frac{|1 + x(\beta^k(t) - \beta^{k+1}(t))|}{|1 + x(\beta^k(t) - \beta^{k+1}(t))| - |y(\beta^k(t) - \beta^{k+1}(t))|}$$

$$\geq 1, \quad t \in [s_0, \infty).$$

Thus,

$$|e_{z,\beta}(t)| = \left| e_{x,\beta}(t) e_{\frac{iy}{1+x(\beta(t)-t)},\beta}(t) \right|$$

$$= |e_{x,\beta}(t)| \left| e_{\frac{iy}{1+x(\beta(t)-t)},\beta}(t) \right|$$

$$\geq |e_{x,\beta}(t)|, \quad t \in [s_0, \infty),$$

and

$$|e_{\ominus_\beta z,\beta}(t)| \leq |e_{\ominus_\beta x,\beta}(t)|$$

$$= e_{\ominus_\beta x,\beta}(t), \quad t \in [s_0, \infty).$$

This completes the proof. $\qquad\qquad\square$

4.2 Definition and properties of the β-Laplace transform

Definition 4.2.1. Let $f \in V([s_0, \infty))$. Then the β-Laplace transform of f is defined as

$$\mathscr{L}_\beta(f)(z) = \int_{s_0}^{\infty} f(t) e_{\ominus_\beta z,\beta}(\beta(t)) d_\beta t \tag{4.1}$$

for $z \in \mathscr{R}_\beta$ for which integral (4.1) exists.

Theorem 4.2.2. *Let $f \in V([s_0, \infty))$ be of exponential order $\lambda \in \mathbb{R}$. Then the β-integral (4.1) converges absolutely for $|z| > \lambda$, provided that*

$$\lim_{t \to \infty} e_{\ominus_\beta z,\beta}(t) = 0.$$

Proof. Let $z = x + iy$, $x, y \in \mathbb{R}$, and $|z| > \lambda$. Then

$$e_{\ominus_\beta x,\beta}(\beta(t)) = e_{\ominus_\beta,x,\beta}(t) + (\beta(t) - t) D_\beta e_{\ominus_\beta x,\beta}(t)$$

$$= \left(1 - \frac{x(\beta(t) - t)}{1 + x(\beta(t) - t)}\right) e_{\ominus \beta x, \beta}(t)$$

$$= \frac{1}{1 + x(\beta(t) - t)} e_{\ominus \beta x, \beta}(t), \quad t \in [s_0, \infty).$$

Since f is of exponential order λ, there is a constant $M > 0$ such that

$$|f(t)| \le M e_{\lambda, \beta}(t), \quad t \in [s_0, \infty).$$

Now applying Theorem 4.1.2, we have

$$\left| \int_{s_0}^{\infty} f(t) e_{\ominus \beta z, \beta}(\beta(t)) d_\beta t \right| \le \int_{s_0}^{\infty} |f(t)| |e_{\ominus \beta z, \beta}(\beta(t))| d_\beta t$$

$$\le \int_{s_0}^{\infty} |f(t)| e_{\ominus \beta x, \beta}(\beta(t)) d_\beta t$$

$$\le M \int_{s_0}^{\infty} e_{\lambda, \beta}(t) e_{\ominus \beta x, \beta}(\beta(t)) d_\beta t$$

$$= M \int_{s_0}^{\infty} \frac{1}{1 + (\beta(t) - t)x} e_{\lambda, \beta}(t) e_{\ominus \beta x, \beta}(t) d_\beta t$$

$$= M \int_{s_0}^{x} \frac{1}{1 + (\beta(t) - t)x} e_{\lambda \ominus \beta x, \beta}(t) d_\beta t$$

$$= \frac{M}{\lambda - x} \int_{s_0}^{\infty} \frac{\lambda - x}{1 + (\beta(t) - t)x} e_{\lambda \ominus \beta x, \beta}(t) d_\beta t$$

$$= \frac{M}{\lambda - x} \int_{s_0}^{\infty} (\lambda \ominus_\beta x)(t) e_{\lambda \ominus \beta x}(t) d_\beta t$$

$$= \frac{M}{\lambda - x} \int_{s_0}^{\infty} D_\beta e_{\lambda \ominus \beta x, \beta}(t) d_\beta t$$

$$= \frac{M}{\lambda - x} \lim_{b \to \infty} \int_{s_0}^{b} D_\beta e_{\lambda \ominus \beta x, \beta}(t) d_\beta t$$

$$= \frac{M}{\lambda - x} \lim_{b \to \infty} e_{\lambda \ominus \beta x, \beta}(t) \Big|_{t=s_0}^{t=b}$$

$$= \frac{M}{\lambda - x} \lim_{b \to \infty} \left(e_{\lambda \ominus \beta x, \beta}(b) - 1 \right)$$

$$= \frac{M}{x - \lambda}.$$

Hence, the β-integral (4.1) converges absolutely. This completes the proof. $\qquad\square$

Example 4.2.3. We will find

$$\mathscr{L}_\beta(1)(z), \quad z \in \mathbb{C}, \quad z \ne 0.$$

For $z \in \mathbb{C}$, $z \ne 0$, we have

$$\mathscr{L}_\beta(1)(z) = \int_{s_0}^{\infty} e_{\ominus \beta z, \beta}(\beta(t)) d_\beta t$$

$$= \int_{s_0}^{\infty} \frac{1}{1+(\beta(t)-t)z} e_{\ominus\beta z,\beta}(t)d_\beta t$$

$$= -\frac{1}{z}\int_{s_0}^{\infty} \frac{-z}{1+(\beta(t)-t)z} e_{\ominus\beta z,\beta}(t)d_\beta t$$

$$= -\frac{1}{z}\int_{s_0}^{\infty} \ominus_\beta z e_{\ominus\beta z,\beta}(t)d_\beta t$$

$$= -\frac{1}{z}\int_{s_0}^{\infty} \ominus_\beta z e_{\ominus\beta z,\beta}(t)d_\beta t$$

$$= -\frac{1}{z}\int_{s_0}^{\infty} D_\beta e_{\ominus\beta z,\beta}(t)d_\beta t$$

$$= -\frac{1}{z}\lim_{b\to\infty}\int_{s_0}^{b} D_\beta e_{\ominus\beta z,\beta}(t)d_\beta t$$

$$= -\frac{1}{z}\lim_{b\to\infty} e_{\ominus\beta z,\beta}(t)\Big|_{t=s_0}^{t=b}$$

$$= -\frac{1}{z}\lim_{b\to\infty}\left(e_{\ominus\beta z,\beta}(b)-1\right)$$

$$= \frac{1}{z},$$

provided that $\lim_{t\to\infty} e_{\ominus\beta z,\beta}(t)=0$. Thus,

$$\mathscr{L}_\beta(1)(z)=\frac{1}{z}, \quad z\neq 0, \text{ provided that } \lim_{t\to\infty} e_{\ominus\beta z,\beta}(t)=0.$$

Example 4.2.4. Let $z,\lambda\in\mathscr{R}_\beta$, $z\in\mathbb{C}$, $z\neq\lambda$. We will find

$$\mathscr{L}_\beta(e_{\lambda,\beta}(t))(z),$$

provided that $\lim_{t\to\infty} e_{\lambda\ominus\beta z,\beta}(t)=0$.

We have

$$\mathscr{L}_\beta(e_{\lambda,\beta}(t))(z)=\int_{s_0}^{\infty} e_{\lambda,\beta}(t)e_{\ominus\beta z}(\beta(t))d_\beta t$$

$$= \int_{s_0}^{\infty} \frac{1}{1+(\beta(t)-t)z} e_{\lambda,\beta}(t)e_{\ominus\beta z,\beta}(t)d_\beta t$$

$$= \int_{s_0}^{\infty} \frac{1}{1+(\beta(t)-t)z} e_{\lambda\ominus\beta z,\beta}(t)d_\beta t$$

$$= \frac{1}{\lambda-z}\int_{s_0}^{\infty} \frac{\lambda-z}{1+(\beta(t)-t)z} e_{\lambda\ominus\beta z,\beta}(t)d_\beta t$$

$$= \frac{1}{\lambda-z}\int_{s_0}^{\infty} (\lambda\ominus_\beta z)(t)e_{\lambda\ominus\beta z,\beta}(t)d_\beta t$$

$$= \frac{1}{\lambda-z}\int_{s_0}^{\infty} D_\beta e_{\lambda\ominus\beta z,\beta}(t)d_\beta t$$

$$= \frac{1}{\lambda - z} \lim_{b \to \infty} \int_{s_0}^{b} D_\beta e_{\lambda \ominus \beta z, \beta}(t) d_\beta t$$

$$= \frac{1}{\lambda - z} \lim_{b \to \infty} e_{\lambda \ominus \beta z, \beta}(t) \Big|_{t=s_0}^{t=b}$$

$$= \frac{1}{\lambda - z} \lim_{b \to \infty} \left(e_{\lambda \ominus \beta z, \beta}(b) - e_{\lambda \ominus \beta z, \beta}(s_0) \right)$$

$$= \frac{1}{z - \lambda}.$$

Hence,

$$\mathscr{L}_\beta(e_{\lambda, \beta}(t))(z) = \frac{1}{z - \lambda}, \quad \text{provided that } \lim_{t \to \infty} e_{\ominus \beta z, \beta}(t) = 0.$$

Example 4.2.5. Let $\lambda, \mu, z \in \mathscr{R}_\beta$, $z \neq \lambda + \mu$. We will find

$$\mathscr{L}_\beta \left(e_{\frac{\lambda}{1 + \mu(\beta(t) - t)}, \beta}(t) e_{\mu, \beta}(t) \right)(z),$$

provided that

$$\lim_{t \to \infty} e_{(\lambda + \mu) \ominus \beta, z, \beta}(t) = 0.$$

We have

$$\frac{\lambda}{1 + \mu(\beta(t) - t)} \oplus_\beta \mu = \frac{\lambda}{1 + \mu(\beta(t) - t)} + \mu + (\beta(t) - t)\frac{\lambda \mu}{1 + \mu(\beta(t) - t)}$$

$$= \frac{\lambda(1 + \mu(\beta(t) - t))}{1 + \mu(\beta(t) - t)} + \mu$$

$$= \lambda + \mu, \quad t \in I.$$

Now using Example 4.2.4, we obtain

$$\mathscr{L}_\beta \left(e_{\frac{\lambda}{1 + \mu(\beta(t) - t)}, \beta}(t) e_{\mu, \beta}(t) \right)(z) = \mathscr{L}_\beta \left(e_{\frac{\lambda}{1 + \mu(\beta(t) - t)} \oplus \mu, \beta}(t) \right)(z)$$

$$= \mathscr{L}_\beta \left(e_{\lambda + \mu, \beta}(t) \right)$$

$$= \frac{1}{z - (\lambda + \mu)}.$$

Thus,

$$\mathscr{L}_\beta \left(e_{\frac{\lambda}{1 + \mu(\beta(t) - t)}, \beta}(t) e_{\mu, \beta}(t) \right)(z) = \frac{1}{z - (\lambda + \mu)}.$$

Theorem 4.2.6. *Let $f, g \in V([s_0, \infty))$ and $c_1, c_2 \in \mathbb{C}$. Then*

$$\mathscr{L}_\beta(c_1 f(t) + c_2 g(t))(z) = c_1 \mathscr{L}_\beta(f(t))(z) + c_2 \mathscr{L}_\beta(g(t))(z).$$

Proof. Using Definition 4.2.1 and the properties of β-integral, we have

$$\mathscr{L}_\beta(c_1 f(t) + c_2 g(t))(z) = \int_{s_0}^{\infty} (c_1 f(t) + c_2 g(t)) e_{\ominus \beta z, \beta}(\beta(t)) d_\beta t$$

$$= \int_{s_0}^{\infty} c_1 f(t) e_{\ominus \beta z, \beta}(\beta(t)) d_\beta t + \int_{s_0}^{\infty} c_2 g(t) e_{\ominus \beta z, \beta}(\beta(t)) d_\beta t$$

$$= c_1 \int_{s_0}^{\infty} f(t) e_{\ominus \beta z, \beta}(\beta(t)) d_\beta t + c_2 \int_{s_0}^{\infty} g(t) e_{\ominus \beta z, \beta}(\beta(t)) d_\beta t$$

$$= c_1 \mathscr{L}_\beta(f(t))(z) + c_2 \mathscr{L}_\beta(g(t))(z).$$

This completes the proof. $\qquad\square$

Example 4.2.7. We will find $\mathscr{L}_\beta(\sin_{\alpha,\beta}(t))(z)$. We have

$$\mathscr{L}_\beta(\sin_{\alpha,\beta}(t))(z) = \mathscr{L}_\beta\left(\frac{e_{i\alpha,\beta}(t) - e_{-i\alpha,\beta}(t)}{2i}\right)$$

$$= \frac{1}{2i}\left(\mathscr{L}_\beta(e_{i\alpha,\beta}(t))(z) - \mathscr{L}_\beta(e_{-i\alpha,\beta}(t))(z)\right)$$

$$= \frac{1}{2i}\left(\frac{1}{z - i\alpha} - \frac{1}{z + i\alpha}\right)$$

$$= \frac{1}{2i}\frac{z + i\alpha - z + i\alpha}{(z - i\alpha)(z + i\alpha)}$$

$$= \frac{1}{2i}\frac{2i\alpha}{z^2 + \alpha^2}$$

$$= \frac{\alpha}{z^2 + \alpha^2}.$$

Hence,

$$\mathscr{L}_\beta(\sin_{\alpha,\beta}(t))(z) = \frac{\alpha}{z^2 + \alpha^2}.$$

Example 4.2.8. We will find $\mathscr{L}_\beta(\cos_{\alpha,\beta}(t))(z)$. We have

$$\mathscr{L}_\beta(\cos_{\alpha,\beta}(t))(z) = \mathscr{L}_\beta\left(\frac{e_{i\alpha,\beta}(t) + e_{-i\alpha,\beta}(t)}{2}\right)$$

$$= \frac{1}{2}\left(\mathscr{L}_\beta(e_{i\alpha,\beta}(t))(z) + \mathscr{L}_\beta(e_{-i\alpha,\beta}(t))(z)\right)$$

$$= \frac{1}{2}\left(\frac{1}{z - i\alpha} + \frac{1}{z + i\alpha}\right)$$

$$= \frac{1}{2}\frac{z + i\alpha + z - i\alpha}{(z - i\alpha)(z + i\alpha)}$$

$$= \frac{1}{2}\frac{2z}{z^2 + \alpha^2}$$

$$= \frac{z}{z^2 + \alpha^2}.$$

Hence,

$$\mathscr{L}_\beta(\cos_{\alpha,\beta}(t))(z) = \frac{z}{z^2 + \alpha^2}.$$

Example 4.2.9. We will find $\mathscr{L}_\beta(\sinh_{\alpha,\beta}(t))(z)$. We have

$$
\begin{aligned}
\mathscr{L}_\beta(\sinh_{\alpha,\beta}(t))(z) &= \mathscr{L}_\beta\left(\frac{e_{\alpha,\beta}(t) - e_{-\alpha,\beta}(t)}{2}\right) \\
&= \frac{1}{2}\left(\mathscr{L}_\beta(e_{\alpha,\beta}(t))(z) - \mathscr{L}_\beta(e_{-\alpha,\beta}(t))(z)\right) \\
&= \frac{1}{2}\left(\frac{1}{z-\alpha} - \frac{1}{z+\alpha}\right) \\
&= \frac{1}{2}\frac{z+\alpha-z+\alpha}{(z-\alpha)(z+\alpha)} \\
&= \frac{1}{2}\frac{2\alpha}{z^2-\alpha^2} \\
&= \frac{\alpha}{z^2-\alpha^2}.
\end{aligned}
$$

Hence,

$$
\mathscr{L}_\beta(\sinh_{\alpha,\beta}(t))(z) = \frac{\alpha}{z^2-\alpha^2}.
$$

Example 4.2.10. We will find $\mathscr{L}_\beta(\cosh_{\alpha,\beta}(t))(z)$. We have

$$
\begin{aligned}
\mathscr{L}_\beta(\cosh_{\alpha,\beta}(t))(z) &= \mathscr{L}_\beta\left(\frac{e_{\alpha,\beta}(t) + e_{-\alpha,\beta}(t)}{2}\right) \\
&= \frac{1}{2}\left(\mathscr{L}_\beta(e_{\alpha,\beta}(t))(z) + \mathscr{L}_\beta(e_{-\alpha,\beta}(t))(z)\right) \\
&= \frac{1}{2}\left(\frac{1}{z-\alpha} + \frac{1}{z+\alpha}\right) \\
&= \frac{1}{2}\frac{z+\alpha+z-\alpha}{(z-\alpha)(z+\alpha)} \\
&= \frac{1}{2}\frac{2z}{z^2-\alpha^2} \\
&= \frac{z}{z^2-\alpha^2}.
\end{aligned}
$$

Hence,

$$
\mathscr{L}_\beta(\cosh_{\alpha,\beta}(t))(z) = \frac{z}{z^2-\alpha^2}.
$$

Exercise 4.2.11. Find

$$
\mathscr{L}_\beta\left(e_{2,\beta}(t) - e_{-1,\beta}(t)\right)(z).
$$

4.3 The β-Laplace transform of β-derivative

In this section, we derive the β-Laplace transform of β-derivatives of arbitrary order.

Theorem 4.3.1. *Let $f \in V([s_0, \infty))$ be a function of exponential order λ. Then for $n \in \mathbb{N}$, we have*

$$\mathscr{L}_\beta\left(D_\beta^n f(t)\right)(z) = z^n \mathscr{L}_\beta(f(t))(z) - \sum_{j=0}^{n-1} z^{n-1-j} D_\beta^j f(s_0), \tag{4.2}$$

provided that

$$\lim_{t\to\infty} D_\beta^k f(t) e_{\ominus\beta,z}(t) = 0, \quad k \in \{0, \ldots, n-1\}.$$

Proof. We will use the principle of mathematical induction. For $n = 1$, we have

$$\mathscr{L}_\beta(D_\beta f(t))(z) = \int_{s_0}^{\infty} D_\beta f(t) e_{\ominus\beta z,\beta}(\beta(t)) d_\beta t.$$

Then using β-integration by parts, we obtain

$$\mathscr{L}_\beta(D_\beta f(t))(z) = \lim_{b\to\infty}\left(f(t)e_{\ominus\beta z,\beta}(t)\Big|_{t=s_0}^{t=b}\right) - \int_{s_0}^{\infty}(\ominus\beta z)(t) f(t) e_{\ominus\beta z,\beta}(t) d_\beta t$$

$$= -f(s_0) + \int_{s_0}^{\infty} \frac{z}{1 + z(\beta(t) - t)} f(t) e_{\ominus\beta z,\beta}(t) d_\beta t$$

$$= -f(s_0) + z \int_{s_0}^{\infty} f(t) e_{\ominus\beta z,\beta}(\beta(t)) d_\beta t$$

$$= -f(s_0) + z\mathscr{L}_\beta(f(t))(z).$$

Thus, (4.2) holds for $n = 1$. Assume that (4.2) holds for some $n \in \mathbb{N}$. We will prove the statement for $n + 1$, that is,

$$\mathscr{L}_\beta\left(D_\beta^{n+1} f(t)\right)(z) = z^{n+1} \mathscr{L}_\beta(f(t))(z) - \sum_{j=0}^{n} z^{n-j} D_\beta^j f(s_0),$$

provided that

$$\lim_{t\to\infty} D_\beta^k f(t) e_{\ominus\beta,z}(t) = 0, \quad k \in \{0, \ldots, n\}.$$

Indeed, using β-integration by parts, we have

$$\mathscr{L}_\beta(D_\beta^{n+1} f(t))(z) = \int_{s_0}^{\infty} D_\beta^{n+1} f(t) e_{\ominus\beta z,\beta}(\beta(t)) d_\beta t$$

$$= \lim_{b\to\infty}\left(D_\beta^n f(t) e_{\ominus\beta z,\beta}(t)\Big|_{t=s_0}^{t=b}\right) - \int_{s_0}^{\infty}(\ominus\beta z)(t) D_\beta^n f(t) e_{\ominus\beta z,\beta}(t) d_\beta t$$

$$= -D_\beta^n f(s_0) + \int_{s_0}^{\infty} \frac{z}{1 + z(\beta(t) - t)} D_\beta^n f(t) e_{\ominus\beta z,\beta}(t) d_\beta t$$

$$= -D_\beta^n f(s_0) + z \int_{s_0}^{\infty} D_\beta^n f(t) e_{\ominus\beta z,\beta}(\beta(t)) d_\beta t$$

$$= -D_\beta^n f(s_0) + z\mathscr{L}_\beta\left(D_\beta^n f(t)\right)(z)$$

$$= -D_\beta^n f(s_0) + z \left(z^n \mathscr{L}_\beta(f(t))(z) - \sum_{j=0}^{n-1} z^{n-1-j} D_\beta^j f(s_0) \right)$$

$$= z^{n+1} \mathscr{L}_\beta(f(t))(z) - \sum_{j=0}^{n} z^{n-j} D_\beta^j f(s_0).$$

Thus, (4.2) holds for $n + 1 \in \mathbb{N}$. Now applying the principle of mathematical induction, we conclude that (4.2) holds for all $n \in \mathbb{N}$. This completes the proof. $\qquad\square$

4.4 **The β-Laplace transform of β-integrals**

In this section, we will find the β Laplace transform for β-integrals. We have the following result.

Theorem 4.4.1. *Let $f \in V([s_0, \infty))$ be of exponential order λ. Then*

$$\mathscr{L}_\beta(F(t))(z) = \frac{1}{z} \mathscr{L}_\beta(f(t))(z),$$

provided that

$$\lim_{t \to \infty} F(t) e_{\ominus\beta z, \beta}(t) = 0,$$

where

$$F(t) = \int_{s_0}^{t} f(\tau) d_\beta \tau.$$

Proof. We have

$$\mathscr{L}_\beta(F(t))(z) = \int_{s_0}^{\infty} F(t) e_{\ominus\beta z, \beta}(\beta(t)) d_\beta t$$

$$= \int_{s_0}^{\infty} \frac{1}{1 + z(\beta(t) - t)} F(t) e_{\ominus\beta z, \beta}(t) d_\beta t$$

$$= -\frac{1}{z} \int_{s_0}^{\infty} \frac{-z}{1 + z(\beta(t) - t)} F(t) e_{\ominus\beta z, \beta}(t) d_\beta t$$

$$= -\frac{1}{z} \int_{s_0}^{\infty} \ominus\beta z F(t) e_{\ominus\beta z, \beta}(t) d_\beta t$$

$$= -\frac{1}{z} \int_{s_0}^{\infty} F(t) D_\beta e_{\ominus\beta z, \beta}(t) d_\beta t$$

$$= -\frac{1}{z} \lim_{b \to \infty} \left(F(t) e_{\ominus\beta z, \beta}(t) \Big|_{t=s_0}^{t=b} \right) + \frac{1}{z} \int_{s_0}^{\infty} f(t) e_{\ominus\beta z, \beta}(\beta(t)) d_\beta t$$

$$= \frac{1}{z} \int_{s_0}^{\infty} f(t) e_{\ominus\beta z, \beta}(\beta(t)) d_\beta t$$

$$= \frac{1}{z} \mathscr{L}_\beta(f(t))(z).$$

This completes the proof. $\qquad\square$

Definition 4.4.2. Let $\lambda \in \mathcal{R}_\beta$. For each $n \in \mathbb{N}_0$, define the functions $\psi_k : I \to \mathbb{C}$ as follows:

$$\psi_0(t) = 1,$$

$$\psi_k(t) = \int_{s_0}^{t} \frac{1}{1 + \lambda(\beta(t) - t)} \psi_{k-1}(t) d_\beta t, \quad k \in \mathbb{N}.$$

Theorem 4.4.3. *Let $\lambda \in \mathcal{R}_\beta$. For $n \in \mathbb{N}_0$, we have*

$$\mathcal{L}_\beta(\psi_n(t)e_{\lambda,\beta}(t)) = \frac{1}{(z-\lambda)^{n+1}}, \tag{4.3}$$

provided that

$$\lim_{t \to \infty} \psi_k(t)e_{\lambda \ominus \beta z, \beta}(t) = 0, \quad k \in \{0, \dots, n\}.$$

Proof. We will use the principle of mathematical induction. Suppose $n = 0$. Then

$$\mathcal{L}_\beta(\psi_0(t)e_{\lambda,\beta}(t))(z) = \mathcal{L}_\beta(e_{\lambda,\beta}(t))(z)$$

$$= \frac{1}{z-\lambda}.$$

Thus, (4.3) holds for $n = 0$. Now assume that (4.3) holds for some $n \in \mathbb{N}_0$. We will prove that

$$\mathcal{L}_\beta(\psi_{n+1}(t)e_{\lambda,\beta}(t)) = \frac{1}{(z-\lambda)^{n+2}},$$

provided that

$$\lim_{t \to \infty} \psi_{n+1}(t)e_{\lambda \ominus \beta z, \beta}(t) = 0.$$

Indeed, we have

$$\mathcal{L}_\beta(\psi_{n+1}(t)e_{\lambda,\beta}(t)) = \int_{s_0}^{\infty} \psi_{n+1}(t)e_{\lambda,\beta}(t)e_{\ominus \beta z, \beta}(\beta(t))d_\beta t$$

$$= \int_{s_0}^{\infty} \psi_{n+1}(t) \frac{1}{1 + z(\beta(t) - t)} e_{\lambda,\beta}(t)e_{\ominus \beta z, \beta}(t)d_\beta t$$

$$= \int_{s_0}^{\infty} \psi_{n+1}(t) \frac{1}{1 + z(\beta(t) - t)} e_{\lambda \ominus \beta z, \beta}(t)d_\beta t$$

$$= \frac{1}{\lambda - z} \int_{s_0}^{\infty} \psi_{n+1}(t) \frac{\lambda - z}{1 + z(\beta(t) - t)} e_{\lambda \ominus \beta z, \beta}(t)d_\beta t$$

$$= \frac{1}{\lambda - z} \int_{s_0}^{\infty} \psi_{n+1}(t)(\lambda \ominus_\beta z)(t)e_{\lambda \ominus \beta z, \beta}(t)d_\beta t$$

$$= \frac{1}{\lambda - z} \int_{s_0}^{\infty} \psi_{n+1}(t)D_\beta e_{\lambda \ominus \beta z, \beta}(t)d_\beta t$$

$$= \frac{1}{\lambda - z} \lim_{b \to \infty} \left(\psi_{n+1}(t)e_{\lambda \ominus \beta z, \beta}(t)d_\beta t \Big|_{t=s_0}^{t=b} \right) + \frac{1}{z - \lambda} \int_{s_0}^{\infty} D_\beta \psi_{n+1}(t)e_{\lambda \ominus \beta z, \beta}(\beta(t))d_\beta t$$

$$= \frac{1}{z-\lambda} \int_{s_0}^{\infty} \psi_n(t) \frac{1}{1+\lambda(\beta(t)-t)} e_{\lambda \ominus \beta z, \beta}(\beta(t)) d_\beta t$$

$$= \frac{1}{z-\lambda} \int_{s_0}^{\infty} \psi_n(t) \frac{1}{1+\lambda(\beta(t)-t)} e_{\lambda,\beta}(\beta(t)) e_{\ominus \beta z, \beta}(\beta(t)) d_\beta t$$

$$= \frac{1}{z-\lambda} \int_{s_0}^{\infty} \psi_n(t) e_{\lambda,\beta}(t) e_{\ominus \beta z, \beta}(\beta(t)) d_\beta t$$

$$= \frac{1}{z-\lambda} \mathscr{L}_\beta(\psi_n(t) e_{\lambda,\beta}(t))$$

$$= \frac{1}{z-\lambda} \frac{1}{(z-\lambda)^{n+1}}$$

$$= \frac{1}{(z-\lambda)^{n+2}}.$$

Thus,

$$\mathscr{L}_\beta(\psi_{n+1}(t) e_{\lambda,\beta}(t)) = \frac{1}{(z-\lambda)^{n+2}}.$$

By the principle of mathematical induction, it follows that (4.3) holds for all $n \in \mathbb{N}_0$. This completes the proof. $\qquad \square$

Definition 4.4.4. The inverse β-Laplace transform of a function F is a function f such that

$$\mathscr{L}_\beta(f) = F,$$

which we denote by $\mathscr{L}_\beta^{-1}(F)$.

Example 4.4.5. Let $I = \mathbb{R}$, and let

$$\beta(t) = \frac{1}{3}t + 2, \quad t \in I.$$

We will find

$$\mathscr{L}_\beta^{-1}\left(\frac{1}{(z-1)^2}\right)(t), \quad t \in I.$$

Here $s_0 = 3$. Now applying Theorem 4.4.3 for $\lambda = 1$, we obtain

$$\mathscr{L}_\beta^{-1}\left(\frac{1}{(z-1)^2}\right)(t) = \psi_1(t) e_{1,\beta}(t), \quad t \in I. \tag{4.4}$$

Also, by Example 1.1.5, we have

$$\beta^k(t) = \left(\frac{1}{3}\right)^k t + 2[k]_{\frac{1}{3}}$$

$$= \left(\frac{1}{3}\right)^k + 2\left(\frac{1 - \left(\frac{1}{3}\right)^k}{1 - \frac{1}{3}}\right)$$

$$= \left(\frac{1}{3}\right)^k t + 2\left(\frac{1 - \left(\frac{1}{3}\right)^k}{\frac{2}{3}}\right)$$

$$= \left(\frac{1}{3}\right)^k t + 3\left(1 - \left(\frac{1}{3}\right)^k\right)$$

$$= \left(\frac{1}{3}\right)^k t - 3\left(\frac{1}{3}\right)^k + 3$$

$$= \left(\frac{1}{3}\right)^k (t - 3) + 3, \quad t \in I, \quad k \in \mathbb{N}.$$

Now by the definition of ψ_1, we get

$$\psi_1(t) = \int_3^t \frac{1}{1 + \left(\frac{1}{3}\tau + 2 - \tau\right)}\, d_\beta \tau$$

$$= \int_3^t \frac{1}{1 + 2 - \frac{2}{3}\tau}\, d_\beta \tau$$

$$= 3\int_3^t \frac{1}{9 - 2\tau}\, d_\beta \tau$$

$$= 3\sum_{k=0}^{\infty} \left(\left(\frac{1}{3}\right)^k (t - 3) + 3 - \left(\frac{1}{3}\right)^{k+1}(t - 3) - 3\right)\frac{1}{9 - 2\left(\left(\frac{1}{3}\right)^k (t - 3) + 3\right)}$$

$$= 3\sum_{k=0}^{\infty} \left(\frac{1}{3}\right)^k (t - 3)\left(1 - \frac{1}{3}\right)\frac{1}{9 - 6 - 2\left(\frac{1}{3}\right)^k (t - 3)}$$

$$= 2\sum_{k=0}^{\infty} \frac{\left(\frac{1}{3}\right)^k (t - 3)}{3 - 2\left(\frac{1}{3}\right)^k (t - 3)}$$

$$= 2(t - 3)\sum_{k=0}^{\infty} \frac{\left(\frac{1}{3}\right)^k}{3 - 2\left(\frac{1}{3}\right)^k (t - 3)},$$

that is,

$$\psi_1(t) = 2(t - 3)\sum_{k=0}^{\infty} \frac{\left(\frac{1}{3}\right)^k}{3 - 2\left(\frac{1}{3}\right)^k (t - 3)}, \quad t \in I.$$

Also, by definition, the β-exponential function $e_{1,\beta}(t)$, $t \in I$, has the representation

$$e_{1,\beta}(t) = \prod_{k=0}^{\infty} \frac{1}{1 - \left(\left(\frac{1}{3}\right)^k (t - 3) + 3 - \left(\frac{1}{3}\right)^{k+1} - 3\right)}$$

$$= \prod_{k=0}^{\infty} \frac{1}{1 - \left(\frac{1}{3}\right)^k (t-3)\left(1 - \frac{1}{3}\right)}$$

$$= \prod_{k=0}^{\infty} \frac{1}{1 - 2\left(\frac{1}{3}\right)^k (t-3)}, \quad t \in I.$$

Therefore, (4.4) becomes

$$\mathscr{L}_\beta\left(\frac{1}{(z-1)^2}\right)(t) = \psi_1(t)e_{1,\beta}(t)$$

$$= 2(t-3)\left(\sum_{k=0}^{\infty} \frac{\left(\frac{1}{3}\right)^k}{3 - 2\left(\frac{1}{3}\right)^k (t-3)}\right) \times \left(\prod_{k=0}^{\infty} \frac{1}{1 - 2\left(\frac{1}{3}\right)^k (t-3)}\right),$$

that is, for $t \in I$,

$$\mathscr{L}_\beta\left(\frac{1}{(z-1)^2}\right)(t) = 2(t-3)\left(\sum_{k=0}^{\infty} \frac{\left(\frac{1}{3}\right)^k}{3 - 2\left(\frac{1}{3}\right)^k (t-3)}\right) \times \left(\prod_{k=0}^{\infty} \frac{1}{1 - 2\left(\frac{1}{3}\right)^k (t-3)}\right).$$

Example 4.4.6. We will find

$$\mathscr{L}_\beta^{-1}\left(\frac{z}{(z^2-1)(z^2+4)}\right)(t), \quad t \in I.$$

First, for $z \in \mathbb{C}$, we write

$$\frac{z}{(z^2-1)(z^2+4)} = \frac{a_1 z + a_2}{z^2 - 1} + \frac{a_3 z + a_4}{z^2 + 4},$$

where $a_1, a_2, a_3, a_4 \in \mathbb{R}$. Then we have

$$\frac{z}{(z^2-1)(z^2+4)} = \frac{(a_1 z + a_2)(z^2 + 4) + (a_3 z + a_4)(z^2 - 1)}{(z^2 - 1)(z^2 + 4)}$$

$$= \frac{a_1 z^3 + a_2 z^2 + 4a_1 z + 4a_2 + a_3 z^3 - a_3 z + a_4 z^2 - a_4}{(z^2 - 1)(z^2 + 4)}$$

$$= \frac{(a_1 + a_3)z^3 + (a_2 + a_4)z^2 + (4a_1 - a_3)z + 4a_2 - a_4}{(z^2 - 1)(z^2 + 4)}$$

$$= \frac{(a_1 + a_3)z^3 + (a_2 + a_4)z^2 + (4a_1 - a_2)z + 4a_2 - a_4}{(z^2 - 1)(z^2 + 4)}.$$

Thus, we get the system of equations

$$a_1 + a_3 = 0,$$

$$a_2 + a_4 = 0,$$

$$4a_1 - a_3 = 1,$$
$$4a_2 - a_4 = 0,$$

whereupon

$$a_1 = \frac{1}{5},$$
$$a_2 = 0,$$
$$a_3 = -\frac{1}{5},$$
$$a_4 = 0.$$

Then

$$\frac{z}{(z^2-1)(z^2+4)} = \frac{1}{5}\frac{z}{z^2-1} - \frac{1}{5}\frac{z}{z^2+4}, \quad z \in \mathbb{C}.$$

Hence,

$$\mathcal{L}_\beta^{-1}\left(\frac{z}{(z^2-1)(z^2+4)}\right) = \mathcal{L}_\beta^{-1}\left(\frac{1}{5}\frac{z}{z^2-1} - \frac{1}{5}\frac{z}{z^2+4}\right)$$
$$= \mathcal{L}_\beta^{-1}\left(\frac{1}{5}\frac{z}{z^2-1}\right) - \mathcal{L}_\beta^{-1}\left(\frac{1}{5}\frac{z}{z^2+4}\right)$$
$$= \frac{1}{5}\mathcal{L}_\beta^{-1}\left(\frac{z}{z^2-1}\right) - \frac{1}{5}\mathcal{L}_\beta^{-1}\left(\frac{z}{z^2+4}\right)$$
$$= \frac{1}{5}\cosh_{1,\beta}(t) - \frac{1}{5}\cos_{2,\beta}(t).$$

Thus,

$$\mathcal{L}_\beta^{-1}\left(\frac{z}{(z^2-1)(z^2+4)}\right) = \frac{1}{5}\cosh_{1,\beta}(t) - \frac{1}{5}\cos_{2,\beta}(t), \quad t \in I.$$

Exercise 4.4.7. Find

$$\mathcal{L}_\beta^{-1}\left(\frac{1}{(z^2+1)^2}\right)(t), \quad t \in I.$$

Answer.

$$\frac{1}{2}\left(\sin_{1,\beta}(t) - \cos_{1,\beta}(t)\int_{s_0}^t \frac{d\beta\tau}{1+(\beta(\tau)-\tau)^2} - \sin_{1,\beta}(t)\int_{s_0}^t \frac{(\beta(\tau)-\tau)d\beta\tau}{1+(\beta(\tau)-\tau)^2}\right), \quad t \in I.$$

4.5 Advanced practical problems

Problem 4.5.1. Find $\mathcal{L}_\beta(f(t))(z)$, where

1. $f(t) = e_{-1,\beta}(t) - e_{-2,\beta}(t), t \in I$;

2. $f(t) = 2\sin_{2,\beta}(t) - 3\cos_{1,\beta}(t), t \in I$;

3. $f(t) = 4e_{1,\beta}(t) - 4\sinh_{3,\beta}(t), t \in I$;
4. $f(t) = \sin_{4,\beta}(t) - e_{3,\beta}(t), t \in I$;
5. $f(t) = \sinh_{2,\beta}(t) - \cosh_{2,\beta}(t), t \in I$.

Answer. 1. $\dfrac{1}{(z+1)(z+2)}$.

2. $\dfrac{-3z^3 + 4z^2 - 12z + 4}{(z^2+1)(z^2+4)}$.

3. $\dfrac{4(z^2 - 2z - 2)}{(z-1)(z^2-4)}$.

4. $\dfrac{-z^2 + 4z - 28}{(z-3)(z^2+16)}$.

5. $-\dfrac{1}{z+2}$.

Problem 4.5.2. Find $\mathscr{L}_\beta^{-1}(g(z))(t), t \in I$, where

1. $g(z) = \dfrac{z^3 - 2z^2 + 1}{z^2(z-1)(z+2)}, z \in \mathbb{C}$;

2. $g(z) = \dfrac{2z}{z^2 + z - 20}, z \in \mathbb{C}$;

3. $g(z) = \dfrac{z}{(z^2+1)^2}, z \in \mathbb{C}$.

Answer. 1. $-\dfrac{1}{4}t + \dfrac{1}{16}e_{2,\beta}(t) + \dfrac{15}{16}e_{-2,\beta}(t), t \in I$.

2. $\dfrac{10}{9}e_{-5,\beta}(t) + \dfrac{8}{9}e_{4,\beta}(t), t \in I$.

3.

$$\frac{1}{2}\left(-\cos_{1,\beta}(t)\int_{s_0}^t \frac{1}{1+(\beta(\tau)-\tau)^2}\,d_\beta\tau + \sin_{1,\beta}(t)\int_{s_0}^t \frac{\beta(\tau)-\tau}{1+(\beta(\tau)-\tau)^2}\,d_\beta\tau\right), \quad t \in I.$$

4.6 **Notes and references**

In this chapter, we have defined the generalized quantum Laplace transform, that is, β-Laplace transform and proved some of its properties. Some special classes of functions ψ_k, $k \in \mathbb{N}_0$, are defined, and the β-Laplace transforms of some β-elementary functions are calculated. Some of the results in this chapter can be found in [22].

First-order linear β-differential equations*

Throughout this chapter, let $I \subseteq \mathbb{R}$, and let $\beta \colon I \to \mathbb{R}$ be a first-kind general quantum operator. By $\mathscr{C}_\beta^k$, where $k \in \mathbb{N}_0$, we denote the set of all functions $f \colon I \to \mathbb{R}$ for which $D_\beta^l f$ exist and are continuous on I for all $l \in \{0, \ldots, k\}$. For convenience, we will denote $\mathscr{C}_\beta^0$ by $\mathscr{C}_\beta$.

5.1 Linear β-differential operators

Definition 5.1.1. Let $f \in \mathscr{R}_\beta$. We define the operator $L_1 \colon \mathscr{C}_\beta^1 \to \mathscr{C}_\beta$ by

$$L_1 x = D_\beta x - f x \quad \text{on} \quad I.$$

The β-adjoint operator $L_{1\beta} \colon \mathscr{C}_\beta^1 \to \mathscr{C}_\beta$ is defined by

$$L_{1\beta} x = D_\beta x + f x^\beta \quad \text{on} \quad I.$$

Theorem 5.1.2 (The β-Lagrange identity). *If $x, y \in \mathscr{C}_\beta^1$, then*

$$x^\beta L_1 y + y L_{1\beta} x = D_\beta(xy).$$

Proof. Using Definition 5.1.1, we have

$$
\begin{aligned}
x^\beta L_1 y + y L_{1\beta} x &= x^\beta (D_\beta y - f y) + y (D_\beta x + f x^\beta) \\
&= x^\beta D_\beta y - f x^\beta y + f x^\beta y + y D_\beta x \\
&= x^\beta D_\beta y + y D_\beta x \\
&= D_\beta(xy).
\end{aligned}
$$

This completes the proof. $\qquad\square$

Corollary 5.1.3. *If x and y are solutions of the equations*

$$L_1 x = 0$$

---

* This book has a companion website hosting complementary materials. Visit this URL to access it: https://www.elsevier.com/books-and-journals/book-companion/9780443328046.

Generalized Quantum Calculus with Applications. https://doi.org/10.1016/B978-0-44-332804-6.00010-8

and

$$L_{1\beta}y = 0,$$

respectively, then

$$x(t)y(t) = x(s_0)y(s_0), \quad t \in I.$$

Proof. From Theorem 5.1.2, it follows that

$$D_\beta(xy)(t) = 0, \quad t \in I.$$

Now applying Theorem 1.5.1, we obtain

$$x(t)y(t) = x(s_0)y(s_0), \quad t \in I.$$

This completes the proof. $\qquad\qquad\square$

5.2 Homogeneous initial value problems

Consider the initial value problem (IVP)

$$\begin{aligned}
D_\beta x &= f(t)x, \quad t > s_0, \quad t \in I, \\
x(s_0) &= x_0,
\end{aligned} \tag{5.1}$$

where $x_0 \in \mathbb{R}$ and $f \in \mathcal{R}_\beta$. We have the following.

Theorem 5.2.1. *Let $x_0 \in \mathbb{R}$ and $f \in \mathcal{R}_\beta$. Then the unique solution of the IVP (5.1) is given by*

$$x(t) = x_0 e_{f,\beta}(t), \quad t \geq s_0.$$

Proof. We have

$$\begin{aligned}
D_\beta x(t) &= x_0 f(t) e_{f,\beta}(t) \\
&= f(t)x(t), \quad t > s_0,
\end{aligned}$$

and

$$\begin{aligned}
x(s_0) &= x_0 e_{f,\beta}(s_0) \\
&= x_0.
\end{aligned}$$

Now let y be another solution of the IVP (5.1). Then

$$\begin{aligned}
D_\beta\left(\frac{y}{x}\right) &= \frac{(D_\beta y(t))x(t) - y(t)D_\beta x(t)}{x(t)x(\beta(t))} \\
&= \frac{f(t)y(t)x(t) - f(t)y(t)x(t)}{x(t)x(\beta(t))} \\
&= 0, \quad t \geq s_0.
\end{aligned}$$

Hence, by Theorem 1.5.1, we conclude that

$$\frac{y(t)}{x(t)} = \frac{y(s_0)}{x(s_0)}$$
$$= \frac{x_0}{x_0}$$
$$= 1, \quad t \geq s_0.$$

Thus,

$$x(t) = y(t), \quad t \geq s_0.$$

This completes the proof. $\qquad\square$

Consider the initial value problem (IVP)

$$D_\beta x = f(t)x^\beta, \quad t > s_0, \quad t \in I,$$
$$x(s_0) = x_0, \tag{5.2}$$

where $x_0 \in \mathbb{R}$ and $-f \in \mathcal{R}_\beta$. We have the following.

Theorem 5.2.2. *Let $x_0 \in \mathbb{R}$ and $-f \in \mathcal{R}_\beta$. Then the unique solution of the IVP (5.2) is given by*

$$x(t) = x_0 E_{f,\beta}(t), \quad t \geq s_0.$$

Proof. We have

$$D_\beta x(t) = x_0 f(t) E_{f,\beta}(\beta(t))$$
$$= f(t)x(\beta(t)), \quad t > s_0,$$

and

$$x(s_0) = x_0 E_{f,\beta}(s_0)$$
$$= x_0.$$

Now let y be another solution of the IVP (5.2). Then

$$D_\beta\left(\frac{y}{x}\right)(t) = \frac{(D_\beta y(t))x(t) - y(t)D_\beta x(t)}{x(t)x(\beta(t))}$$
$$= \frac{f(t)y^\beta(t)x(t) - f(t)y(t)x^\beta(t)}{x(t)x(\beta(t))}$$
$$= f(t)\left(\left(\frac{y}{x}\right)^\beta(t) - \frac{y(t)}{x(t)}\right), \quad t \geq s_0.$$

In view of Theorem 1.3.2, we can write

$$D_\beta\left(\frac{y}{x}\right)(t) = f(t)\left[\left(\frac{y}{x}\right)(t) + (\beta(t) - t)D_\beta\left(\frac{y}{x}\right)(t)\right], \quad t \geq s_0.$$

If $\beta(t) \neq t$, then

$$D_\beta \left(\frac{y}{x} \right)(t) = f(t)(\beta(t) - t)D_\beta \left(\frac{y}{x} \right)(t),$$

whereupon

$$(1 - f(t)(\beta(t) - t))D_\beta \left(\frac{y}{x} \right)(t) = 0,$$

which yields

$$D_\beta \left(\frac{y}{x} \right)(t) = 0. \tag{5.3}$$

Now let $\beta(t) = t$. Then (5.3) holds. Hence by Theorem 1.5.1 we conclude that

$$\begin{aligned}
\frac{y(t)}{x(t)} &= \frac{y(s_0)}{x(s_0)} \\
&= \frac{x_0}{x_0} \\
&= 1, \quad t \geq s_0.
\end{aligned}$$

Thus,

$$x(t) = y(t), \quad t \geq s_0.$$

This completes the proof. $\qquad\square$

5.3 Nonhomogeneous initial value problems

In this section, we investigate the following IVP

$$\begin{aligned}
D_\beta x &= f(t)x + g(t), \quad t > s_0, \quad t \in I, \\
x(s_0) &= x_0,
\end{aligned} \tag{5.4}$$

where $f \in \mathscr{R}_\beta$ and $g \in \mathscr{C}(I)$.

Theorem 5.3.1. *The unique solution of the IVP (5.4) is given by*

$$x(t) = e_{f,\beta}(t) \left(x_0 + \int_{s_0}^{t} g(\tau)E_{-f,\beta}(\beta(\tau))d_\beta\tau \right), \quad t \geq s_0.$$

Proof. First, note that

$$e_{f,\beta}(t)E_{f,\beta}(t) = 1, \quad t \geq s_0,$$

and let

$$x(t) = x_0 e_{f,\beta}(t) + e_{f,\beta}(t) \int_{s_0}^{t} g(\tau)E_{-f,\beta}(\beta(\tau))d_\beta\tau, \quad t \geq s_0.$$

Then

$$D_\beta x(t) = x_0 D_\beta e_{f,\beta}(t) + D_\beta \left(e_{f,\beta}(t) \int_{s_0}^t g(\tau) E_{-f,\beta}(\beta(\tau)) d_\beta \tau \right)$$

$$= x_0 f(t) e_{f,\beta}(t) + \left(D_\beta e_{f,\beta}(t) \right) \int_{s_0}^t g(\tau) E_{-f,\beta}(\beta(\tau)) d_\beta \tau + e_{f,\beta}(\beta(t)) D_\beta \int_{s_0}^t g(\tau) E_{-f,\beta}(\beta(\tau)) d_\beta \tau$$

$$= x_0 f(t) e_{f,\beta}(t) + f(t) e_{f,\beta}(t) \int_{s_0}^t g(\tau) E_{-f,\beta}(\beta(\tau)) d_\beta \tau + e_{f,\beta}(\beta(t)) g(t) E_{-f,\beta}(\beta(t))$$

$$= f(t) \left(x_0 e_{f,\beta}(t) + e_{f,\beta}(t) \int_{s_0}^t g(\tau) E_{-f,\beta}(\beta(\tau)) d_\beta \tau \right) + g(t)$$

$$= f(t) x(t) + g(t), \quad t > s_0,$$

that is,

$$D_\beta x(t) = f(t) x(t) + g(t), \quad t > s_0.$$

Also,

$$x(s_0) = x_0 e_{f,\beta}(s_0) + e_{f,\beta}(s_0) \int_{s_0}^{s_0} g(\tau) E_{-f,\beta}(\beta(\tau)) d_\beta \tau$$

$$= x_0.$$

Hence, x is a solution to the IVP (5.4). Now assume that the IVP (5.4) has two solutions, say x_1 and x_2. Set $y = x_1 - x_2$. Then y is a solution to the IVP

$$D_\beta x = f(t) x, \quad t > s_0,$$
$$x(s_0) = 0.$$

Since the trivial solution is its solution, applying Theorem 5.2.1, we conclude that

$$y(t) = 0, \quad t \geq s_0,$$

that is,

$$x_1(t) = x_2(t), \quad t \geq s_0.$$

This completes the proof. $\qquad\square$

Now we consider the IVP

$$D_\beta x = f(t) x^\beta + g(t), \quad t > s_0, \quad t \in I,$$
$$x(s_0) = x_0,$$

(5.5)

where $-f \in \mathscr{R}_\beta$, $g \in \mathscr{C}(I)$, and $x_0 \in \mathbb{R}$. Since

$$x^\beta = x + (\beta(t) - t) x, \quad t \geq s_0,$$

(5.5) can be rewritten in the form

$$D_\beta x = f(t)x + f(t)(\beta(t) - t)D_\beta x + g(t), \quad t > s_0,$$

or

$$(1 - f(t)(\beta(t) - t))D_\beta x = f(t)x + g(t), \quad t > s_0,$$

whereupon

$$D_\beta x = \frac{f(t)}{1 - f(t)(\beta(t) - t)}x + \frac{g(t)}{1 - f(t)(\beta(t) - t)}, \quad t > s_0,$$

or

$$D_\beta x = \ominus(-f)(t)x + \frac{g(t)}{1 - f(t)(\beta(t) - t)}, \quad t > s_0.$$

Now applying Theorem 5.3.1, we get that the unique solution of the IVP (5.5) can be represented in the form

$$x(t) = e_{\ominus(-f),\beta}(t)\left(x_0 + \int_{s_0}^{t} \frac{g(\tau)}{1 - f(\tau)(\beta(\tau) - \tau)}E_{-(\ominus(-f)),\beta}(\beta(\tau))d_\beta\tau\right), \quad t \geq s_0.$$

5.4 The β-Laplace transform method

Consider the IVP

$$D_\beta x = ax + f(t), \quad t > s_0, \quad t \in I,$$
$$x(s_0) = x_0,$$
$$\tag{5.6}$$

where $a, x_0 \in \mathbb{R}$ and $f \in \mathscr{C}(I)$. Taking the β-Laplace transform of both sides of the first equation of (5.6) and using the initial condition, we obtain

$$z\mathscr{L}_\beta(x)(z) - x_0 = a\mathscr{L}_\beta(x)(z) + \mathscr{L}_\beta(f(t))(z),$$

or

$$(z - a)\mathscr{L}_\beta(x)(z) = x_0 + \mathscr{L}_\beta(f(t))(z),$$

whereupon

$$\mathscr{L}_\beta(x)(z) = \frac{x_0}{z - a} + \frac{1}{z - a}\mathscr{L}_\beta(f(t))(z).$$

Then taking inverse β-Laplace transform, we obtain

$$x(t) = \mathscr{L}_\beta^{-1}\left(\frac{x_0}{z - a} + \frac{1}{z - a}\mathscr{L}_\beta(f(t))(z)\right)$$

$$= x_0\mathscr{L}_\beta^{-1}\left(\frac{1}{z - a}\right) + \mathscr{L}_\beta^{-1}\left(\frac{1}{z - a}\mathscr{L}_\beta(f(t))(z)\right)$$

$$= x_0 e_{a,\beta}(t) + \mathscr{L}_\beta^{-1}\left(\frac{1}{z - a}\mathscr{L}_\beta(f(t))(z)\right), \quad t \geq s_0.$$

Thus, the solution of the IVP (5.6) is given by

$$x(t) = x_0 e_{a,\beta}(t) + \mathscr{L}_\beta^{-1}\left(\frac{1}{z-a}\mathscr{L}_\beta(f(t))(z)\right), \quad t \geq s_0.$$

Example 5.4.1. Consider the IVP

$$D_\beta x = x + e_{2,\beta}(t), \quad t > s_0, \quad t \in I,$$
$$x(s_0) = 1.$$

Taking the β-Laplace transform of both sides of the first equation of the considered IVP, we obtain

$$z\mathscr{L}_\beta(x)(z) - 1 = \mathscr{L}_\beta(x)(z) + \mathscr{L}_\beta(e_{2,\beta}(t))(z)$$
$$= \mathscr{L}_\beta(x)(z) + \frac{1}{z-2},$$

whereupon

$$(z-1)\mathscr{L}_\beta(x)(z) = 1 + \frac{1}{z-2},$$

or

$$\mathscr{L}_\beta(x)(z) = \frac{1}{z-1} + \frac{1}{(z-1)(z-2)}$$
$$= \frac{1}{z-1} + \frac{1}{z-2} - \frac{1}{z-1}$$
$$= \frac{1}{z-2}.$$

This gives

$$x(t) = e_{2,\beta}(t).$$

Hence, the solution of the considered IVP is given by

$$x(t) = e_{2,\beta}(t), \quad t \geq s_0.$$

Example 5.4.2. Consider the IVP

$$D_\beta x = x + e_{3,\beta}(t), \quad t > s_0,$$
$$x(s_0) = 1.$$

Taking the β-Laplace transform of both sides of the first equation of the considered IVP, we obtain

$$z\mathscr{L}_\beta(x)(z) - 1 = \mathscr{L}_\beta(x)(z) + \mathscr{L}_\beta(e_{3,\beta}(t))(z)$$
$$= \mathscr{L}_\beta(x)(z) + \frac{1}{z-3},$$

whereupon

$$(z-1)\mathscr{L}_\beta(x)(z) = 1 + \frac{1}{z-3},$$

or

$$\mathcal{L}_\beta(x)(z) = \frac{1}{z-1} + \frac{1}{(z-1)(z-3)}$$

$$= \frac{1}{z-1} + \frac{1}{2}\left(\frac{1}{z-3} - \frac{1}{z-1}\right)$$

$$= \frac{1}{2(z-1)} + \frac{1}{2(z-3)}.$$

This gives

$$x(t) = \frac{1}{2}(e_{1,\beta}(t) + e_{3,\beta}(t)).$$

Hence, the solution of the considered IVP is given by

$$x(t) = \frac{1}{2}(e_{1,\beta}(t) + e_{3,\beta}(t)) \quad t \geq s_0.$$

Exercise 5.4.3. Find the solution of the IVP

$$D_\beta x = 2x + e_{2,\beta}(t) + \cos_{3,\beta}(t), \quad t > s_0,$$
$$x(s_0) = 1.$$

Answer.

$$x(t) = \frac{13}{10}e_{1,\beta}(t) + e_{2,\beta}(t) - \frac{3}{10}\cos_{3,\beta}(t) - \frac{1}{10}\sin_{3,\beta}(t), \quad t \geq s_0.$$

5.5 Notes and references

This chapter deals with the homogeneous and nonhomogeneous initial value problems for first-order β-differential equations. We give integral representations of their solutions. The β-Laplace transform method is considered. Some of the results in this chapter can be found in [9,13].

Second-order linear β-differential equations[*]

Throughout this chapter, we suppose $I \subseteq \mathbb{R}$, and $\beta : I \to \mathbb{R}$ is a first-kind general quantum operator.

6.1 β-Wronskians

In this section, we investigate the second-order linear β-differential equation

$$D_\beta^2 x + f(t) D_\beta x + g(t) x = h(t), \tag{6.1}$$

where $f, g, h \in \mathscr{C}(I)$. Define the operator $L_2 : \mathscr{C}_\beta^2(I) \to \mathscr{C}(I)$ as follows:

$$L_2 x(t) = D_\beta^2 x(t) + f(t) D_\beta x(t) + g(t) x(t), \quad t \in I.$$

Then, (6.1) can be rewritten in the form

$$L_2 x(t) = h(t), \quad t \in I.$$

Definition 6.1.1. If $h \equiv 0$ on I, then (6.1) is said to be a homogeneous equation. If $h \not\equiv 0$ on I, then (6.1) is said to be a nonhomogeneous equation.

Theorem 6.1.2 (Principle of superposition). *The operator $L_2 : \mathscr{C}_\beta^2(I) \to \mathscr{C}(I)$ is a linear operator. If x_1, x_2 are solutions to the homogeneous equation $L_2 x = 0$, then so does $a x_1 + b x_2$ for all $a, b \in \mathbb{R}$.*

Proof. Let $a, b \in \mathbb{R}$ and $x_1, x_2 \in \mathscr{C}_\beta^2(I)$. Then

$$L_2(a x_1 + b x_2)(t) = D_\beta^2(a x_1 + b x_2)(t) + f(t) D_\beta(a x_1 + b x_2)(t) + g(t)(a x_1 + b x_2)(t)$$

$$= a D_\beta^2 x_1(t) + a f(t) D_\beta x_1(t) + a g(t) x_2(t) + b D_\beta^2 x_2(t) + b f(t) D_\beta x_2(t) + b g(t) x_2(t)$$

$$= a L_2 x_1(t) + b L_2 x_2(t), \quad t \in I.$$

Thus, $L_2 : \mathscr{C}_\beta^2(I) \to \mathscr{C}(I)$ is a linear operator. Now, since

$$L_2 x_1(t) = 0 \quad \text{and}$$

---

[*] This book has a companion website hosting complementary materials. Visit this URL to access it: https://www.elsevier.com/books-and-journals/book-companion/9780443328046.

Generalized Quantum Calculus with Applications. https://doi.org/10.1016/B978-0-44-332804-6.00011-X

$$L_2 x_2(t) = 0, \quad t \in I,$$

we find that

$$L_2(\alpha x_1 + \beta x_2)(t) = \alpha L_2 x_1(t) + \beta L_2 x_2(t)$$
$$= 0, \quad t \in I.$$

This completes the proof. $\qquad\qquad\qquad\qquad\qquad\qquad\qquad\qquad\qquad\qquad\qquad\qquad\square$

Definition 6.1.3. The β-differential equation (6.1) is said to be β-regressive provided

$$1 - (\beta(t) - t)f(t) + (\beta(t) - t)^2 g(t) \neq 0, \quad t \in I.$$

Definition 6.1.4. For any two β-differentiable functions x_1 and x_2, we define the β-Wronskian $W_\beta = W_\beta(x_1, x_2)$ as follows

$$W_\beta(x_1, x_2(t) = \det \begin{pmatrix} x_1(t) & x_2(t) \\ D_\beta x_1(t) & D_\beta x_2(t) \end{pmatrix}, \quad t \in I.$$

We say that two solutions x_1 and x_2 of $L_2 x = 0$ form a fundamental set (or a fundamental system) of solutions of $L_2 x = 0$ provided

$$W_\beta(x_1, x_2)(t) \neq 0, \quad t \in I.$$

Example 6.1.5. Let $\beta(t) - t \neq 0$ for some $t \in I$. Then

$$W_\beta(x_1, x_2)(t) = \det \begin{pmatrix} x_1(t) & x_2(t) \\ D_\beta x_1(t) & D - \beta x_2(t) \end{pmatrix}$$

$$= \det \begin{pmatrix} x_1(t) & x_2(t) \\ \dfrac{x_1(\beta(t)) - x_1(t)}{\beta(t) - t} & \dfrac{x_2(\beta(t)) - x_2(t)}{\beta(t) - t} \end{pmatrix}$$

$$= \frac{1}{\beta(t) - t} \det \begin{pmatrix} x_1(t) & x_2(t) \\ x_1(\beta(t)) & x_2(\beta(t)) \end{pmatrix} - \frac{1}{\beta(t) - t} \det \begin{pmatrix} x_1(t) & x_2(t) \\ x_1(t) & x_2(t) \end{pmatrix}$$

$$= \frac{1}{\beta(t) - t} \det \begin{pmatrix} x_1(t) & x_2(t) \\ x_1(\beta(t)) & x_2(\beta(t)) \end{pmatrix}.$$

Theorem 6.1.6. *If the pair of functions x_1, x_2 forms a fundamental system of solutions for $L_2 x = 0$, then*

$$x(t) = a x_1(t) + b x_2(t), \quad t \in I, \tag{6.2}$$

where $a, b \in \mathbb{R}$, is a general solution of $L_2 x = 0$, that is, every function of the form (6.2) is a solution of $L_2 x = 0$, and every solution is of the form (6.2).

Proof. Assume that the pair of functions x_1, x_2 forms a fundamental system of solutions for $L_2 x = 0$. Then, by Theorem 6.1.2, it follows that any function of the form

$$a x_1(t) + b x_2(t), \quad t \in I,$$

where $a, b \in \mathbb{R}$, is a solution to $L_2 x = 0$. Now let y be a solution of $L_2 x = 0$. Set

$$y(s_0) = y_0 \quad \text{and}$$
$$D_\beta y(s_0) = y_1.$$

Denote

$$u(t) = \alpha x_1(t) + \gamma x_2(t), \quad t \in I,$$

where $\alpha, \gamma \in \mathbb{R}$, which will be determined so that

$$u(s_0) = y_0 \quad \text{and}$$
$$D_\beta u(s_0) = y_1.$$

Then we have

$$\alpha x_1(s_0) + \gamma x_2(s_0) = y_0 \quad \text{and}$$
$$\alpha D_\beta x_1(s_0) + \gamma D_\beta x_2(s_0) = y_1,$$

whereupon

$$\alpha = \frac{y_0 D_\beta x_2(s_0) - y_1 x_2(s_0)}{W_\beta(x_1, x_2)(s_0)}, \quad \text{and}$$
$$\gamma = \frac{y_1 x_1(s_0) - y_0 D_\beta x_0(s_0)}{W_\beta(x_1, x_2)(s_0)}.$$

Therefore,

$$u(t) = \frac{y_0 D_\beta x_2(s_0) - y_1 x_1(s_0)}{W_\beta(x_1, x_2)(s_0)} x_1(t) + \frac{y_1 x_1(s_0) - y_0 D_\beta x_1(s_0)}{W_\beta(x_1, x_2)(s_0)} x_2(t), \quad t \in I.$$

Now, applying the existence and uniqueness theorem (see Chapter 7, Theorem 7.5.2), we get

$$u(t) = y(t), \quad t \in I.$$

This completes the proof. $\qquad\square$

Now we will prove some important properties of the β-Wronskian.

Theorem 6.1.7. *Let x_1 and x_2 be twice β-differentiable functions on I. Then the β-Wronskian $W_\beta(x_1, x_2)$ has the following properties:*

1.

$$W_\beta(x_1, x_2)(t) = \det \begin{pmatrix} x_1(\beta(t)) & x_2(\beta(t)) \\ D_\beta x_1(t) & D_\beta x_2(t) \end{pmatrix};$$

2.

$$D_\beta W_\beta(x_1, x_2)(t) = \det \begin{pmatrix} x_1(\beta(t)) & x_2(\beta(t)) \\ D_\beta^2 x_1(t) & D_\beta^2 x_2(t) \end{pmatrix};$$

3.

$$D_\beta W_\beta(x_1, x_2)(t) = \det \begin{pmatrix} x_1(\beta(t)) & x_2(\beta(t)) \\ L_2 x_1(t) & L_2 x_2(t) \end{pmatrix} + (-f(t) + (\beta(t) - t)g(t))W_\beta(x_1, x_2)(t).$$

Proof. **1.** We have

$$W_\beta(x_1, x_2)(t) = \det \begin{pmatrix} x_1(t) & x_2(t) \\ D_\beta x_1(t) & D_\beta x_2(t) \end{pmatrix}$$

$$= \det \begin{pmatrix} x_1(\beta(t)) - (\beta(t) - t)D_\beta x_1(t) & x_2(\beta(t)) - (\beta(t) - t)D_\beta x_2(t) \\ D_\beta x_1(t) & D_\beta x_2(t) \end{pmatrix}$$

$$= \det \begin{pmatrix} x_1(\beta(t)) & x_2(\beta(t)) \\ D_\beta x_1(t) & D_\beta x_2(t) \end{pmatrix} - (\beta(t) - t)\det \begin{pmatrix} D_\beta x_1(t) & D_\beta x_2(t) \\ D_\beta x_1(t) & D_\beta x_2(t) \end{pmatrix}$$

$$= \det \begin{pmatrix} x_1(\beta(t)) & x_2(\beta(t)) \\ D_\beta x_1(t) & D_\beta x_2(t) \end{pmatrix}.$$

2. We have

$$D_\beta W_\beta(x_1, x_2)(t) = D_\beta \det \begin{pmatrix} x_1(t) & x_2(t) \\ D_\beta x_1(t) & D_\beta x_2(t) \end{pmatrix}$$

$$= D_\beta(x_1 D_\beta x_2 - x_2 D_\beta x_1)(t)$$

$$= D_\beta x_1(t)D_\beta x_2(t) + x_1(\beta(t))D_\beta^2 x_2(t) - D_\beta x_1(t)D_\beta x_2(t) - x_2(\beta(t))D_\beta^2 x_1(t)$$

$$= x_1(\beta(t))D_\beta^2 x_2(t) - x_2(\beta(t))D_\beta^2 x_1(t)$$

$$= \det \begin{pmatrix} x_1(\beta(t)) & x_2(\beta(t)) \\ D_\beta^2 x_1(t) & D_\beta^2 x_2(t) \end{pmatrix}.$$

3. By the previous property, we get

$$D_\beta W_\beta(x_1, x_2)(t) = \det \begin{pmatrix} x_1(\beta(t)) & x_2(\beta(t)) \\ D_\beta x_1(t) & D_\beta x_2(t) \end{pmatrix}$$

$$= \det \begin{pmatrix} x_1(\beta(t)) & x_2(\beta(t)) \\ L_2 x_1(t) - f(t)D_\beta x_1(t) - g(t)x_1(t) & L_2 x_2(t) - f(t)D_\beta x_2(t) - g(t)x_2(t) \end{pmatrix}$$

$$= \det \begin{pmatrix} x_1(\beta(t)) & x_2(\beta(t)) \\ L_2 x_1(t) & L_2 x_2(t) \end{pmatrix}$$

$$\quad + \det \begin{pmatrix} x_1(\beta(t)) & x_2(\beta(t)) \\ -f(t)D_\beta x_1(t) - g(t)x_1(t) & -f(t)D_\beta x_2(t) - g(t)x_2(t) \end{pmatrix}$$

$$= \det \begin{pmatrix} x_1(\beta(t)) & x_2(\beta(t)) \\ L_2 x_1(t) & L_2 x_2(t) \end{pmatrix}$$

$$
\begin{aligned}
&+ \det \begin{pmatrix} x_1(\beta(t)) \\ -f(t)D_\beta x_1(t) - g(t)x_1(\beta(t)) + (\beta(t)-t)g(t)D_\beta x_1(t) \\[4pt] x_2(\beta(t)) \\ -f(t)D_\beta x_2(t) - g(t)x_2(\beta(t)) + (\beta(t)-t)g(t)D_\beta x_2(t) \end{pmatrix} \\[6pt]
&= \det \begin{pmatrix} x_1(\beta(t)) & x_2(\beta(t)) \\ L_2 x_1(t) & L_2 x_2(t) \end{pmatrix} \\[6pt]
&+ \det \begin{pmatrix} x_1(\beta(t)) & x_2(\beta(t)) \\ (-f(t)+g(t)(\beta(t)-t))D_\beta x_1(t) & (-f(t)+g(t)(\beta(t)-t))D_\beta x_2(t) \end{pmatrix} \\[6pt]
&= \det \begin{pmatrix} x_1(\beta(t)) & x_2(\beta(t)) \\ L_2 x_1(t) & L_2 x_2(t) \end{pmatrix} + (-f(t)+g(t)(\beta(t)-t))W_\beta(x_1,x_2)(t).
\end{aligned}
$$

This completes the proof. $\qquad\qquad\qquad\qquad\qquad\qquad\qquad\qquad\qquad\qquad\qquad\square$

6.2 β-Abel theorems

Theorem 6.2.1 (β-Abel theorem I). *Assume that $L_2 x = 0$ is β-regressive and x_1 and x_2 are its solutions. Then*

$$
W_\beta(x_1, x_2)(t) = e_{-f(t)+(\beta(t)-t)g(t),\beta}(t) W_\beta(x_1, x_2)(s_0). \tag{6.3}
$$

Proof. Since x_1 and x_2 are solutions of $L_2 x = 0$, we have

$$
\begin{aligned}
L_2 x_1(t) &= 0 \quad \text{and} \\
L_2 x_2(t) &= 0.
\end{aligned}
$$

This gives

$$
D_\beta W_\beta(x_1, x_2)(t) = (-f(t) + (\beta(t)-t)g(t))W_\beta(x_1, x_2)(t),
$$

whereupon we get (6.3). This completes the proof. $\qquad\qquad\qquad\qquad\qquad\square$

 Alternatively, we will consider the β-differential equation

$$
D_\beta^2 x + f(t)(D_\beta x)^\beta + g(t)x^\beta = 0. \tag{6.4}
$$

Definition 6.2.2. Eq. (6.4) is said to be β-regressive provided $f \in \mathcal{R}_\beta(I)$ and $g \in \mathcal{C}(I)$.

Theorem 6.2.3. *If* (6.4) *is β-regressive, then it is equivalent to a β-regressive equation of the form $L_2 x = 0$. Conversely, if $L_2 x = 0$ is β-regressive, then it is equivalent to a β-regressive equation of the form* (6.4).

Proof. Let (6.4) be β-regressive. Then

$$
1 + f(t)(\beta(t) - t) \neq 0.
$$

Moreover,

$$D_\beta^2 x(t) + f(t)(D_\beta x(t))^\beta + g(t)(x(t))^\beta$$
$$= D_\beta^2 x(t) + f(t)(D_\beta x(t) + (\beta(t) - t)D_\beta^2 x(t)) + g(t)(x(t) + (\beta(t) - t)D_\beta x(t))$$
$$= (1 + f(t)(\beta(t) - t))D_\beta^2 x(t) + (f(t) + g(t)(\beta(t) - t))D_\beta x(t) + g(t)x(t).$$

Now, (6.4) yields

$$D_\beta^2 x(t) + \frac{f(t) + g(t)(\beta(t) - t)}{1 + f(t)(\beta(t) - t)} D_\beta x(t) + \frac{g(t)}{1 + f(t)(\beta(t) - t)} x(t) = 0.$$

We will show that the last equation is a β-regressive equation of the form $L_2 x = 0$. Indeed, we have

$$1 - (\beta(t) - t)\frac{f(t) + g(t)(\beta(t) - t)}{1 + f(t)(\beta(t) - t)} + (\beta(t) - t)^2 \frac{g(t)}{1 + f(t)(\beta(t) - t)} = \frac{1}{1 + f(t)(\beta(t) - t)}$$
$$\neq 0.$$

Next, assume that $L_2 x = 0$ is a β-regressive equation. Then

$$1 - f(t)(\beta(t) - t) + g(t)(\beta(t) - t)^2 \neq 0.$$

Moreover,

$$D_\beta^2 x(t) + f(t)D_\beta x(t) + g(t)x(t) = D_\beta^2 x(t) + f(t)((D_\beta x)(\beta(t)) - (\beta(t) - t)D_\beta^2 x(t))$$
$$+ g(t)(x(\beta(t)) - (\beta(t) - t)D_\beta x(t))$$
$$= (1 - f(t)(\beta(t) - t))D_\beta^2 x(t) + f(t)(D_\beta x)(\beta(t)) + g(t)x(\beta(t))$$
$$- g(t)(\beta(t) - t)((D_\beta x)(\beta(t)) - (\beta(t) - t)D_\beta^2 x(t))$$
$$= (1 - f(t)(\beta(t) - t) + g(t)(\beta(t) - t)^2)D_\beta^2 x(t)$$
$$+ (f(t) - g(t)(\beta(t) - t))(D_\beta x)(\beta(t)) + g(t)x(\beta(t)).$$

Now, $L_2 x = 0$ yields

$$D_\beta^2 x(t) + \frac{f(t) - g(t)(\beta(t) - t)}{1 - f(t)(\beta(t) - t) + g(t)(\beta(t) - t)^2}(D_\beta x)(\beta(t)) + \frac{g(t)}{f(t) - g(t)(\beta(t) - t)} x(\beta(t)) = 0.$$

We see that the last equation is β-regressive. Indeed,

$$1 + (\beta(t) - t)\frac{f(t) - g(t)(\beta(t) - t)}{1 - f(t)(\beta(t) - t) + g(t)(\beta(t) - t)^2} = \frac{1}{1 - f(t)(\beta(t) - t) + g(t)(\beta(t) - t)^2}$$
$$\neq 0.$$

This completes the proof. $\qquad\square$

Theorem 6.2.4 (β-Abel theorem II). *Assume that (6.4) is β-regressive. Suppose that x_1 and x_2 are two solutions of (6.4). Then*

$$W_\beta(x_1, x_2)(t) = e_{\ominus_\beta f, \beta}(t) W_\beta(x_1, x_2)(s_0). \tag{6.5}$$

Proof. By the properties of the β-Wronskian we have

$$D_\beta W_\beta(x_1, x_2)(t) = \det \begin{pmatrix} x_1(\beta(t)) & x_2(\beta(t)) \\ D_\beta^2 x_1(t) & D_\beta^2 x_2(t) \end{pmatrix}$$

$$= \det \begin{pmatrix} x_1(\beta(t)) & x_2(\beta(t)) \\ -f(t)(D_\beta x_1)(\beta(t)) - g(t)x_1(\beta(t)) & -f(t)(D_\beta x_2)(\beta(t)) - g(t)x_2(\beta(t)) \end{pmatrix}$$

$$= \det \begin{pmatrix} x_1(\beta(t)) & x_2(\beta(t)) \\ -f(t)(D_\beta x_1)(\beta(t)) & -f(t)(D_\beta x_2)(\beta(t)) \end{pmatrix}$$

$$= -f(t) \det \begin{pmatrix} x_1(\beta(t)) & x_2(\beta(t)) \\ (D_\beta x_1)(\beta(t)) & (D_\beta x_2)(\beta(t)) \end{pmatrix}$$

$$= -f(t) W_\beta(x_1, x_2)(\beta(t)),$$

that is,

$$D_\beta W_\beta(x_1, x_2)(t) = -f(t) W_\beta(x_1, x_2)(\beta(t)),$$

whereupon we get (6.5). This completes the proof. $\qquad\square$

Corollary 6.2.5. *Assume that $g \in \mathscr{C}(I)$. Then the β-Wronskian of any two solutions of*

$$D_\beta^2 x + g(t)x^\beta = 0 \tag{6.6}$$

is constant.

Proof. Comparing (6.6) with (6.4), we get $f \equiv 0$ on I. Then

$$e_{\ominus_\beta f, \beta}(t) = 1, \quad t \in I.$$

Also, we see that (6.6) is β-regressive. So, applying (6.5), we get

$$W_\beta(x_1, x_2)(t) = W_\beta(x_1, x_2)(s_0), \quad t \in I.$$

This completes the proof. $\qquad\square$

6.3 Homogeneous second-order linear β-differential equations with constant coefficients

In this section, we investigate the homogeneous second-order linear β-differential equation with constant coefficients

$$D_\beta^2 x + a D_\beta x + bx = 0, \tag{6.7}$$

where $a, b \in \mathbb{R}$. Suppose that (6.7) is regressive, that is,

$$1 - a(\beta(t) - t) + b(\beta(t) - t)^2 \neq 0.$$

We will search a solution of (6.7) in the form

$$x(t) = e_{\lambda,\beta}(t), \quad t \in I. \tag{6.8}$$

We have

$$D_\beta x(t) = \lambda e_{\lambda,\beta}(t) \quad \text{and}$$
$$D_\beta^2 x(t) = \lambda^2 e_{\lambda,\beta}(t), \quad t \in I.$$

Then

$$D_\beta^2 x(t) + a D_\beta x(t) + b x(t) = \lambda^2 e_{\lambda,\beta}(t) + a\lambda e_{\lambda,\beta}(t) + b e_{\lambda,\beta}(t)$$
$$= (\lambda^2 + a\lambda + b) e_{\lambda,\beta}(t).$$

In view of (6.7), we obtain

$$(\lambda^2 + a\lambda + b) e_{\lambda,\beta}(t) = 0, \quad t \in I.$$

Since $e_{\lambda,\beta}(\cdot)$ does not vanish, the function given in (6.8) is a solution to (6.7) if and only if

$$\lambda^2 + a\lambda + b = 0. \tag{6.9}$$

Definition 6.3.1. Eq. (6.9) is said to be the characteristic equation of Eq. (6.7).

The roots of Eq. (6.9) are given by

$$\lambda_1 = \frac{-a - \sqrt{a^2 - 4b}}{2} \quad \text{and}$$
$$\lambda_2 = \frac{-a + \sqrt{a^2 - 4b}}{2}.$$

Theorem 6.3.2. *Eq. (6.7) is β-regressive if and only if $\lambda_1, \lambda_2 \in \mathcal{R}_\beta$.*

Proof. Eq. (6.7) is β-regressive if and only if

$$0 \neq 1 - a(\beta(t) - t) + b(\beta(t) - t)^2$$
$$= (\lambda_1(\beta(t) - t) + 1)(\lambda_2(\beta(t) - t) + 1)$$

if and only if

$$1 + \lambda_1(\beta(t) - t) \neq 0,$$
$$1 + \lambda_2(\beta(t) - t) \neq 0.$$

This completes the proof. $\square$

Theorem 6.3.3. *Consider (6.7). Suppose that $a^2 - 4b \neq 0$. If $(\beta(t) - t)b - a \in \mathcal{R}_\beta(I)$, $t \in I$, then*

$$e_{\lambda_1,\beta}(\cdot) \quad \text{and} \quad e_{\lambda_2,\beta}(\cdot) \tag{6.10}$$

form a fundamental system of solutions for (6.7).

Proof. Since λ_1 and λ_2 are solutions to (6.9), (6.10) are solutions to (6.7). Next,

$$W_\beta(e_{\lambda_1,\beta}(\cdot), e_{\lambda_2,\beta}(\cdot))(t) = \det \begin{pmatrix} e_{\lambda_1,\beta}(t) & e_{\lambda_2,\beta}(t) \\ \lambda_1 e_{\lambda_1,\beta}(t) & \lambda_2 e_{\lambda_2,\beta}(t) \end{pmatrix}$$

$$= \lambda_2 e_{\lambda_1,\beta}(t) e_{\lambda_2,\beta}(t) - \lambda_1 e_{\lambda_1,\beta}(t) e_{\lambda_2,\beta}(t)$$

$$= (\lambda_2 - \lambda_1) e_{\lambda_1,\beta}(t) e_{\lambda_2,\beta}(t)$$

$$= \sqrt{a^2 - 4b}\; e_{\lambda_1,\beta}(t) e_{\lambda_2,\beta}(t)$$

$$\neq 0, \quad t \in I.$$

Thus, (6.10) form a fundamental system for (6.7). This completes the proof. $\qquad\square$

Example 6.3.4. Let $\gamma > 0$ and $-\gamma^2(\beta(t) - t) \in \mathscr{R}_\beta(I)$, $t \in I$. Then

$$\cosh_\gamma(\cdot) \quad \text{and} \quad \sinh_\gamma(\cdot)$$

form a fundamental system of

$$D_\beta^2 x - \gamma^2 x = 0.$$

Indeed, we have

$$D_\beta \sinh_{\gamma,\beta}(t) = \gamma \cosh_{\gamma,\beta}(t),$$
$$D_\beta^2 \sinh_{\gamma,\beta}(t) = \gamma^2 \sinh_{\gamma,\beta}(t),$$
$$D_\beta \cosh_{\gamma,\beta}(t) = \gamma \sinh_{\gamma,\beta}(t),$$
$$D_\beta^2 \cosh_{\gamma,\beta}(t) = \gamma^2 \cosh_{\gamma,\beta}(t),$$

and

$$W_\beta(\sinh_{\gamma,\beta}(\cdot), \cosh_{\gamma,\beta}(\cdot))(t) = \det \begin{pmatrix} \cosh_{\gamma,\beta}(t) & \sinh_{\gamma,\beta}(t) \\ \gamma \sinh_{\gamma,\beta}(t) & \gamma \cosh_{\gamma,\beta}(t) \end{pmatrix}$$

$$= \gamma(\cosh_{\gamma,\beta}(t))^2 - \gamma(\sinh_{\gamma,\beta}(t))^2$$

$$= \gamma$$

$$\neq 0.$$

Exercise 6.3.5. Suppose that $a^2 - 4b > 0$. Let

$$p = -\frac{a}{2} \quad \text{and}$$

$$q = \frac{\sqrt{a^2 - 4b}}{2}.$$

If $p, (\beta(t) - t)b - a \in \mathscr{R}_\beta(I)$, then prove that

$$\cosh_{\frac{q}{1+(\beta(t)-t)p},\beta}(\cdot) e_p(\cdot) \quad \text{and} \quad \sinh_{\frac{q}{1+(\beta(t)-t)p},\beta}(\cdot) e_p(\cdot)$$

form a fundamental system of solutions for (6.7).

6.4 Reduction of order

In this section, we investigate the homogeneous second-order linear β-differential equation (6.7) in the case where $a^2 = 4b$. Then

$$\lambda_1 = \lambda_2 = p,$$

where $p = -\dfrac{a}{2}$. Hence

$$a = -2p,$$
$$b = p^2,$$

and (6.7) takes the form

$$D_\beta^2 x - 2pD_\beta x + p^2 x = 0. \tag{6.11}$$

One solution of (6.11) is

$$x_1(t) = e_p(t), \quad t \in I.$$

We will find another linearly independent solution of (6.11). For this, we will use the method of reduction of order. We will search a solution of the form

$$x(t) = v(t)e_{p,\beta}(t),$$

where v is a β-differentiable function such that

$$D_\beta v(s_0) = 1, \quad v(s_0) = 0.$$

We have

$$D_\beta x(t) = D_\beta v(t)e_{p,\beta}(\beta(t)) + v(t)pe_{p,\beta}(t)$$
$$= (1 + p(\beta(t) - t))D_\beta v(t)e_{p,\beta}(t) + pv(t)e_{p,\beta}(t), \quad t \in I,$$

and

$$D_\beta^2 x(t) = D_\beta \left(D_\beta v(t)(1 + p(\beta(t) - t)) \right) e_{p,\beta}(\beta(t))$$
$$+ D_\beta v(t)(1 + p(\beta(t) - t))pe_{p,\beta}(t) + pD_\beta v(t)e_{p,\beta}(\beta(t)) + p^2 v(t)e_{p,\beta}(t)$$
$$= D_\beta \left(D_\beta v(t)(1 + p(\beta(t) - t)) \right) e_{p,\beta}(\beta(t))$$
$$+ 2pD_\beta v(t)(1 + p(\beta(t) - t))e_{p,\beta}(t) + p^2 v(t)e_{p,\beta}(t), \quad t \in I.$$

Now,

$$D_\beta^2 x(t) - 2pD_\beta x(t) + p^2 x(t) = D_\beta \left(D_\beta v(t)(1 + p(\beta(t) - t)) \right) e_{p,\beta}(\beta(t))$$
$$+ 2pD_\beta v(t)(1 + p(\beta(t) - t))e_{p,\beta}(t) + p^2 v(t)e_{p,\beta}(t)$$
$$- 2pD_\beta v(t)(1 + p(\beta(t) - t))e_{p,\beta}(t) - 2p^2 v(t)e_{p,\beta}(t) + p^2 v(t)e_{p,\beta}(t)$$
$$= D_\beta \left(D_\beta v(t)(1 + p(\beta(t) - t)) \right) e_{p,\beta}(\beta(t)).$$

In view of (6.11), we obtain

$$D_\beta \left(D_\beta v(t)(1 + p(\beta(t) - t)) \right) e_{p,\beta}(\beta(t)) = 0, \quad t \in I.$$

This gives

$$D_\beta v(t)(1 + p(\beta(t) - t)) = 1, \quad t \in I,$$

that is,

$$D_\beta v(t) = \frac{1}{1 + p(\beta(t) - t)}, \quad t \in I,$$

Now, β-integrating the last equation from t_0 to t, $t \in I$, we obtain

$$v(t) = \int_{t_0}^{t} \frac{1}{1 + p(\beta(\tau) - \tau)} d_\beta \tau, \quad t \in I.$$

Therefore, another solution of (6.11) is

$$x(t) = e_{p,\beta}(t) \int_{t_0}^{t} \frac{1}{1 + p(\beta(\tau) - \tau)} d_\beta \tau, \quad t \in I.$$

6.5 Method of factoring

In this section, we represent the method of factoring for a class of second-order linear β-differential equations

$$D_\beta \left(D_\beta x - fx \right)(t) - g(t)(D_\beta x(t) - f(t)x(t)) = 0, \quad t \in I, \tag{6.12}$$

where $f, g \in \mathscr{R}_\beta(I)$. We set

$$y(t) = D_\beta x(t) - f(t)x(t), \quad t \in I.$$

Then, (6.12) takes the form

$$D_\beta y(t) = g(t)y(t), \quad t \in I.$$

Hence,

$$y(t) = ce_{g,\beta}(t), \quad t \in I,$$

where $c \in \mathbb{R}$. Therefore,

$$D_\beta x(t) = f(t)x(t) + ce_{g,\beta}(t), \quad t \in I,$$

and

$$x(t) = e_{f,\beta}(t) \left(c_1 + c \int_{s_0}^{t} e_{g,\beta}(\tau) E_{-f,\beta}(\beta(\tau)) d_\beta \tau \right), \quad t \in I,$$

where $c_1 \in \mathbb{R}$.

Example 6.5.1. Let $I = [7, 15]$, and let

$$\beta(t) = \frac{6}{7}t + 1, \quad t \in I.$$

We will find a general solution of

$$D_\beta^2 x - 2(t+1)D_\beta x + 4tx = 0, \quad t \in I,$$

using the method of factoring. Here

$$f(t) = 2 \quad \text{and}$$
$$g(t) = 2t, \quad t \in I.$$

By the computations in Example 1.1.5, we find

$$[k]_{\frac{6}{7}} = \frac{1 - \left(\frac{6}{7}\right)^k}{1 - \frac{6}{7}}$$

$$= 7\left(1 - \left(\frac{6}{7}\right)^k\right), \quad k \in \mathbb{N},$$

and

$$\beta^k(t) = \left(\frac{6}{7}\right)^k t + 7\left(1 - \left(\frac{6}{7}\right)^k\right)$$

$$= \left(\frac{6}{7}\right)^k (t - 7) + 7,$$

$$\beta^{k+1}(t) = \left(\frac{6}{7}\right)^{k+1} (t - 7) + 7, \quad k \in \mathbb{N}, \quad t \in I.$$

Therefore,

$$E_{-f,\beta}(t) = \prod_{k=0}^{\infty}\left(1 - 2\left(\left(\frac{6}{7}\right)^k (t - 7) + 7 - \left(\frac{6}{7}\right)^{k+1}(t - 7) - 7\right)\right)$$

$$= \prod_{k=0}^{\infty}\left(1 - 2\left(1 - \frac{6}{7}\right)\left(\frac{6}{7}\right)^k (t - 7)\right)$$

$$= \prod_{k=0}^{\infty}\left(1 - \frac{2}{7}(t - 7)\left(\frac{6}{7}\right)^k\right), \quad t \in I,$$

and

$$E_{-f,\beta}(\beta(t)) = \prod_{k=0}^{\infty}\left(1 - \frac{2}{7}\left(\frac{6}{7}t + 1 - 7\right)\left(\frac{6}{7}\right)^k\right)$$

$$= \prod_{k=0}^{\infty}\left(1 - \frac{12}{49}(t - 7)\left(\frac{6}{7}\right)^k\right), \quad t \in I.$$

Also,

$$e_{f,\beta}(t) = \frac{1}{\prod\limits_{k=0}^{\infty}\left(1 - 2(t-7)\left(\frac{6}{7}\right)^k\right)}, \quad t \in I,$$

and

$$e_{g,\beta}(t) = \frac{1}{\prod\limits_{k=0}^{\infty}\left(1 - 2\left(\left(\frac{6}{7}\right)^k(t-7) - 7\right)\frac{1}{7}\left(\frac{6}{7}\right)^k(t-7)\right)}$$

$$= \prod\limits_{l=0}^{\infty}\frac{1}{\left(1 - \frac{2}{7}\left(\frac{36}{49}\right)^l(t-7)^2 + 2\left(\frac{6}{7}\right)^l(t-7)\right)}, \quad t \in I.$$

Hence,

$$\int_{s_0}^{t} e_{g,\beta}(t)E_{-f,\beta}(\beta(t))d_\beta t = \sum\limits_{k=0}^{\infty}\left(\beta^k(t) - \beta^{k+1}(t)\right)e_{g,\beta}(t)E_{-f,\beta}(\beta(t))$$

$$= \frac{t-7}{7}\sum\limits_{k=0}^{\infty}\left(\frac{6}{7}\right)^k\prod\limits_{l=0}^{\infty}\frac{1 - \frac{12}{49}(t-7)\left(\frac{6}{7}\right)^k}{\left(1 - \frac{2}{7}\left(\frac{36}{49}\right)^l(t-7)^2 + 2\left(\frac{6}{7}\right)^l(t-7)\right)}, \quad t \in I.$$

Therefore,

$$y(t) = \frac{1}{\prod\limits_{k=0}^{\infty}\left(1 - 2(t-7)\left(\frac{6}{7}\right)^k\right)}\left(c_1 + c\frac{t-7}{7}\sum\limits_{k=0}^{\infty}\left(\frac{6}{7}\right)^k\prod\limits_{l=0}^{\infty}\frac{1 - \frac{12}{49}(t-7)\left(\frac{6}{7}\right)^k}{\left(1 - \frac{2}{7}\left(\frac{36}{49}\right)^l(t-7)^2 + 2\left(\frac{6}{7}\right)^l(t-7)\right)}\right),$$

$t \in I$, where $c, c_1 \in \mathbb{R}$.

Exercise 6.5.2. Let $I = [5, 20]$, and let

$$\beta(t) = \frac{5}{6}t + 1, \quad t \in I.$$

Solve the β-differential equation

$$D_\beta^2 x - 7D_\beta x + 12x = 0.$$

Answer.

$$x(t) = \frac{1}{\prod\limits_{k=0}^{\infty}\left(1 - 3(t-6)\left(\frac{5}{6}\right)^k\right)}\left(c_1 + c\left(\frac{t-6}{6}\right)\sum\limits_{k=0}^{\infty}\left(\frac{5}{6}\right)^k\prod\limits_{l=0}^{\infty}\frac{1 - \frac{1}{2}(t-6)\left(\frac{5}{6}\right)^l}{1 - \frac{2}{3}\left(\frac{5}{6}\right)^l(t-6)}\right), \quad t \in I,$$

where $c, c_1 \in \mathbb{R}$.

Theorem 6.5.3. *Let p and q be constants, and let*

$$f = p + q \quad \text{and}$$
$$g = pq.$$

Then

$$D_\beta^2 x + f D_\beta x + g x = 0 \tag{6.13}$$

can be factored in the form (6.12).

Proof. Since $f = p + q$ and $g = pq$, we have

$$\begin{aligned}
D_\beta^2 x + f D_\beta x + g x &= D_\beta^2 x + (p + q) D_\beta x + pq x \\
&= D_\beta^2 x + p D_\beta x + q D_\beta x + pq x \\
&= D_\beta (D_\beta x + px) + q(D_\beta x + px),
\end{aligned}$$

which completes the proof. $\qquad\square$

Theorem 6.5.4. *Let $p, q \in \mathscr{C}_\beta^1(I)$, and let*

$$f = -p^\beta - q \quad \text{and}$$
$$g = pq - D_\beta p.$$

Then (6.13) *can be factored in the form* (6.12).

Proof. Since $f = -p^\beta - q$ and $g = pq - D_\beta p$, we have

$$\begin{aligned}
D_\beta^2 x + f D_\beta x + g x &= D_\beta^2 x - (p^\beta + q) D_\beta x + (pq - D_\beta p) x \\
&= D_\beta^2 x - p^\beta D_\beta x - q D_\beta x + pq x - D_\beta p x \\
&= D_\beta (D_\beta x - px) - q(D_\beta x - px),
\end{aligned}$$

which completes the proof. $\qquad\square$

Example 6.5.5. We will factor the β-differential equation

$$D_\beta^2 x - (t + \beta(t)) D_\beta x + (t^2 - 1) x = 0, \quad t \in I.$$

Here

$$f(t) = -t - \beta(t) \quad \text{and}$$
$$g(t) = t^2 - 1, \quad t \in I.$$

Let

$$p(t) = t^2 \quad \text{and}$$
$$q(t) = t, \quad t \in I.$$

Then

$$-p^{\beta}(t) - q(t) = -\beta(t) - t$$
$$= f(t), \quad t \in I,$$

and

$$p(t)q(t) - D_{\beta}p(t) = t^2 - 1$$
$$= g(t), \quad t \in I.$$

Therefore, the given β-differential equation can be factored as

$$D_{\beta}(D_{\beta}x - tx) - t(D_{\beta}x - tx) = 0, \quad t \in I.$$

Exercise 6.5.6. Factor the following β-differential equations:

1. $D_{\beta}^2 x - (5 = t)D_{\beta}x + 5tx = 0, t \in I$;
2. $D_{\beta}^2 x - (6 + t)D_{\beta}x + (5 + 5t)x = 0, t \in I.$

Answer. 1. $D_{\beta}(D_{\beta}x - 5x) - t(D_{\beta}x - 5x) = 0, t \in I.$
2. $D_{\beta}(D_{\beta}x - 5x) - (t + 1)(D_{\beta}x - 5x) = 0, t \in I.$

6.6 Nonconstant coefficients

In this section, we investigate the second-order β-differential equation

$$D_{\beta}^2 x - p^{\textcircled{2}\beta}(t)x^{\beta} = 0, \quad t \in I, \tag{6.14}$$

where $p \in \mathscr{R}_{\beta}(I)$ is nonconstant. Note that

$$p^{\textcircled{2}\beta}(t) = \frac{p^2}{1 + p(\beta(t) - t)}, \quad t \in I.$$

Let

$$x_1(t) = e_{p,\beta}(t).$$

Then

$$D_{\beta}x_1(t) = pe_{p,\beta}(t),$$
$$D_{\beta}^2 x_1(t) = p^2 e_{p,\beta}(t),$$
$$x_1(\beta(t)) = e_{p,\beta}(\beta(t))$$
$$= (1 + p(\beta(t) - t))e_{p,\beta}(t), \quad t \in I.$$

This gives

$$D_{\beta}^2 x_1(t) - p^{\textcircled{2}\beta}(t)x_1(\beta(t)) = p^2 e_{p,\beta}(t) - \frac{p^2}{1 + p(\beta(t) - t)}(1 + p(\beta(t) - t))e_{p,\beta}(t)$$

$$= p^2 e_{p,\beta}(t) - p^2 e_{p,\beta}(t)$$
$$= 0, \quad t \in I.$$

Thus, x_1 is a solution to (6.14). Next, suppose that x_2 is a solution to the IVP

$$D_\beta^2 x - p^{\oslash\beta}(t)x^\beta = 0, \quad t \in I,$$
$$x(s_0) = 0,$$
$$D_\beta x(s_0) = 1.$$

Then, by the β-Abel theorem II (Theorem 6.2.4), we have

$$W_\beta(x_1, x_2)(t) = W_\beta(x_1, x_2)(s_0)$$
$$= \det \begin{pmatrix} x_1(s_0) & x_2(s_0) \\ D_\beta x_1(s_0) & D_\beta x_2(s_0) \end{pmatrix}$$
$$= \det \begin{pmatrix} 1 & 0 \\ p & 1 \end{pmatrix}$$
$$= 1, \quad t \in I.$$

Hence,

$$1 = \det \begin{pmatrix} x_1(t) & x_2(t) \\ D_\beta x_1(t) & D_\beta x_2(t) \end{pmatrix}$$
$$= \det \begin{pmatrix} e_{p,\beta}(t) & x_2(t) \\ p e_{p,\beta}(t) & D_\beta x_2(t) \end{pmatrix}$$
$$= e_{p,\beta}(t) D_\beta x_2(t) - p e_{p,\beta}(t) x_2(t), \quad t \in I,$$

whereupon

$$e_{p,\beta}(t) D_\beta x_2(t) = p e_{p,\beta}(t) x_2(t) + 1, \quad t \in I,$$

which gives

$$D_\beta x_2(t) = p x_2(t) + e_{\ominus_\beta p, \beta}(t), \quad t \in I,$$
$$x_2(s_0) = 0.$$

Therefore,

$$x_2(t) = \int_{s_0}^t E_{p,\beta}(\beta(\tau)) e_{\ominus_\beta p, \beta}(\tau) d_\beta \tau$$
$$= \int_{s_0}^t \frac{1}{1 + p(\beta(\tau) - \tau)} E_{-p,\beta}(\tau) e_{\ominus_\beta p, \beta}(\tau) d_\beta \tau$$
$$= \int_{s_0}^t \frac{1}{1 + p(\beta(\tau) - \tau)} \frac{e_{\ominus_\beta p, \beta}(\tau)}{e_{p,\beta}(\tau)} d_\beta \tau$$

$$= \int_{s_0}^{t} \frac{1}{1 + p(\beta(\tau) - \tau)} (e_{\ominus_\beta p, \beta}(\tau))^2 d_\beta \tau, \quad t \in I.$$

Note that x_1 and x_2 form a fundamental system of solutions for (6.14). Now, suppose that $h \in \mathscr{R}_\beta(I)$ solves the β-differential equation

$$D_\beta x + x^{\oslash\beta} + f(t)x^\beta + g(t) = 0, \quad t \in I, \tag{6.15}$$

where f and g are as in (6.4). Let

$$x_3(t) = e_{h,\beta}(t), \quad t \in I.$$

Then

$$D_\beta x_3(t) = h(t)e_{h,\beta}(t),$$
$$(D_\beta x_3)^\beta(t) = h(\beta(t))e_{h,\beta}(\beta(t)), \quad t \in I,$$

and

$$D_\beta^2 x_3(t) = D_\beta h(t)e_{h,\beta}(\beta(t)) + (h(t))^2 e_{h,\beta}(t)$$
$$= D_\beta h(t)e_{h,\beta}(\beta(t)) + \frac{(h(t))^2}{1 + h(t)(\beta(t) - t)} e_{h,\beta}(\beta(t))$$
$$= D_\beta h(t)e_{h,\beta}(\beta(t)) + h^{\oslash\beta}(t)e_{h,\beta}(\beta(t)), \quad t \in I.$$

Hence,

$$D_\beta^2 x_3(t) + f(t)(D_\beta x_3)^\beta(t) + g(t)x_3^\beta(t) = D_\beta h(t)e_{h,\beta}(\beta(t)) + h^{\oslash\beta}(t)e_{h,\beta}(\beta(t))$$
$$+ f(t)(h(\beta(t))e_{h,\beta}(\beta(t)) + g(t)e_{h,\beta}(\beta(t))$$
$$= (D_\beta h(t) + h^{\oslash\beta}(t) + f(t)h(\beta(t)) + g(t))e_{h,\beta}(\beta(t)).$$

Since h is a solution to (6.15), we get

$$D_\beta^2 x_3(t) + f(t)(D_\beta x_3)^\beta(t) + g(t)x_3^\beta(t) = 0, \quad t \in I,$$

that is, x_3 is a solution to (6.4). Now, let x_4 be a solution to (6.4) such that

$$x_4(s_0) = 0,$$
$$D_\beta x_4(s_0) = 0.$$

Then

$$e_{h,\beta}(t)(D_\beta x_4(t) - h(t)x_4(t)) = W_\beta(x_3, x_4)(t)$$
$$= e_{\ominus_\beta f, \beta}(t)W_\beta(x_3, x_4)(s_0).$$

Since $W_\beta(x_3, x_4)(s_0) = 1$, we have

$$e_{h,\beta}(t)(D_\beta x_4(t) - h(t)x_4(t)) = e_{\ominus_\beta f, \beta}(t), \quad t \in I.$$

Therefore,

$$D_\beta\left(\frac{x_4}{e_{h,\beta}}\right)(t) = \frac{D_\beta x_4(t)e_{h,\beta}(t) - h(t)e_{h,\beta}(t)x_4(t)}{e_{h,\beta}(t)e_{h,\beta}(\beta(t))}$$

$$= \frac{e_{\ominus_\beta f,\beta}(t)}{e_{h,\beta}(t)e_{h,\beta}(\beta(t))}, \quad t \in I,$$

whereupon

$$\frac{x_4(t)}{e_{h,\beta}(t)} = \int_{s_0}^{t} \frac{e_{\ominus_\beta f,\beta}(\tau)}{e_{h,\beta}(\tau)e_{h,\beta}(\beta(\tau))} d_\beta\tau$$

$$= \int_{s_0}^{t} \frac{e_{\ominus_\beta f,\beta}(\tau)}{(1+h(\tau)(\beta(\tau)-\tau))(e_{h,\beta}(\tau))^2} d_\beta\tau, \quad t \in I,$$

that is,

$$x_4(t) = e_{h,\beta}(t)\int_{s_0}^{t} \frac{e_{\ominus_\beta f,\beta}(\tau)}{(1+h(\tau)(\beta(\tau)-\tau))(e_{h,\beta}(\tau))^2} d_\beta\tau, \quad t \in I.$$

Note that x_3 and x_4 form a fundamental system of solutions for (6.4).

Example 6.6.1. Consider the β-differential equation

$$D_\beta^2 x = (D_\beta p_1 - (p_1(t))^2 + q_1^2)x^\beta, \quad t \in I, \tag{6.16}$$

where $p_1: I \to \mathbb{R}$ is a function, and q_1 is a nonzero constant such that

$$2p_1(t) + (\beta(t) - t)((p_1(t))^2 - q_1^2) = 0, \quad t \in I. \tag{6.17}$$

We will prove that

$$x_1(t) = e_{p_1-q_1,\beta}(t) \quad \text{and}$$
$$x_2(t) = e_{p_1+q_1,\beta}(t), \quad t \in I,$$

form a fundamental system of solutions for (6.16). Using condition (6.17), we get

$$D_\beta x_1(t) = (p_1(t) - q_1)e_{p_1-q_1,\beta}(t)$$
$$= (p_1(t) - q_1)x_1(t),$$
$$x_1(\beta(t)) = x_1(t) + (\beta(t) - t)D_\beta x_1(t)$$
$$= (1 + (\beta(t) - t)(p_1(t) - q_1))x_1(t), \quad t \in I,$$

and

$$D_\beta^2 x_1(t) = D_\beta(p_1 - q_1)(t)x_1(\beta(t)) + (p_1(t) - q_1)D_\beta x_1(t)$$
$$= D_\beta p_1(t)x_1(\beta(t)) + (p_1(t) - q_1)^2 x_1(t)$$
$$= \left(D_\beta p_1(t) + \frac{(p_1(t) - q_1)^2}{1 + (\beta(t) - t)(p_1(t) - q_1)}\right)x_1(\beta(t))$$

$$= \left(D_\beta p_1(t) + q_1^2 - (p_1(t))^2 + (p_1(t))^2 - q_1^2 + \frac{(p_1(t) - q_1)^2}{1 + (\beta(t) - t)(p_1(t) - q_1)} \right) x_1(\beta(t))$$

$$= \left(D_\beta p_1(t) + q_1^2 - (p_1(t))^2 + (p_1(t) - q_1)\left(p_1(t) + q_1 + \frac{p_1(t) - q_1}{1 + (\beta(t) - t)(p_1(t) - q_1)} \right) \right) x_1(\beta(t))$$

$$= \left(D_\beta p_1(t) + q_1^2 - (p_1(t))^2 \right.$$

$$\left. + (p_1(t) - q_1)\left(\frac{p_1(t) + q_1 + p_1(t) - q_1 + (\beta(t) - t)((p_1(t))^2 - q_1^2)}{1 + (\beta(t) - t)(p_1(t) - q_1)} \right) \right) x_1(\beta(t))$$

$$= (D_\beta p_1(t) + q_1^2 - (p_1(t))^2)x_1(\beta(t)), \quad t \in I,$$

that is, x_1 is a solution to (6.16). Moreover,

$$D_\beta x_2(t) = (p_1(t) + q_1)e_{p_1 + q_1, \beta}(t)$$
$$= (p_1(t) + q_1)x_2(t),$$
$$x_2(\beta(t)) = x_2(t) + (\beta(t) - t)D_\beta x_2(t)$$
$$= (1 + (\beta(t) - t)(p_1(t) + q_1))x_2(t), \quad t \in I,$$

and

$$D_\beta^2 x_2(t) = D_\beta(p_1 + q_1)(t)x_2(\beta(t)) + (p_1(t) + q_1)D_\beta x_1(t)$$

$$= D_\beta p_1(t)x_2(\beta(t)) + (p_1(t) + q_1)^2 x_2(t)$$

$$= \left(D_\beta p_1(t) + \frac{(p_1(t) + q_1)^2}{1 + (\beta(t) - t)(p_1(t) + q_1)} \right) x_2(\beta(t))$$

$$= \left(D_\beta p_1(t) + q_1^2 - (p_1(t))^2 + (p_1(t))^2 - q_1^2 + \frac{(p_1(t) + q_1)^2}{1 + (\beta(t) - t)(p_1(t) + q_1)} \right) x_1(\beta(t))$$

$$= \left(D_\beta p_1(t) + q_1^2 - (p_1(t))^2 + (p_1(t) + q_1)\left(p_1(t) - q_1 + \frac{p_1(t) + q_1}{1 + (\beta(t) - t)(p_1(t) + q_1)} \right) \right) x_1(\beta(t))$$

$$= \left(D_\beta p_1(t) + q_1^2 - (p_1(t))^2 \right.$$

$$\left. + (p_1(t) + q_1)\left(\frac{p_1(t) - q_1 + p_1(t) + q_1 + (\beta(t) - t)((p_1(t))^2 - q_1^2)}{1 + (\beta(t) - t)(p_1(t) + q_1)} \right) \right) x_2(\beta(t))$$

$$= (D_\beta p_1(t) + q_1^2 - (p_1(t))^2)x_2(\beta(t)), \quad t \in I,$$

that is, x_2 is a solution to (6.16). Next,

$$W_\beta(x_1, x_2)(t) = \det \begin{pmatrix} x_1(t) & x_2(t) \\ D_\beta x_1(t) & D_\beta x_2(t) \end{pmatrix}$$

$$= \det \begin{pmatrix} x_1(t) & x_2(t) \\ (p_1(t) - q_1)x_1(t) & (p_1(t) + q_1)x_2(t) \end{pmatrix}$$

$$= (p_1(t) + q_1)x_1(t)x_2(t) - (p_1(t) - q_1)x_1(t)x_2(t)$$

$$= 2qx_1(t)x_2(t)$$
$$\neq 0, \quad t \in I.$$

Thus, x_1 and x_2 form a fundamental system of solutions for (6.16).

Exercise 6.6.2. Let $\alpha \in \mathscr{R}_\beta(I)$ be such that α and $\ominus_\beta\alpha$ are β-differentiable on I. Suppose that $\alpha \neq \ominus_\beta\alpha$ and $D_\beta(\alpha) = D_\beta(\ominus_\beta\alpha)$. Prove that the functions

$$x_1(t) = \frac{e_{\alpha,\beta}(t) + e_{\ominus_\beta\alpha,\beta}(t)}{2} \quad \text{and}$$
$$x_2(t) = \frac{e_{\alpha,\beta}(t) - e_{\ominus_\beta\alpha,\beta}(t)}{2}, \quad t \in I,$$

form a fundamental system of solutions for the β-differential equation

$$D_\beta^2 x = (D_\beta\alpha + \alpha^{\textcircled{2}\beta}(t))x^\beta, \quad t \in I.$$

6.7 β-Euler–Cauchy equations

In this section, we investigate the second-order β-differential equation

$$t\beta(t)D_\beta^2 x + at D_\beta x + bx = 0, \tag{6.18}$$

where $a, b \in \mathbb{R}$ are such that

$$t\beta(t) - at(\beta(t) - t) + b(\beta(t))^2 \neq 0, \quad t \in I.$$

Definition 6.7.1. The β-differential equation (6.18) is called the β-Euler–Cauchy equation.

The characteristic equation of (6.18) is

$$\lambda^2 + (a - 1)\lambda + b = 0. \tag{6.19}$$

Suppose that

$$(a - 1)^2 - 4b > 0.$$

Then (6.19) has two roots, say λ_1 and λ_2. Let

$$x_1(t) = e_{\frac{\lambda_1}{t},\beta}(t),$$
$$x_2(t) = e_{\frac{\lambda_2}{t},\beta}(t), \quad t \in I.$$

Then we have

$$D_\beta x_1(t) = \frac{\lambda_1}{t} e_{\frac{\lambda_1}{t},\beta}(t),$$

$$D_\beta^2 x_1(t) = -\frac{\lambda_1}{t\beta(t)}e_{\frac{\lambda_1}{t},\beta}(t) + \frac{\lambda_1^2}{t\beta(t)}e_{\frac{\lambda_1}{t},\beta}(t), \quad t \in I,$$

and

$$D_\beta x_2(t) = \frac{\lambda_2}{t}e_{\frac{\lambda_2}{t},\beta}(t),$$

$$D_\beta^2 x_2(t) = -\frac{\lambda_2}{t\beta(t)}e_{\frac{\lambda_2}{t},\beta}(t) + \frac{\lambda_2^2}{t\beta(t)}e_{\frac{\lambda_2}{t},\beta}(t), \quad t \in I.$$

Hence,

$$t\beta(t)D_\beta^2 x_1(t) + at D_\beta x_1(t) + bx_1(t) = t\beta(t)\left(-\frac{\lambda_1}{t\beta(t)}e_{\frac{\lambda_1}{t},\beta}(t) + \frac{\lambda_1^2}{t\beta(t)}e_{\frac{\lambda_1}{t},\beta}(t)\right) + at\frac{\lambda_1}{t}e_{\frac{\lambda_1}{t},\beta}(t) + be_{\frac{\lambda_1}{t},\beta}(t)$$

$$= (-\lambda_1 + \lambda_1^2 + a\lambda_1 + b)e_{\frac{\lambda_1}{t},\beta}(t)$$

$$= (\lambda_1^2 + (a-1)\lambda_1 + b)e_{\frac{\lambda_1}{t},\beta}(t)$$

$$= 0, \quad t \in I.$$

As above,

$$t\beta(t)D_\beta^2 x_2(t) + at D_\beta x_2(t) + bx_2(t) = 0, \quad t \in I.$$

Thus, x_1 and x_2 are solutions to (6.18). Next,

$$W_\beta(x_1, x_2)(t) = \det\begin{pmatrix} x_1(t) & x_2(t) \\ D_\beta x_1(t) & D_\beta x_2(t) \end{pmatrix}$$

$$= \det\begin{pmatrix} x_1(t) & x_2(t) \\ \frac{\lambda_1}{t}x_1(t) & \frac{\lambda_2}{t}x_2(t) \end{pmatrix}$$

$$= \frac{\lambda_2}{t}x_1(t)x_2(t) - \frac{\lambda_1}{t}x_1(t)x_2(t)$$

$$= \frac{\lambda_2 - \lambda_1}{t}x_1(t)x_2(t)$$

$$\neq 0, \quad t \in I.$$

Hence, x_1 and x_2 form a fundamental system of solutions for (6.18).

Example 6.7.2. Consider the β-differential equation

$$t\beta(t)D_\beta^2 x - 4t D_\beta x + 6x = 0, \quad t \in I.$$

The corresponding characteristic equation is

$$\lambda^2 - 5\lambda + 6 = 0$$

with roots

$$\lambda_1 = 2 \quad \text{and}$$
$$\lambda_2 = 3.$$

Therefore

$$e_{\frac{2}{t},\beta}(t), \quad e_{\frac{3}{t},\beta}(t), \quad t \in I,$$

form a fundamental system of solutions for the considered β-differential equation.

Exercise 6.7.3. Find a fundamental system of solutions for the second-order β-differential equation

$$t\beta(t)D_\beta^2 x + 5t D_\beta x + 3x = 0, \quad t \in I.$$

Answer.

$$e_{-\frac{1}{t},\beta}(t), \quad e_{-\frac{3}{t},\beta}(t), \quad t \in I.$$

Now we consider the second-order β-differential equation

$$t\beta(t)D_\beta^2 x + (1-2\alpha)t D_\beta x + \alpha^2 x = 0, \quad t \in I, \tag{6.20}$$

where $\alpha \in \mathbb{R}$ is such that

$$t\beta(t) + (2\alpha-1)t(\beta(t)-t) + \alpha^2(\beta(t))^2 \neq 0, \quad t \in I.$$

The characteristic equation of (6.20) is

$$\lambda^2 - 2\alpha\lambda + \alpha^2 = 0.$$

Then we see that $\lambda = \alpha$ is a root of this characteristic equation, and hence

$$x_3(t) = e_{\frac{\alpha}{t},\beta}(t), \quad t \in I,$$

is a solution to (6.20). Let

$$x_4(t) = e_{\frac{\alpha}{t},\beta}(t) \int_{s_0}^t \frac{1}{1+\tau+\alpha(\beta(\tau)-\tau)} d_\beta\tau, \quad t \in I.$$

Then

$$D_\beta x_4(t) = \frac{\alpha}{t} e_{\frac{\alpha}{t},\beta}(t) \int_{s_0}^t \frac{1}{\tau+\alpha(\beta(\tau)-\tau)} d_\beta\tau + e_{\frac{\alpha}{t},\beta}(\beta(t)) \frac{1}{t+\alpha(\beta(t)-t)}$$

$$= \frac{\alpha}{t} x_4(t) + \frac{1}{t} x_3(t),$$

and

$$D_\beta^2 x_4(t) = -\frac{\alpha}{t\beta(t)} x_4(t) + \frac{\alpha}{\beta(t)} D_\beta x_4(t) - \frac{1}{t\beta(t)} x_3(t) + \frac{\alpha}{t\beta(t)} x_3(t)$$

$$= -\frac{\alpha}{t\beta(t)} x_4(t) + \frac{\alpha^2}{t\beta(t)} x_4(t) + \frac{\alpha}{t\beta(t)} x_3(t) - \frac{1}{t\beta(t)} x_3(t) + \frac{\alpha}{t\beta(t)} x_3(t)$$

$$= \frac{\alpha(\alpha-1)}{t\beta(t)} x_4(t) + \frac{2\alpha-1}{t\beta(t)} x_3(t), \quad t \in I.$$

Hence,

$$t\beta(t)D_\beta^2 x_4(t) + (1 - 2\alpha)t D_\beta x_4(t) + \alpha^2 x_4(t) = t\beta(t)\left(\frac{\alpha(\alpha-1)}{t\beta(t)}x_4(t) + \frac{2\alpha-1}{t\beta(t)}x_3(t)\right)$$
$$+ (1 - 2\alpha)t\left(\frac{\alpha}{t}x_4(t) + \frac{1}{t}x_3(t)\right) + \alpha^2 x_4(t)$$
$$= (\alpha(\alpha-1) + \alpha(1-2\alpha) + \alpha^2)x_4(t)$$
$$+ (2\alpha - 1 + 1 - 2\alpha)x_3(t)$$
$$= 0, \quad t \in I.$$

Thus, it follows that x_4 is a solution to (6.20). Next,

$$W_\beta(x_3, x_4)(t) = \det\begin{pmatrix} x_3(t) & x_4(t) \\ D_\beta x_3(t) & D_\beta x_4(t) \end{pmatrix}$$
$$= \det\begin{pmatrix} x_3(t) & x_4(t) \\ \dfrac{\alpha}{t}x_3(t) & \dfrac{\alpha}{t}x_4(t) + \dfrac{1}{t}x_3(t) \end{pmatrix}$$
$$= \left(\frac{\alpha}{t}x_4(t) + \frac{1}{t}x_3(t)\right)x_3(t) - \frac{\alpha}{t}x_3(t)x_4(t)$$
$$= \frac{1}{t}(x_3(t))^2$$
$$\neq 0, \quad t \in I.$$

Therefore, x_3 and x_4 form a fundamental system of solutions for (6.20).

Example 6.7.4. Consider the β-differential equation

$$t\beta(t)D_\beta^2 x - 9t D_\beta x + 25x = 0, \quad t \in I.$$

The corresponding characteristic equation is

$$\lambda^2 - 10\lambda + 25 = 0.$$

We see that $\lambda = 5$ is one of the roots of this characteristic equation and the functions

$$e_{\frac{5}{t},\beta}(t) \quad \text{and} \quad e_{\frac{5}{t},\beta}(t)\int_{s_0}^{t}\frac{1}{\tau + 5(\beta(\tau) - \tau)}d_\beta\tau, \quad t \in I,$$

form a fundamental system of solutions for the considered β-differential equation.

Exercise 6.7.5. Find a fundamental system of solutions for the second-order β-differential equation

$$t\beta(t)D_\beta^2 x - 5t D_\beta x + 9x = 0, \quad t \in I.$$

Answer.

$$e_{\frac{3}{t},\beta}(t) \quad \text{and} \quad e_{\frac{3}{t},\beta}(t)\int_{s_0}^{t}\frac{1}{\tau + 3(\beta(\tau) - \tau)}d_\beta\tau, \quad t \in I.$$

6.8 Variation of parameters

In this section, we investigate the nonhomogeneous second-order β-differential equation (6.1) by employing the method of variation of parameters. Suppose that x_1 and x_2 form a fundamental system of solutions for the homogeneous second-order β-differential equation $L_2 x = 0$. We will search a particular solution of (6.1) in the form

$$x_p(t) = \gamma(t)x_1(t) + \delta(t)x_2(t), \quad t \in I,$$

where $\gamma, \delta \in \mathscr{C}_\beta^1(I)$ will be determined further. We have

$$D_\beta x_p(t) = D_\beta \gamma(t)x_1(\beta(t)) + \gamma(t)D_\beta x_1(t) + D_\beta \delta(t)x_2(\beta(t)) + \delta(t)D_\beta x_2(t), \quad t \in I.$$

Assume that

$$D_\beta x\gamma(t)x_1(\beta(t)) + D_\beta \delta(t)x_2(\beta(t)) = 0, \quad t \in I. \tag{6.21}$$

Then

$$D_\beta x_p(t) = \gamma(t)D_\beta x_1(t) + \delta(t)D_\beta x_2(t), \quad t \in I,$$

and

$$D_\beta^2 x_p(t) = D_\beta \gamma(t)(D_\beta x_1)(\beta(t)) + \gamma(t)D_\beta^2 x_1(t) + D_\beta \gamma(t)(D_\beta x_2)(\beta(t)) + \delta(t)D_\beta^2 x_2(t), \quad t \in I.$$

Hence,

$$
\begin{aligned}
h(t) &= L_2 x_p(t) \\
&= D_\beta^2 x_p(t) + f(t)D_\beta x_p(t) + g(t)x_p(t) \\
&= D_\beta \gamma(t)(D_\beta x_1)(\beta(t)) + \gamma(t)D_\beta^2 x_1(t) + D_\beta \delta(t)(D_\beta x_2)(\beta(t)) + \delta(t)D_\beta^2 x_2(t) \\
&\quad + f(t)\gamma(t)D_\beta x_1(t) + f(t)\delta(t)D_\beta x_2(t) + g(t)\gamma(t)x_1(t) + g(t)\delta(t)x_2(t) \\
&= \gamma(t)(D_\beta^2 x_1(t) + f(t)D_\beta x_1(t) + g(t)x_1(t)) + \delta(t)(D_\beta^2 x_2(t) + f(t)D_\beta x_2(t) + g(t)x_2(t)) \\
&\quad + D_\beta \gamma(t)(D_\beta x_1)(\beta(t)) + D_\beta \delta(t)(D_\beta x_2)(\beta(t)) \\
&= D_\beta \gamma(t)(D_\beta x_1)(\beta(t)) + D_\beta \delta(t)(D_\beta x_2)(\beta(t)), \quad t \in I.
\end{aligned}
$$

From the last equation and (6.21), we get the following system

$$
\begin{aligned}
D_\beta \gamma(t)x_1(\beta(t)) + D_\beta \delta(t)x_2(\beta(t)) &= 0, \\
D_\beta \gamma(t)(D_\beta x_1)(\beta(t)) + D_\beta \delta(t)(D_\beta x_2)(\beta(t)) &= h(t), \quad t \in I,
\end{aligned}
$$

whereupon

$$D_\beta \gamma(t) = -\frac{(D_\beta x_2)(\beta(t))h(t)}{W_\beta(x_1, x_2)(\beta(t))} \quad \text{and}$$

$$D_\beta \delta(t) = \frac{(D_\beta x_1)(\beta(t))h(t)}{W_\beta(x_1, x_2)(\beta(t))}, \quad t \in I.$$

Now, suppose that

$$\gamma(s_0) = 0 \quad \text{and}$$
$$\delta(s_0) = 0.$$

Then

$$\gamma(t) = -\int_{s_0}^{t} \frac{(D_\beta x_2)(\beta(\tau))h(\tau)}{W_\beta(x_1, x_2)(\beta(\tau))} d_\beta\tau, \quad \text{and}$$

$$\delta(t) = \int_{s_0}^{t} \frac{(D_\beta x_1)(\beta(\tau))h(\tau)}{W_\beta(x_1, x_2)(\beta(\tau))} d_\beta\tau, \quad t \in I.$$

Therefore, a particular solution of (6.1) is given by

$$x_p(t) = \gamma(t)x_1(t) + \delta(t)x_2(t)$$
$$= -x_1(t)\int_{s_0}^{t} \frac{(D_\beta x_2)(\beta(\tau))h(\tau)}{W_\beta(x_1, x_2)(\beta(\tau))} d_\beta\tau + x_2(t)\int_{s_0}^{t} \frac{(D_\beta x_1)(\beta(\tau))h(\tau)}{W_\beta(x_1, x_2)(\beta(\tau))} d_\beta\tau$$
$$= \int_{s_0}^{t} \frac{(D_\beta x_1)(\beta(\tau))x_2(t) - (D_\beta x_2)(\beta(\tau))x_1(t)}{W_\beta(x_1, x_2)(\beta(\tau))} h(\tau)d_\beta\tau, \quad t \in I.$$

Hence, a general solution of (6.1) is

$$x(t) = c_1 x_1(t) + c_2 x_2(t) + \int_{s_0}^{t} \frac{(D_\beta x_1)(\beta(\tau))x_2(t) - (D_\beta x_2)(\beta(\tau))x_1(t)}{W_\beta(x_1, x_2)(\beta(\tau))} h(\tau)d_\beta\tau, \quad t \in I.$$

Example 6.8.1. We will find a general solution of the nonhomogeneous second-order β-differential equation

$$D_\beta^2 x - 5D_\beta x + 6x = e_{4,\beta}(t), \quad t \in I.$$

The corresponding homogeneous equation is

$$D_\beta^2 x(t) - 5D_\beta x(t) + 6x(t) = 0, \quad t \in I,$$

and its characteristic equation is

$$\lambda^2 - 5\lambda + 6 = 0.$$

Then the roots of this characteristic equation are

$$\lambda_1 = 2 \quad \text{and}$$
$$\lambda_2 = 3.$$

Hence

$$x_1(t) = e_{2,\beta}(t),$$
$$x_2(t) = e_{3,\beta}(t), \quad t \in I,$$

form a fundamental system of solutions of the corresponding homogeneous equation. We have

$$W_\beta(x_1, x_2)(t) = \det \begin{pmatrix} x_1(t) & x_2(t) \\ D_\beta x_1(t) & D_\beta x_2(t) \end{pmatrix}$$

$$= \det \begin{pmatrix} e_{2,\beta}(t) & e_{3,\beta}(t) \\ 2e_{2,\beta}(t) & 3e_{3,\beta}(t) \end{pmatrix}$$

$$= 3e_{2,\beta}(t)e_{3,\beta}(t) - 2e_{2,\beta}(t)e_{3,\beta}(t)$$

$$= e_{2,\beta}(t)e_{3,\beta}(t), \quad t \in I,$$

and

$$\int_{s_0}^t \frac{(D_\beta x_1)(\beta(\tau))x_2(t) - (D_\beta x_2)(\beta(\tau))x_1(t)}{W_\beta(x_1, x_2)(\beta(\tau))} h(\tau)d_\beta\tau$$

$$= \int_{s_0}^t \frac{2e_{2,\beta}(\beta(\tau))e_{3,\beta}(t) - 3e_{3,\beta}(\beta(\tau))e_{2,\beta}(t)}{e_{2,\beta}(\beta(\tau))e_{3,\beta}(\beta(\tau))} d_\beta\tau$$

$$= 2e_{3,\beta}(t) \int_{s_0}^t \frac{e_{4,\beta}(\tau)}{e_{3,\beta}(\beta(\tau))} d_\beta\tau - 3e_{2,\beta}(t) \int_{s_0}^t \frac{e_{4,\beta}(\tau)}{e_{2,\beta}(\beta(\tau))} d_\beta\tau$$

$$= 2e_{3,\beta}(t) \int_{s_0}^t \frac{e_{4,\beta}(\tau)}{(1 + 3(\beta(\tau) - \tau))e_{3,\beta}(\tau)} d_\beta\tau - 3e_{2,\beta}(t) \int_{s_0}^t \frac{e_{4,\beta}(\tau)}{(1 + 2(\beta(\tau) - \tau))e_{2,\beta}(\tau)} d_\beta\tau$$

$$= 2e_{3,\beta}(t) \int_{s_0}^t \frac{e_{\frac{1}{1+3(\beta(\tau)-\tau)},\beta}(\tau)}{1 + 3(\beta(\tau) - \tau)} d_\beta\tau - 3e_{2,\beta}(t) \int_{s_0}^t \frac{e_{\frac{2}{1+2(\beta(\tau)-\tau)},\beta}(\tau)}{1 + 2(\beta(\tau) - \tau)} d_\beta\tau$$

$$= 2e_{3,\beta}(t) \int_{s_0}^t D_\beta e_{\frac{1}{1+3(\beta(\tau)-\tau)},\beta}(\tau)d_\beta\tau - \frac{3}{2}e_{2,\beta}(t) \int_{s_0}^t D_\beta e_{\frac{2}{1+2(\beta(\tau)-\tau)},\beta}(\tau)d_\beta\tau$$

$$= 2e_{3,\beta}(t) \left(e_{\frac{1}{1+3(\beta(t)-t)},\beta}(t) - 1 \right) - \frac{3}{2} \left(e_{\frac{2}{1+2(\beta(t)-t)},\beta}(t) - 1 \right), \quad t \in I.$$

Thus, a general solution of the considered β-differential equation is

$$x(t) = c_1 e_{2,\beta}(t) + c_2 e_{3,\beta}(t) + 2e_{3,\beta}(t)e_{\frac{1}{1+3(\beta(t)-t)},\beta}(t) - \frac{3}{2}e_{2,\beta}(t)e_{\frac{2}{1+2(\beta(t)-t)},\beta}(t)$$

$$= c_1 e_{2,\beta}(t) + c_2 e_{3,\beta}(t) + 2e_{4,\beta}(t) - \frac{3}{2}e_{4,\beta}(t)$$

$$= c_1 e_{2,\beta}(t) + c_2 e_{3,\beta}(t) + \frac{1}{2}e_{4,\beta}(t), \quad t \in I,$$

where $c_1, c_2 \in \mathbb{R}$.

Exercise 6.8.2. Find a general solution to the second-order β-differential equation

$$t\beta(t)D_\beta^2 x - 2t D_\beta x + 2x = e_{\frac{3}{t},\beta}(t), \quad t \in I.$$

Answer.

$$x(t) = c_1 e_{\frac{1}{t},\beta}(t) + c_2 e_{\frac{2}{t},\beta}(t) + \frac{1}{2}e_{\frac{3}{t},\beta}(t), \quad t \in I,$$

where $c_1, c_2 \in \mathbb{R}$.

6.9 **The annihilator method**

In this section, we introduce the annihilator method for β-differential equations.

Definition 6.9.1. We say that a function $f : I \to \mathbb{R}$ can be annihilated provided there is an operator of the form

$$a_0 D_\beta^n + a_1 D_\beta^{n-1} + \cdots + a_{n-1} D_\beta + a_n \mathscr{I}$$

such that

$$(a_0 D_\beta^n + a_1 D_\beta^{n-1} + \cdots + a_{n-1} D_\beta + a_n \mathscr{I}) f(t) = 0, \quad t \in I.$$

Example 6.9.2. Let

$$f(t) = e_{2,\beta}(t), \quad t \in I.$$

Then

$$\begin{aligned}
(D_\beta - 2\mathscr{I}) f(t) &= (D_\beta - 2\mathscr{I}) e_{2,\beta}(t) \\
&= D_\beta e_{2,\beta}(t) - 2 e_{2,\beta}(t) \\
&= 2 e_{2,\beta}(t) - 2 e_{2,\beta}(t) \\
&= 0, \quad t \in I.
\end{aligned}$$

Thus, $D_\beta - 2\mathscr{I}$ is an annihilator for the function f.

Example 6.9.3. Consider the β-differential equation

$$D_\beta^2 x - 5 D_\beta x + 6x = e_{5,\beta}(t), \quad t \in I.$$

It can be rewritten in the form

$$(D_\beta - 2\mathscr{I})(D_\beta - 3\mathscr{I}) x(t) = e_{5,\beta}(t), \quad t \in I. \tag{6.22}$$

Note that $D_\beta - 5\mathscr{I}$ is an annihilator for $e_{5,\beta}(t)$, $t \in I$. Multiplying both sides of (6.22) by $D_\beta - 5\mathscr{I}$, we get

$$(D_\beta - 5\mathscr{I})(D_\beta - 2\mathscr{I})(D_\beta - 3\mathscr{I}) x(t) = 0, \quad t \in I.$$

Hence,

$$x(t) = a e_{2,\beta}(t) + b e_{3,\beta}(t) + c e_{5,\beta}(t), \quad t \in I,$$

where $a, b, c \in \mathbb{R}$. Note that $e_{2,\beta}(t)$ and $e_{3,\beta}(t)$, $t \in I$, form a fundamental system of solutions for the homogeneous β-differential equation

$$D_\beta^2 x - 5 D_\beta x + 6x = 0, \quad t \in I.$$

Then

$$x_p(t) = c e_{5,\beta}(t), \quad t \in I,$$

is a particular solution to the considered β-differential equation. Thus, we have

$$
\begin{aligned}
e_{5,\beta}(t) &= D_\beta^2 x_p(t) - 5 D_\beta x_p(t) + 6 x_p(t) \\
&= (25c - 5c + 6c) e_{5,\beta}(t) \\
&= 26 c e_{5,\beta}(t), \quad t \in I,
\end{aligned}
$$

whereupon

$$
c = \frac{1}{26}.
$$

Consequently, a general solution of the considered β-differential equation is

$$
x(t) = a e_{2,\beta}(t) + b e_{3,\beta}(t) + \frac{1}{26} e_{5,\beta}(t), \quad t \in I.
$$

Exercise 6.9.4. Use the method of annihilator to find a general solution to the second-order β-differential equation

$$
D_\beta^2 x + D_\beta x - 2x = 2 + t, \quad t \in I.
$$

Answer.

$$
x(t) = c_1 e_{1,\beta}(t) + c_2 e_{-2,\beta}(t) - \frac{5}{4} - \frac{1}{2} t, \quad t \in I.
$$

6.10 The β-Laplace transform method

Consider the second-order β-differential equation

$$
D_\beta^2 x + a D_\beta x + b x = f(t), \quad t \in I, \tag{6.23}
$$

where $a, b \in \mathbb{R}$, $f \in \mathscr{C}(I)$, subject to the initial conditions

$$
x(s_0) = x_0 \quad \text{and} \quad D_\beta x(s_0) = x_1, \tag{6.24}
$$

where $x_0, x_1 \in \mathbb{R}$. Taking the β-Laplace transform of both sides of (6.23), we obtain, for $z \in \mathbb{C}$,

$$
\begin{aligned}
\mathscr{L}_\beta(f(t))(z) &= z^2 \mathscr{L}_\beta(x)(z) - z x_0 - x_1 + a z \mathscr{L}_\beta(x)(z) - a x_0 + b \mathscr{L}_\beta(x)(z) \\
&= (z^2 + a z + b) \mathscr{L}_\beta(x)(z) - (a + z) x_0 - x_1,
\end{aligned}
$$

whereupon

$$
\mathscr{L}_\beta(x)(z) = \frac{1}{z^2 + a z + b} \mathscr{L}_\beta(f(t))(z) + \frac{a + z}{z^2 + a z + b} x_0 + \frac{1}{z^2 + a z + b} x_1, \quad z \in \mathbb{C},
$$

and hence the solution of (6.23) subject to (6.24) is given by

$$x(t) = \mathscr{L}_\beta^{-1}\left(\frac{1}{z^2 + az + b}\mathscr{L}_\beta(f(t))(z)\right)(t) + \mathscr{L}_\beta^{-1}\left(\frac{a+z}{z^2 + az + b}\right)(t)x_0$$
$$+ \mathscr{L}_\beta^{-1}\left(\frac{1}{z^2 + az + b}\right)(t)x_1, \quad t \in I,$$

provided that both $\mathscr{L}_\beta$ and $\mathscr{L}_\beta^{-1}$ exist.

Example 6.10.1. Consider the IVP

$$D_\beta^2 x + 3D_\beta x + 2x = 1, \quad t \in I,$$
$$x(s_0) = 1,$$
$$D_\beta x(s_0) = 1.$$

Then its solution is given by

$$x(t) = \mathscr{L}_\beta^{-1}\left(\frac{1}{z^2 + 3z + 2}\mathscr{L}_\beta(1)(z)\right)(t) + \mathscr{L}_\beta^{-1}\left(\frac{z+3}{z^2 + 3z + 2}\right)(t) + \mathscr{L}_\beta^{-1}\left(\frac{1}{z^2 + 3z + 2}\right)(t)$$

$$= \mathscr{L}_\beta^{-1}\left(\frac{1}{z(z+1)(z+2)}\right)(t) + \mathscr{L}_\beta^{-1}\left(\frac{z+3}{(z+1)(z+2)}\right)(t) + \mathscr{L}_\beta^{-1}\left(\frac{1}{(z+1)(z+2)}\right)(t)$$

$$= \mathscr{L}_\beta^{-1}\left(\frac{1}{z(z+1)}\right)(t) - \mathscr{L}_\beta^{-1}\left(\frac{1}{z(z+2)}\right)(t) + \mathscr{L}_\beta^{-1}\left(\frac{z+3}{z+1}\right)(t) - \mathscr{L}_\beta^{-1}\left(\frac{z+3}{z+2}\right)(t)$$

$$= \mathscr{L}_\beta^{-1}\left(\frac{1}{z}\right)(t) - \mathscr{L}_\beta^{-1}\left(\frac{1}{z+1}\right)(t) - \frac{1}{2}\mathscr{L}_\beta^{-1}\left(\frac{1}{z}\right)(t)$$

$$+ \frac{1}{2}\mathscr{L}_\beta^{-1}\left(\frac{1}{z+2}\right)(t) + \mathscr{L}_\beta^{-1}(1)(t) + 2\mathscr{L}_\beta^{-1}\left(\frac{1}{z+1}\right)(t) - \mathscr{L}_\beta^{-1}(1)(t) - \mathscr{L}_\beta^{-1}\left(\frac{1}{z+2}\right)(t)$$

$$= 1 - e_{-1,\beta}(t) - \frac{1}{2} + \frac{1}{2}e_{2,\beta}(t) + 2e_{-1,\beta}(t) - e_{-2,\beta}(t)$$

$$= \frac{1}{2} + \frac{1}{2}e_{-2,\beta}(t) + e_{-1,\beta}(t), \quad t \in I.$$

Exercise 6.10.2. Using the β-Laplace transform method, solve the IVP

$$D_\beta^2 x + D_\beta x - 2x = 0, \quad t \in I,$$
$$x(s_0) = 1,$$
$$D_\beta x(s_0) = 1.$$

Answer.

$$x(t) = e_{1,\beta}(t), \quad t \in I.$$

6.11 Advanced practical problems

Problem 6.11.1. Suppose that $a^2 - 4b < 0$. Let

$$p = -\frac{a}{2},$$

$$q = \frac{\sqrt{4b - a^2}}{2}.$$

If $p, (\beta(t) - t)b - a \in \mathscr{R}_\beta(I)$, then prove that

$$\cos{\frac{q}{1+(\beta(t)-t)p}},\beta(\cdot)e_p(\cdot), \quad \sin{\frac{q}{1+(\beta(t)-t)p}},\beta(\cdot)e_p(\cdot)$$

form a fundamental system of solutions for the homogeneous β-differential equation (6.7).

Problem 6.11.2. Let $I = [4, 30]$, and let

$$\beta(t) = \frac{3}{4}t + 1, \quad t \in I.$$

Solve the β-differential equation

$$D_\beta^2 x - 5D_\beta x + 4x = 0.$$

Answer.

$$x(t) = \frac{1}{\displaystyle\prod_{k=0}^{\infty}\left(1 - 4(t - 4)\left(\frac{3}{4}\right)^k\right)}\left(c_1 + c\,\frac{t-4}{4}\sum_{k=0}^{\infty}\left(\frac{3}{4}\right)^k\prod_{l=0}^{\infty}\frac{1 - \frac{3}{2}(t - 4)\left(\frac{3}{4}\right)^l}{1 - \frac{1}{4}\left(\frac{3}{4}\right)^l(t - 4)}\right),$$

where $t \in I$ and $c, c_1 \in \mathbb{R}$.

Problem 6.11.3. Factorize the following β-differential equations:

1. $D_\beta^2 x - 6D_\beta x + 5x = 0, t \in I$;
2. $D_\beta^2 x - 8D_\beta x + 7x = 0, t \in I$.

Answer. 1. $D_\beta(D_\beta x - 5x) - (D_\beta x - 5x) = 0, t \in I$.
2. $D_\beta(D_\beta x - 7x) - (D_\beta x - 7x) = 0, t \in I$.

Problem 6.11.4. Find a fundamental system of solutions for the β-differential equation

$$t\beta(t)D_\beta^2 x - 5tD_\beta x + 8x = 0, \quad t \in I.$$

Answer.

$$e_{\frac{2}{t},\beta}(t), \quad e_{\frac{4}{t},\beta}(t), \quad t \in I.$$

Problem 6.11.5. Find a fundamental system of solutions for the β-differential equation

$$t\beta(t)D_\beta^2 x - 7tD_\beta x + 16x = 0, \quad t \in I.$$

Answer.

$$e_{\frac{4}{7},\beta}(t), \quad e_{\frac{4}{7},\beta}(t)\int_{s_0}^{t}\frac{1}{\tau+4(\beta(\tau)-\tau)}d_\beta\tau, \quad t\in I.$$

Problem 6.11.6. Find a general solution to the β-differential equation

$$D_\beta^2 x - 3D_\beta x + 2x = e_{2,\beta}(t), \quad t\in I.$$

Answer.

$$x(t) = c_1 e_{1,\beta}(t) + c_2 e_{2,\beta}(t) + e_{2,\beta}(t)\int_{s_0}^{t}\frac{1}{1+2(\beta(\tau)-\tau)}d_\beta\tau, \quad t\in I,$$

where $c_1, c_2 \in \mathbb{R}$.

Problem 6.11.7. Use the method of annihilator to find a general solution to the equation

$$D_\beta^2 x + x = e_{3,\beta}(t), \quad t\in I.$$

Answer.

$$x(t) = c_1 \sin_{1,\beta}(t) + c_2 \cos_{1,\beta}(t) + \frac{1}{10}e_{3,\beta}(t), \quad t\in I.$$

Problem 6.11.8. Using the β-Laplace transform method, solve the IVP

$$D_\beta^2 x - 9x = 0, \quad t\in I,$$
$$x(s_0) = 0,$$
$$D_\beta x(s_0) = 1.$$

Answer.

$$x(t) = \frac{1}{3}\sinh_{3,\beta}(t), \quad t\in I.$$

6.12 Notes and references

In this chapter, we investigated homogeneous and nonhomogeneous second-order linear β-differential equations. The β-Wronskians are defined, and some of their properties are deduced. The β-analogues of the Abel theorem are proved. We define fundamental systems of solutions, and using them, we give certain representations of the general solutions of the considered classes β-differential equations. Further, we investigate β-Euler–Cauchy equations. We introduce the annihilator method and the β-Laplace transform method. A good percentage of the contents of this chapter is new and belongs to the authors.

β-differential systems[★]

Let $I \subseteq \mathbb{R}$, and let $\beta \colon I \to \mathbb{R}$ be a first-kind general quantum operator. Let s_0 be the unique fixed point of the operator β.

7.1 Structure of β-differential systems

Let A be an $m \times n$-matrix on I, $A = (a_{ij})_{1 \leq i \leq m, 1 \leq j \leq n}$, shortly $A = (a_{ij})$, $a_{ij} \colon I \to \mathbb{R}$, $1 \leq i \leq m$, $1 \leq j \leq n$. It is also known as an $m \times n$-matrix-valued function defined on I. The $m \times n$-identity matrix will be denoted by $\mathscr{I}$, and the $m \times n$-zero matrix by O.

Definition 7.1.1. We say that an $m \times n$-matrix A is β-differentiable on I provided each entry of A is β-differentiable on I, and we write

$$D_\beta A = \left(D_\beta a_{ij} \right).$$

Example 7.1.2. Let $I = \mathbb{R}$. Let $\beta(t) = \dfrac{1}{2}t + 1$, $t \in I$, and

$$A(t) = \begin{pmatrix} t+1 & t^2 + t \\ 2t - 3 & 2t^2 - 3t + 2 \end{pmatrix}, \quad t \in I.$$

We will find $D_\beta A(t)$, $t \in I$. We have

$$
\begin{aligned}
a_{11}(t) &= t + 1, \\
a_{12}(t) &= t^2 + t, \\
a_{21}(t) &= 2t - 3, \\
a_{22}(t) &= 2t^2 - 3t + 2, \quad t \in I.
\end{aligned}
$$

Then

$$
\begin{aligned}
D_\beta a_{11}(t) &= 1, \\
D_\beta a_{12}(t) &= \beta(t) + t + 1
\end{aligned}
$$

★ This book has a companion website hosting complementary materials. Visit this URL to access it: https://www.elsevier.com/books-and-journals/book-companion/9780443328046.

Generalized Quantum Calculus with Applications. https://doi.org/10.1016/B978-0-44-332804-6.00012-1

$$= \frac{1}{2}t + 1 + t + 1$$

$$= \frac{3}{2}t + 2,$$

$$D_\beta a_{21}(t) = 2,$$

$$D_\beta a_{22}(t) = 2(\beta(t) + t) - 3$$

$$= 2\left(\frac{1}{2}t + 1 + t\right) - 3$$

$$= 2\left(\frac{3}{2}t + 1\right) - 3$$

$$= 3t + 1 - 3$$

$$= 3t - 2, \quad t \in I.$$

Therefore,

$$D_\beta A(t) = \begin{pmatrix} 1 & \frac{3}{2}t + 2 \\ 2 & 3t - 1 \end{pmatrix}, \quad t \in I.$$

Example 7.1.3. Let $I = \mathbb{R}$, and let

$$\beta(t) = \frac{2t}{3} + 2, \quad t \in I, \quad \text{and}$$

$$A(t) = \begin{pmatrix} t^3 + t & \dfrac{t+1}{t+2} & t \\ 1 & -t^2 & t^2 + t \\ 1 & 2 & t^3 \end{pmatrix}, \quad t \in I.$$

We will find $D_\beta A(t)$, $t \in I$. We have

$$a_{11}(t) = t^3 + t,$$

$$a_{12}(t) = \frac{t+1}{t+2},$$

$$a_{13}(t) = t,$$

$$a_{21}(t) = 1,$$

$$a_{22}(t) = -t^2,$$

$$a_{23}(t) = t^2 + t,$$

$$a_{31}(t) = 1,$$

$$a_{32}(t) = 2,$$

$$a_{33}(t) = t^3, \quad t \in I.$$

Then we obtain

$$D_\beta a_{11}(t) = (\beta(t))^2 + t\beta(t) + t^2 + 1$$

$$= \left(\frac{2}{3}t + 2\right)^2 + t\left(\frac{2}{3}t + 2\right) + t^2 + 1$$

$$= \frac{4}{9}t^2 + \frac{8}{3}t + 4 + \frac{2}{3}t^2 + 2t + t^2 + 1$$

$$= \frac{19}{9}t^2 + \frac{17}{3}t + 5,$$

$$D_\beta a_{12}(t) = \frac{t + 2 - (t + 1)}{(t + 2)(\beta(t) + 2)}$$

$$= \frac{1}{(t + 2)\left(\frac{2}{3}t + 2 + 2\right)}$$

$$= \frac{3}{2(t + 2)(t + 6)},$$

$$D_\beta a_{13}(t) = 1,$$

$$D_\beta a_{21}(t) = 0,$$

$$D_\beta a_{22}(t) = -(\beta(t) + t)$$

$$= -\left(\frac{2}{3}t + 2 + t\right)$$

$$= -\left(\frac{5}{3}t + 2\right),$$

$$D_\beta a_{23}(t) = \beta(t) + t + 1$$

$$= \frac{2}{3}t + 2 + t + 1$$

$$= \frac{5}{3}t + 3,$$

$$D_\beta a_{31}(t) = 0,$$

$$D_\beta a_{32}(t) = 0,$$

$$D_\beta a_{33}(t) = (\beta(t))^2 + t\beta(t) + t^2$$

$$= \left(\frac{2}{3}t + 2\right)^2 + t\left(\frac{2}{3}t + 2\right) + t^2$$

$$= \frac{4}{9}t^2 + \frac{8}{3}t + 4 + \frac{2}{3}t^2 + 2t + t^2$$

$$= \frac{19}{9}t^2 + \frac{14}{3}t + 4, \quad t \in I.$$

Therefore,

$$D_\beta A(t) = \begin{pmatrix} \frac{19}{9}t^2 + \frac{17}{3}t + 5 & \frac{3}{2(t+2)(t+6)} & 1 \\ 0 & -\left(\frac{5}{3}t + 2\right) & \frac{5}{3}t + 3 \\ 0 & 0 & \frac{19}{9}t^2 + \frac{14}{3}t + 4 \end{pmatrix}, \quad t \in I.$$

Example 7.1.4. Let $I = \mathbb{R}$. Define $\beta(t) = \dfrac{3}{4}t + 1$, $t \in I$, and

$$A(t) = \begin{pmatrix} t^2 + 1 & \dfrac{1}{t+1} \\ 2 & 3t \end{pmatrix}, \quad t \in I.$$

We will find $D_\beta A(t)$, $t \in I$. We have

$$a_{11}(t) = t^2 + 1,$$
$$a_{12}(t) = \frac{1}{t+1},$$
$$a_{21}(t) = 2,$$
$$a_{22}(t) = 3t, \quad t \in I.$$

Then we obtain

$$D_\beta a_{11}(t) = \beta(t) + t$$
$$= \frac{3}{4}t + 1 + t$$
$$= \frac{7}{4}t + 1,$$
$$D_\beta a_{12}(t) = -\frac{1}{(t+1)(\beta(t)+1)}$$
$$= -\frac{1}{(t+1)\left(\frac{3}{4}t + 1 + 1\right)}$$
$$= -\frac{4}{(t+1)(3t+8)},$$
$$D_\beta a_{21}(t) = 0,$$
$$D_\beta a_{22}(t) = 3, \quad t \in I.$$

Therefore,

$$D_\beta A(t) = \begin{pmatrix} \dfrac{7}{4}t + 1 & -\dfrac{4}{(t+1)(3t+8)} \\ 0 & 3 \end{pmatrix}, \quad t \in I.$$

Exercise 7.1.5. Let $I = \mathbb{R}$. Define $\beta(t) = \dfrac{1}{2}t + 3$, $t \in I$, and

$$A(t) = \begin{pmatrix} t^3 & t^2 \\ 2t+4 & t-1 \end{pmatrix}, \quad t \in I.$$

Find $D_\beta A(t)$, $t \in I$.

Answer.

$$D_\beta A(t) = \begin{pmatrix} \dfrac{7}{4}t^2 + 6t + 9 & \dfrac{3}{2}t + 3 \\ 2 & 1 \end{pmatrix}, \quad t \in I.$$

Definition 7.1.6. For a β-differentiable $m \times n$-matrix A, we define

$$A^\beta = \left(a_{ij}^\beta \right).$$

Theorem 7.1.7. *If A is a β-differentiable $m \times n$-matrix-valued function on I, then*

$$A^\beta(t) = A(t) + (\beta(t) - t)D_\beta A(t), \quad t \in I.$$

Proof. We have

$$\begin{aligned}
A^\beta(t) &= \left(a_{ij}^\beta(t) \right) \\
&= \left(a_{ij}(t) + (\beta(t) - t)D_\beta a_{ij}(t) \right) \\
&= (a_{ij}(t)) + (\beta(t) - t)\left(D_\beta a_{ij}(t) \right) \\
&= A(t) + (\beta(t) - t)D_\beta A(t), \quad t \in I.
\end{aligned}$$

This completes the proof. $\square$

We further suppose that $B = (b_{ij})_{1 \le i \le m, 1 \le j \le n}$, $b_{ij} : I \to \mathbb{R}$, $1 \le i \le m$, $1 \le j \le n$.

Theorem 7.1.8. *Let A and B be β-differentiable $m \times n$-matrix-valued functions on I. Then*

$$D_\beta(A + B)(t) = D_\beta A(t) + D_\beta B(t), \quad t \in I.$$

Proof. We have

$$(A + B)(t) = (a_{ij}(t) + b_{ij}(t)).$$

Then

$$\begin{aligned}
D_\beta(A + B)(t) &= \left(D_\beta a_{ij}(t) + D_\beta b_{ij}(t) \right) \\
&= \left(D_\beta a_{ij}(t) \right) + \left(D_\beta b_{ij}(t) \right) \\
&= D_\beta A(t) + D_\beta B(t), \quad t \in I.
\end{aligned}$$

This completes the proof. $\square$

Theorem 7.1.9. *Let $\alpha \in \mathbb{R}$, and let A be a β-differentiable $m \times n$-matrix-valued function on I. Then*

$$D_\beta(\alpha A)(t) = \alpha D_\beta A(t), \quad t \in I.$$

Proof. We have

$$\begin{aligned}
D_\beta(\alpha A)(t) &= \left(D_\beta(\alpha a_{ij})(t) \right) \\
&= \left(\alpha D_\beta a_{ij}(t) \right)
\end{aligned}$$

$$= \alpha \left(D_\beta a_{ij}(t) \right)$$
$$= \alpha D_\beta A(t), \quad t \in I.$$

This completes the proof. $\qquad\qquad\qquad\qquad\qquad\qquad\qquad\qquad\qquad\qquad\quad\square$

Theorem 7.1.10. *Let A and B be β-differentiable $n \times n$-matrix-valued functions on I. Then*

$$D_\beta(AB)(t) = D_\beta A(t)B(t) + A^\beta(t)D_\beta B(t)$$
$$= D_\beta A(t)B^\beta(t) + A(t)D_\beta B(t), \quad t \in I.$$

Proof. We have

$$(AB)(t) = \left(\sum_{k=1}^{n} a_{ik}(t)b_{kj}(t) \right), \quad t \in I.$$

Then

$$D_\beta(AB)(t) = \left(D_\beta \left(\sum_{k=1}^{n} a_{ik}b_{kj} \right)(t) \right)$$

$$= \left(\sum_{k=1}^{n} D_\beta(a_{ik}b_{kj})(t) \right)$$

$$= \left(\sum_{k=1}^{n} \left(D_\beta a_{ik}(t)b_{kj}(t) + a_{ik}^\beta(t)D_\beta b_{kj}(t) \right) \right)$$

$$= \left(\sum_{k=1}^{n} D_\beta a_{ik}(t)b_{kj}(t) \right) + \left(\sum_{k=1}^{n} a_{ik}^\beta(t)D_\beta b_{kj}(t) \right)$$

$$= D_\beta A(t)B(t) + A^\beta(t)D_\beta B(t)$$

$$= \left(\sum_{k=1}^{n} \left(D_\beta a_{ik}(t)b_{kj}^\beta(t) + a_{ik}(t)D_\beta b_{kj}(t) \right) \right)$$

$$= \left(\sum_{k=1}^{n} D_\beta a_{ik}(t)b_{kj}^\beta(t) \right) + \left(\sum_{k=1}^{n} a_{ik}(t)D_\beta b_{kj}(t) \right)$$

$$= D_\beta A(t)B^\beta(t) + A(t)D_\beta B(t).$$

Thus,

$$D_\beta(AB)(t) = D_\beta A(t)B^\beta(t) + A(t)D_\beta B(t), \quad t \in I.$$

This completes the proof. $\qquad\qquad\qquad\qquad\qquad\qquad\qquad\qquad\qquad\qquad\quad\square$

Example 7.1.11. Let $I = [0, \infty)$. Define $\beta(t) = \dfrac{1}{2}t, t \in I$,

$$A(t) = \begin{pmatrix} t & t-1 \\ 2 & 3t+1 \end{pmatrix}, \quad \text{and} \quad B(t) = \begin{pmatrix} 1 & t \\ t+1 & t-1 \end{pmatrix}, \quad t \in I.$$

Then

$$(AB)(t) = \begin{pmatrix} t & t-1 \\ 2 & 3t+1 \end{pmatrix} \begin{pmatrix} 1 & t \\ t+1 & t-1 \end{pmatrix}$$

$$= \begin{pmatrix} t^2+t-1 & 2t^2-2t+1 \\ 3t^2+4t+3 & 3t^2-1 \end{pmatrix}$$

$$= C(t)$$

$$= (c_{ij}(t)), \quad t \in I.$$

We have

$$a_{11}(t) = t,$$
$$a_{12}(t) = t-1,$$
$$a_{21}(t) = 2,$$
$$a_{22}(t) = 3t+1,$$
$$b_{11}(t) = 1,$$
$$b_{12}(t) = t,$$
$$b_{21}(t) = t+1,$$
$$b_{22}(t) = t-1,$$
$$c_{11}(t) = t^2+t-1,$$
$$c_{12}(t) = 2t^2-2t+1,$$
$$c_{21}(t) = 3t^2+4t+3,$$
$$c_{22}(t) = 3t^2-1, \quad t \in I.$$

Then

$$D_\beta a_{11}(t) = 1,$$
$$D_\beta a_{12}(t) = 1,$$
$$D_\beta a_{21}(t) = 0,$$
$$D_\beta a_{22}(t) = 3,$$
$$D_\beta b_{11}(t) = 0,$$
$$D_\beta b_{12}(t) = 1,$$
$$D_\beta b_{21}(t) = 1,$$
$$D_\beta b_{22}(t) = 1,$$
$$D_\beta c_{11}(t) = \beta(t)+t+1$$
$$= \frac{1}{2}t+3+t+1$$
$$= \frac{3}{2}t+4,$$

$$D_\beta c_{12}(t) = 2(\beta(t) + t) - 2$$
$$= 2\left(\frac{1}{2}t + 3 + t\right) - 2$$
$$= 3t + 4,$$
$$D_\beta c_{21}(t) = 3(\beta(t) + t) + 4$$
$$= 3\left(\frac{1}{2}t + 3 + t\right) + 4$$
$$= 9t + 13,$$
$$D_\beta c_{22}(t) = 3(\beta(t) + t)$$
$$= 3\left(\frac{1}{2}t + 3 + t\right)$$
$$= \frac{9}{2}t + 3, \quad t \in I.$$

Therefore,

$$D_\beta(AB)(t) = \begin{pmatrix} \frac{3}{2}t + 4 & 3t + 4 \\ \frac{9}{2}t + 13 & \frac{9}{2}t + 9 \end{pmatrix}, \quad t \in I.$$

Also,

$$D_\beta A(t)B(t) = \begin{pmatrix} 1 & 1 \\ 0 & 3 \end{pmatrix} \begin{pmatrix} 1 & t \\ t+1 & t-1 \end{pmatrix}$$
$$= \begin{pmatrix} t+2 & 2t-1 \\ 3t+3 & 3t-3 \end{pmatrix},$$
$$A^\beta(t) = \begin{pmatrix} \beta(t) & \beta(t) - 1 \\ 2 & 3\beta(t) + 1 \end{pmatrix}$$
$$= \begin{pmatrix} \frac{1}{2}t + 3 & \frac{1}{2}t + 2 \\ 2 & \frac{3}{2}t + 10 \end{pmatrix},$$
$$A^\beta(t)D_\beta B(t) = \begin{pmatrix} \frac{1}{2}t + 3 & \frac{1}{2}t + 2 \\ 2 & \frac{3}{2}t + 10 \end{pmatrix} \begin{pmatrix} 0 & 1 \\ 1 & 1 \end{pmatrix}$$
$$= \begin{pmatrix} \frac{1}{2}t + 2 & t + 5 \\ \frac{3}{2}t + 10 & \frac{3}{2}t + 12 \end{pmatrix},$$

$$D_\beta A(t)B(t) + A^\beta(t)D_\beta B(t) = \begin{pmatrix} t+2 & 2t-1 \\ 3t+3 & 3t-3 \end{pmatrix} + \begin{pmatrix} \dfrac{1}{2}t+2 & t+5 \\ \dfrac{3}{2}t+10 & \dfrac{3}{2}t+12 \end{pmatrix}$$

$$= \begin{pmatrix} \dfrac{3}{2}t+4 & 3t+4 \\ \dfrac{9}{2}t+13 & \dfrac{9}{2}t+9 \end{pmatrix}, \quad t \in I.$$

Consequently,

$$D_\beta(AB)(t) = D_\beta A(t)B(t) + A^\beta(t)D_\beta B(t), \quad t \in I.$$

Next,

$$B^\beta(t) = \begin{pmatrix} 1 & \beta(t) \\ \beta(t)+1 & \beta(t)-1 \end{pmatrix}$$

$$= \begin{pmatrix} 1 & \dfrac{1}{2}t+3 \\ \dfrac{1}{2}t+4 & \dfrac{1}{2}t+2 \end{pmatrix},$$

$$D_\beta A(t)B^\beta(t) = \begin{pmatrix} 1 & 1 \\ 0 & 3 \end{pmatrix}\begin{pmatrix} 1 & \dfrac{1}{2}t+3 \\ \dfrac{1}{2}t+4 & \dfrac{1}{2}t+2 \end{pmatrix}$$

$$= \begin{pmatrix} \dfrac{1}{2}t+5 & t+5 \\ \dfrac{3}{2}t+12 & \dfrac{3}{2}t+6 \end{pmatrix},$$

$$A(t)D_\beta B(t) = \begin{pmatrix} t & t-1 \\ 2 & 3t+1 \end{pmatrix}\begin{pmatrix} 0 & 1 \\ 1 & 1 \end{pmatrix}$$

$$= \begin{pmatrix} t-1 & 2t-1 \\ 3t+1 & 3t+3 \end{pmatrix},$$

$$D_\beta A(t)B^\beta(t) + A(t)D_\beta B(t) = \begin{pmatrix} \dfrac{1}{2}t+5 & t+5 \\ \dfrac{3}{2}t+12 & \dfrac{3}{2}t+6 \end{pmatrix} + \begin{pmatrix} t-1 & 2t-1 \\ 3t+1 & 3t+3 \end{pmatrix}$$

$$= \begin{pmatrix} \dfrac{3}{2}t+4 & 3t+4 \\ \dfrac{9}{2}t+13 & \dfrac{9}{2}t+9 \end{pmatrix}, \quad t \in I.$$

Thus,

$$D_\beta(AB)(t) = D_\beta A(t)B^\beta(t) + A(t)D_\beta B(t), \quad t \in I.$$

Exercise 7.1.12. Let $I = \mathbb{R}$. Define $\beta(t) = \dfrac{1}{5}t, t \in I$,

$$A(t) = \begin{pmatrix} t^2 + 1 & t - 2 \\ 2t - 1 & t + 1 \end{pmatrix}, \quad \text{and} \quad B(t) = \begin{pmatrix} t & 2t + 1 \\ t & t - 1 \end{pmatrix}, \quad t \in I.$$

Prove that

$$D_\beta(AB)(t) = D_\beta A(t) B^\beta(t) + A(t) D_\beta B(t), \quad t \in I.$$

Theorem 7.1.13. *Let A be an $n \times n$-matrix-valued function such that A^{-1} exists on I. Then*

$$\left(A^\beta \right)^{-1} (t) = \left(A^{-1} \right)^\beta (t), \quad t \in I.$$

Proof. For any $t \in I$, we have

$$A(t) A^{-1}(t) = \mathscr{I}, \quad t \in I.$$

Then

$$A^\beta(t) \left(A^{-1} \right)^\beta (t) = \mathscr{I},$$

whereupon

$$\left(A^\beta \right)^{-1} (t) = \left(A^{-1} \right)^\beta (t), \quad t \in I.$$

This completes the proof. $\qquad\square$

Example 7.1.14. Let $I = \mathbb{R}$. Define $\beta(t) = lt + l, l \in (0, 1)$, and

$$A(t) = \begin{pmatrix} t + 1 & t + 2 \\ 1 & t + 3 \end{pmatrix}, \quad t \in I.$$

Then

$$A^\beta(t) = \begin{pmatrix} \beta(t) + 1 & \beta(t) + 2 \\ 1 & \beta(t) + 3 \end{pmatrix}$$

$$= \begin{pmatrix} lt + l + 1 & lt + l + 2 \\ 1 & lt + l + 3 \end{pmatrix},$$

$$\left(A^\beta \right)^{-1} (t) = \frac{1}{(lt + l)(lt + l + 3) + 1} \begin{pmatrix} lt + l + 3 & -lt - l - 2 \\ -1 & lt + l + 1 \end{pmatrix}, \quad t \in I.$$

Next,

$$A^{-1}(t) = \frac{1}{t(t + 3) + 1} \begin{pmatrix} t + 3 & -t - 2 \\ -1 & t + 1 \end{pmatrix}, \quad t \in I,$$

whereupon

$$\left(A^{-1}\right)^{\beta}(t) = \frac{1}{\beta(t)(\beta(t)+3)+1}\begin{pmatrix} \beta(t)+3 & -\beta(t)-2 \\ -1 & \beta(t)+1 \end{pmatrix}$$

$$= \frac{1}{(lt+l)(lt+l+3)+1}\begin{pmatrix} lt+l+3 & -lt-l-2 \\ -1 & lt+l+1 \end{pmatrix}, \quad t \in I.$$

Consequently,

$$\left(A^{\beta}\right)^{-1}(t) = \left(A^{-1}\right)^{\beta}(t), \quad t \in I.$$

Exercise 7.1.15. Let $I = \mathbb{R}$. Define $\beta(t) = \dfrac{2}{7}t + 4$, $t \in I$, and

$$A(t) = \begin{pmatrix} t+2 & \dfrac{1}{t+1} \\ t^2+1 & \dfrac{1}{t+2} \end{pmatrix}, \quad t \in I.$$

Prove that

$$\left(A^{\beta}\right)^{-1}(t) = \left(A^{-1}\right)^{\beta}(t), \quad t \in I.$$

Theorem 7.1.16. *Let A be a β-differentiable $n \times n$-matrix-valued function on I with A^{-1} such that $(A^{\beta})^{-1}$ exists on I. Then*

$$D_{\beta}\left(A^{-1}\right)(t) = -A^{-1}(t)\, D_{\beta}A \left(A^{\beta}\right)^{-1}(t)$$

$$= -\left(A^{\beta}\right)^{-1}(t)\, D_{\beta}AA^{-1}(t), \quad t \in I.$$

Proof. We have

$$\mathscr{I} = AA^{-1} \quad \text{on} \quad I,$$

and

$$D_{\beta}(\mathscr{I}) = O,$$

whereupon, using Theorem 7.1.10, we get

$$D_{\beta}(\mathscr{I})(t) = D_{\beta}\left(AA^{-1}\right)(t)$$

$$= D_{\beta}A\left(A^{-1}\right)^{\beta}(t) + A(t)\, D_{\beta}(A^{-1})(t)$$

$$= D_{\beta}A\left(A^{\beta}\right)^{-1}(t) + A(t)\, D_{\beta}\left(A^{-1}\right)(t)$$

$$= D_{\beta}AA^{-1}(t) + A^{\beta}(t)\, D_{\beta}\left(A^{-1}\right)(t), \quad t \in I.$$

Hence

$$A(t)\, D_\beta\left(A^{-1}\right)(t) = -D_\beta A\left(A^\beta\right)^{-1}(t),$$

$$A^\beta D_\beta\left(A^{-1}\right)(t) = -D_\beta A A^{-1}(t), \quad t \in I,$$

and

$$D_\beta\left(A^{-1}\right)(t) = -A^{-1}(t)\, D_\beta A\left(A^\beta\right)^{-1}(t),$$

$$D_\beta\left(A^{-1}\right)(t) = -\left(A^\beta\right)^{-1}(t)\, D_\beta A A^{-1}(t), \quad t \in I.$$

This completes the proof. $\qquad\qquad\square$

Exercise 7.1.17. Let A and B be β-differentiable $n \times n$-matrix-valued functions on I such that B^{-1} and $\left(B^\beta\right)^{-1}$ exist on I. Prove that

$$D_\beta\left(AB^{-1}\right)(t) = \left(D_\beta A - AB^{-1} D_\beta B\right)\left(B^\beta\right)^{-1}(t)$$

$$= \left(D_\beta A - \left(AB^{-1}\right)^\beta D_\beta B\right) B^{-1}(t), \quad t \in I.$$

Definition 7.1.18. We say that a matrix-valued function A is continuous on I provided each entry of A is continuous. The class of such continuous $m \times n$-matrix-valued functions on I is denoted by

$$\mathscr{C} = \mathscr{C}(I) = \mathscr{C}(I, \mathbb{R}^{m \times n}).$$

We further suppose that A and B are $n \times n$-matrix-valued functions.

Definition 7.1.19. We say that an $n \times n$-matrix-valued function A on I is β-regressive with respect to I provided

$$\mathscr{I} + (\beta(t) - t)A(t) \quad \text{is invertible for all} \quad t \in I.$$

The class of such β-regressive and continuous functions is denoted, similarly to the scalar case, by

$$\mathscr{R}_\beta = \mathscr{R}_\beta(I) = \mathscr{R}_\beta(I, \mathbb{R}^{n \times n}).$$

Theorem 7.1.20. *An $n \times n$-matrix-valued function A is β-regressive if and only if the eigenvalues λ_i of A are β-regressive for all $1 \le i \le n$.*

Proof. Let $i \in \{1, \dots, n\}$, and let $\lambda_i(t)$ be the eigenvalue corresponding to an eigenvector $x(t)$, $t \in I$. Then

$$(1 + (\beta(t) - t)\lambda_i(t))\, x(t) = \mathscr{I} y(t) + (\beta(t) - t)\lambda_i(t)x(t)$$

$$= \mathscr{I} x(t) + (\beta(t) - t)A(t)x(t)$$

$$= (\mathscr{I} + (\beta(t) - t)A(t))\, x(t),$$

whereupon the statement follows. This completes the proof. $\qquad\qquad\square$

Example 7.1.21. Let $I = 3\mathbb{Z}$. Define $\beta(t) = \dfrac{t}{2} + 3$, $t \in I$, and

$$A(t) = \begin{pmatrix} 5t & t \\ 4t & 2t \end{pmatrix}, \qquad t \in I.$$

Consider the equation

$$\det \begin{pmatrix} 5t - \lambda(t) & t \\ 4t & 2t - \lambda(t) \end{pmatrix} = 0, \qquad t \in I.$$

Then the corresponding characteristic equation is

$$(\lambda(t) - 5t)(\lambda(t) - 2t) - 4t^2 = 0, \qquad t \in I,$$

that is,

$$(\lambda(t))^2 - 7t\lambda(t) + 6t^2 = 0, \qquad t \in I.$$

The roots of this characteristic equation are

$$\lambda_{1,2}(t) = \frac{7t \pm \sqrt{49t^2 - 24t^2}}{2}$$
$$= \frac{7t \pm 5t}{2}, \qquad t \in I,$$

or

$$\lambda_1(t) = 6t, \quad \text{and}$$
$$\lambda_2(t) = t, \qquad t \in I.$$

Now,

$$1 + (\beta(t) - t)\lambda_{1,2}(t) = 0$$

implies

$$1 + \left(3 - \frac{t}{2}\right)6t = 0 \quad \text{and}$$
$$1 + \left(3 - \frac{t}{2}\right)t = 0, \qquad t \in I,$$

that is,

$$3t^2 - 18t - 1 = 0 \quad \text{and}$$
$$t^2 - 6t - 2 = 0, \qquad t \in I.$$

The last two equations have no roots in I. Consequently,

$$1 + (\beta(t) - t)\lambda_{1,2}(t) \neq 0, \qquad t \in I,$$

that is, the matrix A is β-regressive.

Theorem 7.1.22. *Let A be a 2×2-matrix-valued function. Then, A is β-regressive if and only if*

$$\operatorname{tr} A + (\beta(t) - t) \det A$$

is β-regressive. Here $\operatorname{tr} A$ *denotes the trace of A.*

Proof. Let

$$A(t) = \begin{pmatrix} a_{11}(t) & a_{12}(t) \\ a_{21}(t) & a_{22}(t) \end{pmatrix}, \quad t \in I.$$

Then

$$\mathscr{I} + (\beta(t) - t) A(t) = \begin{pmatrix} 1 & 0 \\ 0 & 1 \end{pmatrix} + \begin{pmatrix} (\beta(t) - t)a_{11}(t) & (\beta(t) - t)a_{12}(t) \\ (\beta(t) - t)a_{21}(t) & (\beta(t) - t)a_{22}(t) \end{pmatrix}$$

$$= \begin{pmatrix} 1 + (\beta(t) - t)a_{11}(t) & (\beta(t) - t)a_{12}(t) \\ (\beta(t) - t)a_{21}(t) & 1 + (\beta(t) - t)a_{22}(t) \end{pmatrix}, \quad t \in I.$$

Therefore, we obtain

$$\det(\mathscr{I} + (\beta(t) - t)A(t)) = (1 + (\beta(t) - t)a_{11}(t))(1 + (\beta(t) - t)a_{22}(t)) - ((\beta(t) - t))^2 a_{12}(t)a_{21}(t)$$

$$= 1 + (\beta(t) - t)a_{22}(t) + (\beta(t) - t)a_{11}(t)$$

$$+ ((\beta(t) - t))^2 a_{11}(t)a_{22}(t) - ((\beta(t) - t))^2 a_{12}(t)a_{21}(t)$$

$$= 1 + (\beta(t) - t)(\operatorname{tr} A)(t) + ((\beta(t) - t))^2 (\det A)(t),$$

that is,

$$\det(\mathscr{I} + (\beta(t) - t)A(t)) = 1 + (\beta(t) - t)\left((\operatorname{tr} A)(t) + (\beta(t) - t)(\det A)(t)\right). \tag{7.1}$$

Suppose A is β-regressive. Then

$$\det(\mathscr{I} + (\beta(t) - t)A(t)) \neq 0, \quad t \in I.$$

In view of (7.1), we obtain

$$1 + (\beta(t) - t)\left((\operatorname{tr} A)(t) + (\beta(t) - t)(\det A)(t)\right) \neq 0, \quad t \in I, \tag{7.2}$$

that is,

$$\operatorname{tr} A + (\beta(t) - t) \det A$$

is β-regressive. Conversely, suppose

$$\operatorname{tr} A + (\beta(t) - t) \det A$$

is β-regressive. Then (7.2) holds. From (7.1), we conclude that A is β-regressive. This completes the proof. $\qquad\square$

Definition 7.1.23. Assume that A and B are β-regressive matrix-valued functions on I. Then we define $A \oplus_\beta B$, $\ominus_\beta A$, and $A \ominus_\beta B$ by

$$(A \oplus_\beta B)(t) = A(t) + B(t) + (\beta(t) - t)A(t)B(t),$$

$$(\ominus_\beta A)(t) = -(I + (\beta(t) - t)A(t))^{-1}A(t), \quad \text{and}$$

$$(A \ominus_\beta B)(t) = (A \oplus_\beta (\ominus_\beta B))(t), \quad t \in I,$$

respectively.

Example 7.1.24. Let $I = \mathbb{R}$. Define $\beta(t) = \dfrac{2}{3}t + 1$, $t \in I$, and

$$A(t) = \begin{pmatrix} 1 & t \\ 2 & 3t \end{pmatrix} \quad \text{and}$$

$$B(t) = \begin{pmatrix} t & 1 \\ 2t & 3 \end{pmatrix}, \quad t \in I.$$

Here

$$\beta(t) - t = \frac{2}{3}t + 1 - t$$

$$= 1 - \frac{t}{3}, \quad t \in I.$$

Then

$$(A \oplus_\beta B)(t) = A(t) + B(t) + (\beta(t) - t)A(t)B(t)$$

$$= \begin{pmatrix} 1 & t \\ 2 & 3t \end{pmatrix} + \begin{pmatrix} t & 1 \\ 2t & 3 \end{pmatrix} + \left(1 - \frac{1}{3}t\right)\begin{pmatrix} 1 & t \\ 2 & 3t \end{pmatrix}\begin{pmatrix} t & 1 \\ 2t & 3 \end{pmatrix}$$

$$= \begin{pmatrix} 1 + t & 1 + t \\ 2(1 + t) & 3(1 + t) \end{pmatrix} + \left(1 - \frac{1}{3}t\right)\begin{pmatrix} t + 2t^2 & 1 + 3t \\ 2t + 6t^2 & 2 + 9t \end{pmatrix}$$

$$= \begin{pmatrix} -\dfrac{2}{3}t^3 + \dfrac{5}{3}t^2 + 2t + 1 & -t^2 + \dfrac{11}{3}t + 2 \\[2mm] -2t^3 + \dfrac{16}{3}t^2 + 4t + 2 & -3t^2 + \dfrac{34}{3}t + 5 \end{pmatrix},$$

and

$$(B \oplus_\beta A)(t) = A(t) + B(t) + (\beta(t) - t)B(t)A(t)$$

$$= \begin{pmatrix} 1 & t \\ 2 & 3t \end{pmatrix} + \begin{pmatrix} t & 1 \\ 2t & 3 \end{pmatrix} + \left(1 - \frac{1}{3}t\right)\begin{pmatrix} t & 1 \\ 2t & 3 \end{pmatrix}\begin{pmatrix} 1 & t \\ 2 & 3t \end{pmatrix}$$

$$= \begin{pmatrix} 1 + t & 1 + t \\ 2(1 + t) & 3(1 + t) \end{pmatrix} + \left(1 - \frac{1}{3}t\right)\begin{pmatrix} t + 2 & t^2 + 3t \\ 2t + 6 & 2t^2 + 9t \end{pmatrix}$$

$$= \begin{pmatrix} -\dfrac{1}{3}t^2 + \dfrac{4}{3}t + 3 & -\dfrac{1}{3}t^3 + 4t + 1 \\[2ex] -\dfrac{2}{3}t^2 + 2t + 8 & -\dfrac{2}{3}t^3 - t^2 + 12t + 3 \end{pmatrix}.$$

Also,

$$\mathscr{I} + (\beta(t) - t)B(t) = \begin{pmatrix} 1 & 0 \\ 0 & 1 \end{pmatrix} + \left(1 - \dfrac{1}{3}t\right)\begin{pmatrix} t & 1 \\ 2t & 3 \end{pmatrix}$$

$$= \begin{pmatrix} -\dfrac{1}{3}t^2 + t + 1 & -\dfrac{1}{3}t + 1 \\[2ex] -\dfrac{2}{3}t^2 + 2t & -t + 4 \end{pmatrix},$$

$$\det(\mathscr{I} + (\beta(t) - t)B(t)) = \left(\dfrac{1}{3}t^2 - t - 1\right)(t - 4) - \left(\dfrac{1}{3}t - 1\right)\left(\dfrac{2}{3}t^2 - 2t\right)$$

$$= \dfrac{1}{3}t^3 - \dfrac{4}{3}t^2 - t^2 + 4t - t + 4 - \left(\dfrac{2}{9}t^3 - \dfrac{2}{3}t^2 - \dfrac{2}{3}t^2 + 2t\right)$$

$$= \dfrac{1}{3}t^3 - \dfrac{7}{3}t^2 + 3t - 4 - \dfrac{2}{9}t^3 + \dfrac{4}{3}t^2 - 2t$$

$$= \dfrac{1}{9}t^3 - t^2 + t - 4,$$

and

$$(\mathscr{I} + (\beta(t) - t)B(t))^{-1} = \dfrac{1}{\frac{1}{9}t^3 - t^2 + t - 4} \begin{pmatrix} -t + 4 & -1 + \dfrac{1}{3}t \\[2ex] -2t + \dfrac{2}{3}t^2 & -\dfrac{1}{3}t^2 + t + 1 \end{pmatrix},$$

$$(\ominus_\beta B)(t) = -(\mathscr{I} + (\beta(t) - t)B(t))^{-1}B(t)$$

$$= -\dfrac{1}{\frac{1}{9}t^3 - t^2 + t - 4} \begin{pmatrix} -t + 4 & -1 + \dfrac{1}{3}t \\[2ex] -2t + \dfrac{2}{3}t^2 & -\dfrac{1}{3}t^2 + t + 1 \end{pmatrix}\begin{pmatrix} t & 1 \\ 2t & 3 \end{pmatrix}$$

$$= -\dfrac{1}{\frac{1}{9}t^3 - t^2 + t - 4} \begin{pmatrix} -\dfrac{1}{3}t^2 + 2t & 1 \\[2ex] 2t & -\dfrac{1}{3}t^2 + t + 3 \end{pmatrix}, \qquad t \in I.$$

Exercise 7.1.25. Let $I = \mathbb{R}$. Define $\beta(t) = 2t$, $t \in I$,

$$A(t) = \begin{pmatrix} 1 & 1 \\ 2 & -1 \end{pmatrix} \quad \text{and}$$

$$B(t) = \begin{pmatrix} 3 & 4 \\ 1 & 0 \end{pmatrix}, \qquad t \in I.$$

Find

1. $(\ominus_\beta A)(t), t \in I$;

2. $(A \oplus_\beta B)(t), t \in I$.

Answer.

$$(\ominus_\beta A)(t) = \begin{pmatrix} \dfrac{1-4t}{1-t-4t^2} & \dfrac{1-t}{1-t-4t^2} \\ \dfrac{2}{1-t-4t^2} & \dfrac{-1-3t}{1-t-4t^2} \end{pmatrix},$$

$$(A \oplus_\beta B)(t) = \begin{pmatrix} 4+4t & 5+4t \\ 3+5t & -1+8t \end{pmatrix}, \quad t \in I.$$

Theorem 7.1.26. *The set $\mathscr{R}_\beta$ under the operation $\oplus_\beta$ is a group.*

Proof. Let $A, B, C \in \mathscr{R}_\beta$. Then

$$(\mathscr{I} + (\beta(t) - t)A)^{-1}, \quad (\mathscr{I} + (\beta(t) - t)B)^{-1}, \quad (\mathscr{I} + (\beta(t) - t)C)^{-1}$$

exist, and

$$\begin{aligned}
\mathscr{I} + (\beta(t) - t)(A \oplus_\beta B) &= \mathscr{I} + (\beta(t) - t)(A + B + (\beta(t) - t)AB) \\
&= \mathscr{I} + (\beta(t) - t)A + (\beta(t) - t)B + (\beta(t) - t)^2 AB \\
&= \mathscr{I} + (\beta(t) - t)A + (\mathscr{I} + (\beta(t) - t)A)(\beta(t) - t)B \\
&= (\mathscr{I} + (\beta(t) - t)A)(\mathscr{I} + (\beta(t) - t)B).
\end{aligned}$$

Therefore,

$$(\mathscr{I} + (\beta(t) - t)(A \oplus_\beta B))^{-1}$$

exists. Hence, $A \oplus_\beta B \in \mathscr{R}_\beta$. Also,

$$\begin{aligned}
(A \oplus_\beta B) \oplus_\beta C &= (A \oplus_\beta B) + C + (\beta(t) - t)(A \oplus_\beta B)C \\
&= A + B + (\beta(t) - t)AB + C + (\beta(t) - t)(A + B + (\beta(t) - t)AB)C \\
&= A + B + (\beta(t) - t)AB + C + (\beta(t) - t)AC + (\beta(t) - t)BC + (\beta(t) - t)^2 ABC,
\end{aligned}$$

and

$$\begin{aligned}
A \oplus_\beta (B \oplus_\beta C) &= A + (B \oplus_\beta C) + (\beta(t) - t)A(B \oplus_\beta C) \\
&= A + B + C + (\beta(t) - t)BC + (\beta(t) - t)A(B + C + (\beta(t) - t)BC) \\
&= A + B + C + (\beta(t) - t)BC + (\beta(t) - t)AB + (\beta(t) - t)AC + (\beta(t) - t)^2 ABC.
\end{aligned}$$

Consequently,

$$(A \oplus_\beta B) \oplus_\beta C = A \oplus_\beta (B \oplus_\beta C),$$

that is, in $\mathscr{R}_\beta$, the associative law holds with respect to $\oplus_\beta$. Also,

$$\begin{aligned}
O \oplus_\beta A &= A \oplus_\beta O \\
&= A,
\end{aligned}$$

that is, O is the identity element in $\mathcal{R}_\beta$ with respect to $\oplus_\beta$. Next,

$$A \oplus_\beta \left(-(\mathcal{I} + (\beta(t) - t)A)^{-1}A\right) = A - (\mathcal{I} + (\beta(t) - t)A)^{-1}A - (\beta(t) - t)(\mathcal{I} + (\beta(t) - t)A)^{-1}A^2$$
$$= A - (\mathcal{I} + (\beta(t) - t)A)^{-1}(\mathcal{I} + (\beta(t) - t)A)A$$
$$= A - A$$
$$= O,$$

that is, the additive inverse of A under the addition $\oplus_\beta$ is $-(\mathcal{I} + (\beta(t) - t)A)^{-1}A$. Note that

$$\mathcal{I} + (\beta(t) - t)\left(-(\mathcal{I} + (\beta(t) - t)A)^{-1}A\right)$$
$$= (\mathcal{I} + (\beta(t) - t)A)^{-1}(\mathcal{I} + (\beta(t) - t)A) - (\mathcal{I} + (\beta(t) - t)A)^{-1}(\beta(t) - t)A$$
$$= (\mathcal{I} + (\beta(t) - t)A)^{-1},$$

and then $-(\mathcal{I} + (\beta(t) - t)A)^{-1}A \in \mathcal{R}_\beta$. This completes the proof. $\qquad\square$

Henceforth, the conjugate and transpose of a matrix A will be denoted by $\overline{A}$ and A^T, respectively. The conjugate transpose of a matrix A will be denoted by $A^* = \left(\overline{A}\right)^T$.

Theorem 7.1.27. *Let A and B be β-regressive matrix-valued functions. Then we have*

1. A^* *is β-regressive;*
2. $\ominus_\beta A^* = \left(\ominus_\beta A\right)^*$;
3. $(A \oplus_\beta B)^* = B^* \oplus_\beta A^*$.

Proof. Since A is β-regressive, $(\mathcal{I} + (\beta(t) - t)A)^{-1}$ exists.

1. We have

$$\mathcal{I} = (\mathcal{I} + (\beta(t) - t)A)(\mathcal{I} + (\beta(t) - t)A)^{-1}$$
$$= \overline{(\mathcal{I} + (\beta(t) - t)\overline{A})(\mathcal{I} + (\beta(t) - t)A)^{-1}}$$
$$= \left((\mathcal{I} + (\beta(t) - t)A)^{-1}\right)^* (\mathcal{I} + (\beta(t) - t)A^*).$$

Therefore both

$$\left(\mathcal{I} + (\beta(t) - t)A^*\right)^{-1} \quad \text{and} \quad \left(\mathcal{I} + (\beta(t) - t)A^T\right)^{-1}$$

exist, and

$$\left(\mathcal{I} + (\beta(t) - t)A^*\right)^{-1} = \left((\mathcal{I} + (\beta(t) - t)A)^{-1}\right)^*,$$
$$\left(\mathcal{I} + (\beta(t) - t)A^T\right)^{-1} = \left((\mathcal{I} + (\beta(t) - t)A)^{-1}\right)^T.$$

Consequently, A^* is β-regressive.

2. We have

$$(\ominus_\beta A)^* = -\left((\mathcal{I} + (\beta(t) - t)A)^{-1}A\right)^*$$

$$= -A^* \left((\mathscr{I} + (\beta(t) - t)A)^{-1} \right)^*$$
$$= -A^* \left(\mathscr{I} + (\beta(t) - t)A^* \right)^{-1}$$
$$= \ominus_\beta A^*.$$

Thus,

$$(\ominus_\beta A)^* = \ominus_\beta A^*.$$

3. We have

$$(A \oplus B)^* = (A + B + (\beta(t) - t)AB)^*$$
$$= A^* + B^* + (\beta(t) - t)B^* A^*$$
$$= B^* + A^* + (\beta(t) - t)B^* A^*$$
$$= B^* \oplus_\beta A^*.$$

Thus,

$$(A \oplus B)^* = B^* \oplus_\beta A^*.$$

This completes the proof. $\square$

7.2 β-matrix exponential function

Definition 7.2.1 (β-Matrix exponential function). Let $A \in \mathscr{R}_\beta$ be a matrix-valued function, and let $s_0 \in I$. The unique solution of the IVP

$$D_\beta X = A(t)X, \quad t \in I,$$
$$X(s_0) = \mathscr{I},$$

is called the β-matrix exponential function. It is denoted by $e_{A,\beta}(\cdot)$.

Theorem 7.2.2. *Let $A, B \in \mathscr{R}_\beta$ be matrix-valued functions, and let $t, s, r \in I$. Then we have*

1. $e_{O,\beta}(t) = \mathscr{I}$, $e_{A_\beta}(s_0) = \mathscr{I}$;
2. $e_{A,\beta}(\beta(t)) = (\mathscr{I} + (\beta(t) - t)A(t))e_{A,\beta}(t)$;
3. $e_{A,\beta}(t)e_{B,\beta}(t) = e_{A\oplus_\beta B,\beta}(t)$ *if $e_{A,\beta}(t)$ and B commute.*

Proof. **1.** Consider the IVP

$$D_\beta X = O, \quad t \in I,$$
$$X(s_0) = \mathscr{I}.$$

Then its unique solution is

$$e_{O,\beta}(t) = \mathscr{I}, \quad t \in I.$$

Now, consider the IVP

$$D_\beta X = A(t)X, \quad t \in I,$$
$$X(s_0) = \mathscr{I}.$$

Using the definition of $e_{A,\beta}(\cdot)$, we obtain

$$e_{A,\beta}(s_0) = \mathscr{I}.$$

2. In view of Theorem 7.1.7 and the definition of $e_{A,\beta}(\cdot)$, we have

$$\begin{aligned}
e_{A,\beta}(\beta(t)) &= e_{A,\beta}(t) + (\beta(t) - t)D_\beta e_{A,\beta}(t,s) \\
&= e_{A,\beta}(t) + (\beta(t) - t)A(t)e_{A,\beta}(t) \\
&= (\mathscr{I} + (\beta(t) - t)A(t))e_{A,\beta}(t).
\end{aligned}$$

3. Let

$$Y(t) = e_{A,\beta}(t)e_{B,\beta}(t).$$

Then

$$\begin{aligned}
D_\beta Y(t) &= D_\beta e_{A,\beta}(t)e^\beta_{B,\beta}(t) + e_{A,\beta}(t)D_\beta e_{B,\beta}(t) \\
&= A(t)e_{A,\beta}(t)(\mathscr{I} + (\beta(t) - t)B(t))e_{B,\beta}(t) + B(t)e_{A,\beta}(t)e_{B,\beta}(t) \\
&= A(t)(\mathscr{I} + (\beta(t) - t)B(t))e_{A,\beta}(t)e_{B,\beta}(t) + B(t)e_{A,\beta}(t)e_{B,\beta}(t) \\
&= (A(t) + B(t) + (\beta(t) - t)A(t)B(t))e_{A,\beta}(t)e_{B,\beta}(t) \\
&= (A \oplus_\beta B)(t)Y(t).
\end{aligned}$$

Also,

$$\begin{aligned}
Y(s_0) &= e_{A,\beta}(s_0)e_{B,\beta}(s_0) \\
&= \mathscr{I}.
\end{aligned}$$

Consequently,

$$e_{A \oplus_\beta B,\beta}(t) = e_{A,\beta}(t)e_{B,\beta}(t).$$

This completes the proof. $\qquad\Box$

7.3 The β-Liouville theorem

Theorem 7.3.1 (β-Liouville formula). *Let $A \in \mathscr{R}_\beta$ be a 2×2-matrix-valued function, and let X be a solution of*

$$D_\beta X = A(t)X.$$

Then X satisfies the β-Liouville formula

$$\det X(t) = e_{\operatorname{tr} A + (\beta(t) - t)\det A, \beta}(t) \det X(s_0), \quad t \in I.$$

Proof. In view of Theorem 7.1.22, it follows that

$$\operatorname{tr} A + (\beta(t) - t) \det A$$

is β-regressive. Let

$$A(t) = \begin{pmatrix} a_{11}(t) & a_{12}(t) \\ a_{21}(t) & a_{22}(t) \end{pmatrix} \quad \text{and}$$

$$X(t) = \begin{pmatrix} x_{11}(t) & x_{12}(t) \\ x_{21}(t) & x_{22}(t) \end{pmatrix}.$$

Then

$$\begin{pmatrix} D_\beta x_{11}(t) & D_\beta x_{12}(t) \\ D_\beta x_{21}(t) & D_\beta x_{22}(t) \end{pmatrix} = \begin{pmatrix} a_{11}(t) & a_{12}(t) \\ a_{21}(t) & a_{22}(t) \end{pmatrix} \begin{pmatrix} x_{11}(t) & x_{12}(t) \\ x_{21}(t) & x_{22}(t) \end{pmatrix}$$

$$= \begin{pmatrix} a_{11}(t)x_{11}(t) + a_{12}(t)x_{12}(t) & a_{11}(t)x_{12}(t) + a_{12}(t)x_{22}(t) \\ a_{21}(t)x_{11}(t) + a_{22}(t)x_{21}(t) & a_{21}(t)x_{12}(t) + a_{22}(t)x_{22}(t) \end{pmatrix},$$

whereupon

$$\begin{cases} D_\beta x_{11}(t) = a_{11}(t)x_{11}(t) + a_{12}(t)x_{21}(t), \\ D_\beta x_{12}(t) = a_{11}(t)x_{12}(t) + a_{12}(t)x_{22}(t), \\ D_\beta x_{21}(t) = a_{21}(t)x_{11}(t) + a_{22}(t)x_{21}(t), \\ D_\beta x_{22}(t) = a_{21}(t)x_{12}(t) + a_{22}(t)x_{22}(t). \end{cases}$$

Then

$$\det X(t) = x_{11}(t)x_{22}(t) - x_{12}(t)x_{21}(t),$$

and

$$D_\beta(\det X)(t) = D_\beta x_{11}(t)x_{22}(t) + x_{11}^\beta(t)D_\beta x_{22}(t) - D_\beta x_{12}(t)x_{21}(t) - x_{12}^\beta(t)D_\beta x_{21}(t)$$

$$= \det \begin{pmatrix} D_\beta x_{11}(t) & D_\beta x_{12}(t) \\ x_{21}(t) & x_{22}(t) \end{pmatrix} + \det \begin{pmatrix} x_{11}^\beta(t) & x_{12}^\beta(t) \\ D_\beta x_{21}(t) & D_\beta x_{22}(t) \end{pmatrix}$$

$$= \det \begin{pmatrix} a_{11}(t)x_{11}(t) + a_{12}(t)x_{21}(t) & a_{11}(t)x_{12}(t) + a_{12}(t)x_{22}(t) \\ x_{21}(t) & x_{22}(t) \end{pmatrix}$$

$$+ \det \begin{pmatrix} x_{11}(t) + (\beta(t) - t)D_\beta x_{11}(t) & x_{12}(t) + (\beta(t) - t)D_\beta x_{12}(t) \\ D_\beta x_{21}(t) & D_\beta x_{22}(t) \end{pmatrix}$$

$$= [a_{11}(t)x_{11}(t)x_{22}(t) + a_{12}(t)x_{21}(t)x_{22}(t) - a_{11}(t)x_{12}(t)x_{21}(t) - a_{12}(t)x_{21}(t)x_{22}(t)]$$

$$+ \det \begin{pmatrix} x_{11}(t) & x_{12}(t) \\ D_\beta x_{21}(t) & x_{22}(t) \end{pmatrix} + (\beta(t) - t)(t)\det \begin{pmatrix} D_\beta x_{11}(t) & D_\beta x_{12}(t) \\ D_\beta x_{21}(t) & D_\beta x_{22}(t) \end{pmatrix}$$

$$= a_{11}(t) \det X(t) + \det \begin{pmatrix} x_{11}(t) & x_{12}(t) \\ D_\beta x_{21}(t) & D_\beta x_{22}(t) \end{pmatrix} + (\beta(t) - t) \det \begin{pmatrix} D_\beta x_{11}(t) & D_\beta x_{12}(t) \\ D_\beta x_{21}(t) & D_\beta x_{22}(t) \end{pmatrix}$$

$$= a_{11}(t) \det X(t) + \det \begin{pmatrix} x_{11}(t) & x_{12}(t) \\ a_{21}(t)x_{11}(t) + a_{22}(t)x_{21}(t) & a_{21}(t)x_{12}(t) + a_{22}(t)x_{22}(t) \end{pmatrix}$$

$$+ (\beta(t) - t) \det \begin{pmatrix} a_{11}(t)x_{11}(t) + a_{12}(t)x_{21}(t) & a_{11}(t)x_{12}(t) + a_{12}(t)x_{22}(t) \\ D_\beta x_{21}(t) & D_\beta x_{22}(t) \end{pmatrix}$$

$$= a_{11}(t) \det X(t) + [a_{21}(t)x_{11}(t)x_{12}(t) + a_{22}(t)x_{11}(t)x_{22}(t)$$
$$- a_{21}(t)x_{11}(t)x_{12}(t) - a_{22}(t)x_{21}(t)x_{12}(t)]$$

$$+ (\beta(t) - t)\left[a_{11}(t) \det \begin{pmatrix} x_{11}(t) & x_{12}(t) \\ D_\beta x_{21}(t) & D_\beta x_{22}(t) \end{pmatrix} + a_{12}(t) \det \begin{pmatrix} x_{21}(t) & x_{22}(t) \\ D_\beta x_{21}(t) & D_\beta x_{22}(t) \end{pmatrix} \right]$$

$$= a_{11}(t) \det X(t) + a_{22}(t) \det X(t)$$

$$+ (\beta(t) - t)\left[a_{11}(t) \det \begin{pmatrix} x_{11}(t) & x_{12}(t) \\ a_{21}(t)x_{11}(t) + a_{22}(t)x_{21}(t) & a_{21}(t)x_{12}(t) + a_{22}(t)x_{22}(t) \end{pmatrix} \right.$$

$$\left. + a_{12}(t) \det \begin{pmatrix} x_{21}(t) & x_{22}(t) \\ a_{21}(t)x_{11}(t) + a_{22}(t)x_{21}(t) & a_{21}(t)x_{12}(t) + a_{22}(t)x_{22}(t) \end{pmatrix} \right]$$

$$= \operatorname{tr} A(t) \det X(t) + (\beta(t) - t)\,[a_{11}(t)(a_{21}(t)x_{12}(t)x_{11}(t) + a_{22}(t)x_{22}(t)x_{11}(t)$$
$$- a_{21}(t)x_{11}(t)x_{12}(t) - a_{22}(t)x_{12}(t)x_{21}(t)) + a_{12}(t)(a_{21}(t)x_{21}(t)x_{12}(t) + a_{22}(t)x_{22}(t)x_{21}(t)$$
$$- a_{21}(t)x_{11}(t)x_{22}(t) - a_{22}(t)x_{21}(t)x_{22}(t))]$$

$$= \operatorname{tr} A(t) \det X(t) + (\beta(t) - t)\,[a_{11}(t)a_{22}(t) \det X(t) - a_{12}(t)a_{21}(t) \det X(t)]$$

$$= \operatorname{tr} A(t) \det X(t) + (\beta(t) - t) \det A(t) \det X(t)$$

$$= [\operatorname{tr} A(t) + (\beta(t) - t) \det A(t)] \det X(t),$$

that is,

$$D_\beta (\det X)(t) = [\operatorname{tr} A(t) + (\beta(t) - t) \det A(t)] \det X(t).$$

Thus,

$$\det X(t) = e_{\operatorname{tr} A + (\beta(t) - t) \det A}(t) \det X(s_0).$$

This completes the proof. $\square$

Example 7.3.2. Let $I = \mathbb{R}$, and let $\beta(t) = \dfrac{1}{2}t + 1$, $t \in I$. Then $s_0 = 2$. Consider the IVP

$$D_\beta X(t) = \begin{pmatrix} 3t & 4t + 1 \\ 3 & 2 + t \end{pmatrix} X(t), \quad t \in I,$$

$$X(2) = \begin{pmatrix} 1 & 0 \\ 1 & 1 \end{pmatrix}.$$

Here

$$A(t) = \begin{pmatrix} 3t & 4t+1 \\ 3 & 2+t \end{pmatrix}$$

$$\beta(t) - t = 1 - \frac{1}{2}t, \quad t \in I.$$

Then

$$\det A(t) = 3t(2+t) - 3(4t+1)$$
$$= 6t + 3\,t^2 - 12t - 3$$
$$= 3t^2 - 6t - 3,$$
$$\operatorname{tr} A(t) = 3t + 2 + t$$
$$= 2 + 4t,$$

$$\operatorname{tr} A(t) + (\beta(t) - t)\det A(t) = 2 + 4t + \left(1 - \frac{1}{2}t\right)(3t^2 - 6t - 3)$$
$$= 2 + 4t + 3t^2 - 6t - 3 - \frac{3}{2}t^3 + 3t^2 + \frac{3}{2}t$$
$$= -\frac{3}{2}t^3 + 6t^2 - \frac{1}{2}t - 1, \quad t \in I.$$

Let

$$f(t) = -\frac{3}{2}t^3 + 6t^2 - \frac{1}{2}t - 1, \quad t \in I.$$

Note that

$$\beta^k(t) = \left(\frac{1}{2}\right)^k t + [k]_{\frac{1}{2}}, \quad k \in \mathbb{N}.$$

Then, using the β-Liouville formula, we get

$$\det X(t) = e_{f,\beta}(t)\det X(2)$$
$$= e_{f,\beta}(t)$$
$$= \frac{1}{\displaystyle\prod_{k=0}^{\infty} \left(1 - f\left(\beta^k(t)\right)\left(\beta^k(t) - \beta^{k+1}(t)\right)\right)}, \quad t \in I.$$

Exercise 7.3.3. Let $A \in \mathscr{R}_\beta$ be an $n \times n$-matrix-valued function, and let X be a solution of

$$D_\beta X = A(t)X.$$

Prove that X satisfies the β-Liouville formula

$$\det X(t) = e_{\operatorname{tr} A + (\beta(t) - t)\det A, \beta}(t)\det X(s_0), \quad t \in I.$$

Hint. Use Theorem 7.3.1 and the principle of mathematical induction.

7.4 Constant coefficients

In this section, we suppose that A is an $n \times n$-matrix-valued constant function such that $A \in \mathscr{R}_\beta$. Consider the β-differential system

$$D_\beta X = AX. \tag{7.3}$$

Theorem 7.4.1. *Let λ and ξ be an eigenpair of A. Then*

$$x(t) = e_{\lambda,\beta}(t)\xi, \quad t \in I,$$

is a solution of (7.3).

Proof. Since

$$A\xi = \lambda\xi,$$

for $t \in I$, we have

$$\begin{aligned}
D_\beta X(t) &= D_\beta e_{\lambda,\beta}(t)\xi \\
&= \lambda e_{\lambda,\beta}(t)\xi \\
&= e_{\lambda,\beta}(t)(\lambda\xi) \\
&= e_{\lambda,\beta}(t)A\xi \\
&= A\left(e_\lambda(t,t_0)\xi\right) \\
&= AX(t).
\end{aligned}$$

Thus,

$$D_\beta X(t) = AX(t), \quad t \in I,$$

that is, X is a solution of (7.3). This completes the proof. $\qquad\square$

Example 7.4.2. Let $I \subseteq \mathbb{R}$ and consider the β-differential system

$$\begin{aligned}
D_\beta x_1(t) &= -3x_1 - 2x_2, \\
D_\beta x_2(t) &= 3x_1 + 4x_2, \quad t \in I.
\end{aligned}$$

Here

$$A = \begin{pmatrix} -3 & -2 \\ 3 & 4 \end{pmatrix}.$$

Then

$$\det \begin{pmatrix} -3 - \lambda & -2 \\ 3 & 4 - \lambda \end{pmatrix} = 0$$

implies

$$\lambda^2 - \lambda - 6 = 0,$$

which yields

$$\lambda_1 = 3 \quad \text{and} \quad \lambda_2 = -2.$$

Thus, the considered β-differential system is β-regressive provided $3, -2 \in \mathscr{R}_\beta$. Note that

$$\xi_1 = \begin{pmatrix} 1 \\ -3 \end{pmatrix} \quad \text{and} \quad \xi_2 = \begin{pmatrix} -2 \\ 1 \end{pmatrix}$$

are eigenvalues corresponding to λ_1 and λ_2, respectively. Therefore

$$x(t) = c_1 e_{3,\beta}(t)\xi_1 + c_2 e_{-2,\beta}(t)\xi_2$$

$$= c_1 e_{3,\beta}(t)\begin{pmatrix} 1 \\ -3 \end{pmatrix} + c_2 e_{-2,\beta}(t)\begin{pmatrix} -2 \\ 1 \end{pmatrix}, \quad t \in I,$$

where $c_1, c_2 \in \mathbb{R}$, is a general solution of the considered β-differential system provided $3, -2 \in \mathscr{R}_\beta$.

Example 7.4.3. Let $I \subseteq \mathbb{R}$ and consider the β-differential system

$$D_\beta x_1(t) = x_1(t) - x_2(t),$$
$$D_\beta x_2(t) = -x_1(t) + 2x_2(t) - x_3(t),$$
$$D_\beta x_3(t) = -x_2(t) + x_3(t), \quad t \in I.$$

Here

$$A = \begin{pmatrix} 1 & -1 & 0 \\ -1 & 2 & -1 \\ 0 & -1 & 1 \end{pmatrix}.$$

Then

$$\det(A - \lambda\mathscr{I}) = \det \begin{pmatrix} 1-\lambda & -1 & 0 \\ -1 & 2-\lambda & -1 \\ 0 & -1 & 1-\lambda \end{pmatrix}$$

$$= -(\lambda - 1)^2(\lambda - 2) + (\lambda - 1) + (\lambda - 1)$$
$$= (\lambda - 1)\left(-(\lambda - 1)(\lambda - 2) + 2\right)$$
$$= (\lambda - 1)\left(-\lambda^2 + 3\lambda\right)$$
$$= -\lambda(\lambda - 1)(\lambda - 3),$$

whereupon

$$\det(A - \lambda\mathscr{I}) = 0$$

yields

$$\lambda_1 = 0, \quad \lambda_2 = 1, \quad \text{and} \quad \lambda_3 = 3.$$

Note that the matrix A is β-regressive, provided $1, 3 \in \mathcal{R}_\beta$ and

$$\xi_1 = \begin{pmatrix} 1 \\ 1 \\ 1 \end{pmatrix}, \xi_2 = \begin{pmatrix} 1 \\ 0 \\ -1 \end{pmatrix}, \quad \text{and} \quad \xi_3 = \begin{pmatrix} 1 \\ -2 \\ 1 \end{pmatrix}$$

are eigenvalues corresponding to λ_1, λ_2, and λ_3, respectively. Consequently,

$$x(t) = c_1 \xi_1 + c_2 e_{1,\beta}(t)\xi_2 + c_3 e_{3,\beta}(t)\xi_3$$

$$= c_1 \begin{pmatrix} 1 \\ 1 \\ 1 \end{pmatrix} + c_2 e_{1,\beta}(t) \begin{pmatrix} 1 \\ 0 \\ -1 \end{pmatrix} + c_3 e_{3,\beta}(t) \begin{pmatrix} 1 \\ -2 \\ 1 \end{pmatrix}, \quad t \in I,$$

where $c_1, c_2, c_3 \in \mathbb{R}$, is a general solution of the considered β-differential system provided $1, 3 \in \mathcal{R}_\beta$.

Example 7.4.4. Let $I \subseteq \mathbb{R}$ and consider the β-differential system

$$D_\beta x_1(t) = -x_1(t) + x_2(t) + x_3(t),$$
$$D_\beta x_2(t) = x_2(t) - x_3(t) + x_4(t),$$
$$D_\beta x_3(t) = 2x_3(t) - 2x_4(t),$$
$$D_\beta x_4(t) = 3x_4(t), \quad t \in I.$$

Here

$$A = \begin{pmatrix} -1 & 1 & 1 & 0 \\ 0 & 1 & -1 & 1 \\ 0 & 0 & 2 & -2 \\ 0 & 0 & 0 & 3 \end{pmatrix}.$$

Then

$$\det(A - \lambda \mathscr{I}) = \det \begin{pmatrix} -1-\lambda & 1 & 1 & 0 \\ 0 & 1-\lambda & -1 & 1 \\ 0 & 0 & 2-\lambda & -2 \\ 0 & 0 & 0 & 3-\lambda \end{pmatrix}$$

$$= (\lambda+1)(\lambda-1)(\lambda-2)(\lambda-3),$$

whereupon

$$\det(A - \lambda \mathscr{I}) = 0$$

yields

$$\lambda_1 = -1, \quad \lambda_2 = 1, \quad \lambda_3 = 2, \quad \text{and} \quad \lambda_4 = 3.$$

The matrix A is β-regressive provided $-1, 1, 2, 5 \in \mathscr{R}_\beta$. Note that

$$\xi_1 = \begin{pmatrix} 1 \\ 0 \\ 0 \\ 0 \end{pmatrix}, \quad \xi_2 = \begin{pmatrix} 0 \\ 1 \\ 0 \\ 0 \end{pmatrix}, \quad \xi_3 = \begin{pmatrix} 0 \\ 0 \\ 1 \\ 0 \end{pmatrix}, \quad \text{and} \quad \xi_4 = \begin{pmatrix} 0 \\ 0 \\ 0 \\ 1 \end{pmatrix}$$

are eigenvectors corresponding to λ_1, λ_2, λ_3, and λ_4, respectively. Consequently,

$$x(t) = c_1 e_{-1,\beta}(t)\xi_1 + c_2 e_{1,\beta}(t)\xi_2 + c_3 e_{2,\beta}(t)\xi_3 + c_4 e_{5,\beta}(t)\xi_4$$

$$= c_1 e_{-1,\beta}(t) \begin{pmatrix} 1 \\ 0 \\ 0 \\ 0 \end{pmatrix} + c_2 e_{1,\beta}(t) \begin{pmatrix} 0 \\ 1 \\ 0 \\ 0 \end{pmatrix} + c_3 e_{2,\beta}(t) \begin{pmatrix} 0 \\ 0 \\ 1 \\ 0 \end{pmatrix} + c_4 e_{3,\beta}(t) \begin{pmatrix} 0 \\ 0 \\ 0 \\ 1 \end{pmatrix},$$

$t \in I$, where $c_1, c_2, c_3, c_4 \in \mathbb{R}$ is a general solution of the considered β-differential system provided $-1, 1, 2, 5 \in \mathscr{R}_\beta$.

Exercise 7.4.5. Find a general solution of the β-differential system

$$\begin{aligned} D_\beta x_1(t) &= x_2(t), \\ D_\beta x_2(t) &= x_1(t), \quad t \in I. \end{aligned}$$

Answer.

$$x(t) = c_1 e_{1,\beta}(t) \begin{pmatrix} 1 \\ 1 \end{pmatrix} + c_2 e_{-1,\beta}(t) \begin{pmatrix} 1 \\ -1 \end{pmatrix}, \quad t \in I,$$

where $c_1, c_2 \in \mathbb{R}$, for which $-1 \in \mathscr{R}_\beta$.

Theorem 7.4.6. *Assume that $A \in \mathscr{R}_\beta$. If*

$$x(t) = u(t) + iv(t), \quad t \in I,$$

is a complex vector-valued solution of (7.3), where u and v are real vector-valued functions on I, then u and v are real vector-valued solutions of (7.3) on I.

Proof. We have

$$\begin{aligned} D_\beta x(t) &= A(t)x(t) \\ &= A(t)\left(u(t) + iv(t)\right) \\ &= A(t)u(t) + iA(t)v(t) \\ &= D_\beta u(t) + iD_\beta v(t), \quad t \in I. \end{aligned}$$

Equating real and imaginary parts, we get

$$D_\beta u(t) = A(t)u(t),$$

$$D_\beta v(t) = A(t)v(t), \quad t \in I.$$

This means that u and v are solutions of (7.3) on I. This completes the proof. $\qquad\square$

Example 7.4.7. Consider the β-differential system

$$D_\beta x_1(t) = x_1(t) + x_2(t),$$
$$D_\beta x_2(t) = -x_1(t) + x_2(t), \quad t \in I.$$

Here

$$A = \begin{pmatrix} 1 & 1 \\ -1 & 1 \end{pmatrix}.$$

Then

$$\det(A - \lambda \mathscr{I}) = \det \begin{pmatrix} 1-\lambda & 1 \\ -1 & 1-\lambda \end{pmatrix}$$
$$= (\lambda - 1)^2 + 1$$
$$= \lambda^2 - 2\lambda + 1 + 1$$
$$= \lambda^2 - 2\lambda + 2,$$

whereupon

$$\det(A - \lambda \mathscr{I}) = 0$$

yields

$$\lambda_{1,2} = 1 \pm i.$$

The matrix A is β-regressive provided $1 \pm i \in \mathscr{R}_\beta$. Note that

$$\xi = \begin{pmatrix} 1 \\ i \end{pmatrix}$$

is an eigenvector corresponding to the eigenvalue $\lambda = 1 + i$. We have

$$x(t) = e_{1+i,\beta}(t) \begin{pmatrix} 1 \\ i \end{pmatrix}$$

$$= e_{1,\beta}(t) \left(\cos_{\frac{1}{1+(\beta(t)-t)},\beta}(t) + i \sin_{\frac{1}{1+(\beta(t)-t)},\beta}(t) \right) \begin{pmatrix} 1 \\ i \end{pmatrix}$$

$$= e_{1,\beta}(t) \left(\begin{pmatrix} \cos_{\frac{1}{1+(\beta(t)-t)},\beta}(t) \\ i\cos_{\frac{1}{1+(\beta(t)-t)},\beta}(t) \end{pmatrix} + \begin{pmatrix} i\sin_{\frac{1}{1+(\beta(t)-t)},\beta}(t) \\ -\sin_{\frac{1}{1+(\beta(t)-t)},\beta}(t) \end{pmatrix} \right)$$

$$= e_{1,\beta}(t) \begin{pmatrix} \cos_{\frac{1}{1(\beta(t)-t)},\beta}(t) \\ -\sin_{\frac{1}{1(\beta(t)-t)},\beta}(t) \end{pmatrix} + ie_{1,\beta}(t) \begin{pmatrix} \sin_{\frac{1}{1+(\beta(t)-t)},\beta}(t) \\ \cos_{\frac{1}{1+(\beta(t)-t)},\beta}(t) \end{pmatrix}.$$

Consequently,

$$e_{1,\beta}(t)\begin{pmatrix} \cos_{\frac{1}{1+(\beta(t)-t)},\beta}(t) \\ -\sin_{\frac{1}{1+(\beta(t)-t)},\beta}(t) \end{pmatrix} \quad \text{and} \quad e_{1,\beta}(t)\begin{pmatrix} \sin_{\frac{1}{1+(\beta(t)-t)},\beta}(t) \\ \cos_{\frac{1}{1+(\beta(t)-t)},\beta}(t) \end{pmatrix}$$

are solutions of the considered β-differential system. Therefore

$$x(t) = c_1 e_{1,\beta}(t)\begin{pmatrix} \cos_{\frac{1}{1+(\beta(t)-t)},\beta}(t) \\ -\sin_{\frac{1}{1+(\beta(t)-t)},\beta}(t) \end{pmatrix} + c_2 e_{1,\beta}(t)\begin{pmatrix} \sin_{\frac{1}{1+(\beta(t)-t)},\beta}(t) \\ \cos_{\frac{1}{1+(\beta(t)-t)},\beta}(t) \end{pmatrix}, \quad t \in I,$$

where $c_1, c_2 \in \mathbb{R}$, is a general solution of the considered β-differential system provided $1 \pm i \in \mathscr{R}_\beta$.

Example 7.4.8. Consider the β-differential system

$$D_\beta x_1(t) = x_2(t),$$
$$D_\beta x_2(t) = x_3(t),$$
$$D_\beta x_3(t) = 2x_1(t) - 4x_2(t) + 3x_3(t), \quad t \in I.$$

Here

$$A = \begin{pmatrix} 0 & 1 & 0 \\ 0 & 0 & 1 \\ 2 & -4 & 3 \end{pmatrix}.$$

Then

$$\det(A - \lambda \mathscr{I}) = \det\begin{pmatrix} -\lambda & 1 & 0 \\ 0 & -\lambda & 1 \\ 2 & -4 & 3-\lambda \end{pmatrix}$$
$$= -\lambda^2(\lambda - 3) + 2 - 4\lambda$$
$$= -\left(\lambda^3 - 3\lambda^2 + 4\lambda - 2\right)$$
$$= -(\lambda - 1)(\lambda^2 - 2\lambda + 2),$$

whereupon

$$\det(A - \lambda \mathscr{I}) = 0$$

yields

$$\lambda_1 = 1 \quad \text{and} \quad \lambda_{2,3} = 1 \pm i.$$

The matrix A is β-regressive provided $1, 1 \pm i \in \mathscr{R}_\beta$. Note that

$$\xi_1 = \begin{pmatrix} 1 \\ 1 \\ 1 \end{pmatrix} \quad \text{and} \quad \xi_2 = \begin{pmatrix} 1 \\ 1+i \\ 2i \end{pmatrix}$$

are eigenvectors corresponding to the eigenvalues $\lambda_1 = 1$ and $\lambda_2 = 1+i$, respectively. Note that

$$e_{1+i,\beta}(t)\begin{pmatrix}1\\1+i\\2i\end{pmatrix} = e_{1,\beta}(t)\left(\cos_{\frac{1}{1+(\beta(t)-t)},\beta}(t) + i\sin_{\frac{1}{1+(\beta(t)-t)},\beta}(t)\right)\begin{pmatrix}1\\1+i\\2i\end{pmatrix}$$

$$= e_{1,\beta}(t)\left(\begin{pmatrix}\cos_{\frac{1}{1+(\beta(t)-t)},\beta}(t)\\(1+i)\cos_{\frac{1}{1+(\beta(t)-t)},\beta}(t)\\2i\cos_{\frac{1}{1+(\beta(t)-t)},\beta}(t)\end{pmatrix} + i\begin{pmatrix}\sin_{\frac{1}{1+(\beta(t)-t)},\beta}(t)\\(1+i)\sin_{\frac{1}{1+(\beta(t)-t)},\beta}(t)\\2i\sin_{\frac{1}{1+(\beta(t)-t)},\beta}(t)\end{pmatrix}\right)$$

$$= e_{1,\beta}(t)\left(\begin{pmatrix}\cos_{\frac{1}{1+(\beta(t)-t)},\beta}(t)\\(1+i)\cos_{\frac{1}{1+(\beta(t)-t)},\beta}(t)\\2i\cos_{\frac{1}{1+(\beta(t)-t)},\beta}(t)\end{pmatrix} + \begin{pmatrix}i\sin_{\frac{1}{1+(\beta(t)-t)},\beta}(t)\\(-1+i)\sin_{\frac{1}{1+(\beta(t)-t)},\beta}(t)\\-2\sin_{\frac{1}{1+(\beta(t)-t)},\beta}(t)\end{pmatrix}\right)$$

$$= e_{1,\beta}(t)\left(\begin{pmatrix}\cos_{\frac{1}{1+(\beta(t)-t)},\beta}(t)\\\cos_{\frac{1}{1+(\beta(t)-t)},\beta}(t) - \sin_{\frac{1}{1+(\beta(t)-t)},\beta}(t)\\-2\sin_{\frac{1}{1+(\beta(t)-t)},\beta}(t)\end{pmatrix}\right.$$

$$\left. + i\begin{pmatrix}\sin_{\frac{1}{1+(\beta(t)-t)},\beta}(t)\\\cos_{\frac{1}{1+(\beta(t)-t)},\beta}(t) + \sin_{\frac{1}{1+(\beta(t)-t)},\beta}(t)\\2\cos_{\frac{1}{1+(\beta(t)-t)},\beta}(t)\end{pmatrix}\right).$$

Thus,

$$x(t) = e_{1,\beta}(t)\left(c_1\begin{pmatrix}1\\1\\1\end{pmatrix} + c_2\begin{pmatrix}\cos_{\frac{1}{1+(\beta(t)-t)},\beta}(t)\\\cos_{\frac{1}{1+(\beta(t)-t)},\beta}(t) - \sin_{\frac{1}{1+(\beta(t)-t)},\beta}(t)\\-2\sin_{\frac{1}{1+(\beta(t)-t)},\beta}(t)\end{pmatrix}\right.$$

$$\left. + c_3\begin{pmatrix}\sin_{\frac{1}{1+(\beta(t)-t)},\beta}(t)\\\cos_{\frac{1}{1+(\beta(t)-t)},\beta}(t) + \sin_{\frac{1}{1+(\beta(t)-t)},\beta}(t)\\2\cos_{\frac{1}{1+(\beta(t)-t)},\beta}(t)\end{pmatrix}\right), \quad t \in I,$$

where $c_1, c_2, c_3 \in \mathbb{R}$, is a general solution of the considered β-differential system provided $1, 1\pm i \in \mathscr{R}_\beta$.

Exercise 7.4.9. Find a general solution of the β-differential system

$$D_\beta x_1(t) = x_1(t) - 2x_2(t) + x_3(t),$$
$$D_\beta x_2(t) = -x_1(t) + x_3(t),$$
$$D_\beta x_3(t) = x_1(t) - 2x_2(t) + x_3(t), \quad t \in I.$$

Theorem 7.4.10 (The β-Putzer algorithm). *Let $A \in \mathcal{R}_\beta$ be a constant $n \times n$-matrix. If $\lambda_1, \lambda_2, \ldots, \lambda_n$ are the eigenvalues of A, then*

$$e_{A,\beta}(t) = \sum_{k=0}^{n-1} r_{k+1}(t) P_k,$$

where

$$r(t) = \begin{pmatrix} r_1(t) \\ \vdots \\ r_n(t) \end{pmatrix}$$

is the solution of the IVP

$$D_\beta r = \begin{pmatrix} \lambda_1 & 0 & 0 & \cdots & 0 \\ 1 & \lambda_2 & 0 & \cdots & 0 \\ 0 & 1 & \lambda_3 & \cdots & 0 \\ \vdots & \vdots & \vdots & \vdots & \vdots \\ 0 & 0 & 0 & \cdots & \lambda_n \end{pmatrix} r, \quad r(t_0) = \begin{pmatrix} 1 \\ 0 \\ 0 \\ \vdots \\ 0 \end{pmatrix}, \tag{7.4}$$

and the P-matrices are recursively defined by

$$P_0 = I \quad and$$
$$P_{k+1} = (A - \lambda_{k+1}\mathcal{I}) P_k, \quad 0 \le k \le n-1.$$

Proof. Since A is β-regressive, we have that all eigenvalues of A are β-regressive. Therefore the IVP (7.4) has a unique solution. We set

$$X(t) = \sum_{k=0}^{n-1} r_{k+1}(t) P_k. \tag{7.5}$$

We have

$$P_1 = (A - \lambda_1\mathcal{I}) P_0$$
$$= (A - \lambda_1\mathcal{I}),$$
$$P_2 = (A - \lambda_2\mathcal{I}) P_1$$
$$= (A - \lambda_2 I)(A - \lambda_1\mathcal{I}),$$
$$\vdots$$
$$P_n = (A - \lambda_n\mathcal{I}) P_{n-1}$$
$$= (A - \lambda_n\mathcal{I})\ldots(A - \lambda_1\mathcal{I})$$
$$= 0.$$

Therefore

$$D_\beta X(t) = \sum_{k=0}^{n-1} D_\beta r_{k+1}(t) P_k,$$

and

$$D_\beta X(t) - AX(t) = \sum_{k=0}^{n-1} D_\beta r_{k+1}(t) P_k - A \sum_{k=0}^{n-1} r_{k+1}(t) P_k$$

$$= D_\beta r_1(t) P_0 + \sum_{k=1}^{n-1} D_\beta r_{k+1}(t) P_k - A \sum_{k=0}^{n-1} r_{k+1}(t) P_k$$

$$= \lambda_1 r_1(t) P_0 + \sum_{k=1}^{n-1} (r_k(t) + \lambda_{k+1} r_{k+1}(t)) P_k - \sum_{k=1}^{n-1} r_{k+1}(t) A P_k$$

$$= \sum_{k=1}^{n-1} r_k(t) P_k + \lambda_1 r_1(t) P_0 + \sum_{k=1}^{n-1} \lambda_{k+1} r_{k+1}(t) P_k - \sum_{k=0}^{n-1} r_{k+1}(t) A P_k$$

$$= \sum_{k=1}^{n-1} r_k(t) P_k + \sum_{k=0}^{n-1} \lambda_{k+1} r_{k+1}(t) P_k - \sum_{k=0}^{n-1} r_{k+1}(t) A P_k$$

$$= \sum_{k=1}^{n-1} r_k(t) P_k - \sum_{k=0}^{n-1} (A - \lambda_{k+1} I) r_{k+1}(t) P_k$$

$$= \sum_{k=1}^{n-1} r_k(t) P_k - \sum_{k=0}^{n-1} r_{k+1}(t) P_{k+1}$$

$$= -r_n(t) P_n$$

$$= 0, \quad t \in I.$$

Thus,

$$D_\beta X(t) = AX(t), \quad t \in I.$$

Also,

$$X(t_0) = \sum_{k=0}^{n-1} r_{k+1}(t_0) P_k$$

$$= r_1(t_0) P_0$$

$$= \mathscr{I}.$$

Hence, X defined in (7.5) is indeed a solution of the IVP (7.4). This completes the proof. $\quad\square$

Example 7.4.11. Consider the β-differential system

$$D_\beta x_1(t) = 2x_1(t) + x_2(t) + 2x_3(t),$$
$$D_\beta x_2(t) = 4x_1(t) + 2x_2(t) + 4x_3(t),$$

$$D_\beta x_3(t) = 2x_1(t) + x_2(t) + 2x_3(t), \quad t \in I.$$

Here

$$A = \begin{pmatrix} 2 & 1 & 2 \\ 4 & 2 & 4 \\ 2 & 1 & 2 \end{pmatrix}.$$

Then

$$\det(A - \lambda \mathscr{I}) = \det \begin{pmatrix} 2-\lambda & 1 & 2 \\ 4 & 2-\lambda & 4 \\ 2 & 1 & 2-\lambda \end{pmatrix}$$

$$= -(\lambda - 2)^3 + 8 + 8 + 4(\lambda - 2) + 4(\lambda - 2) + 4(\lambda - 2)$$

$$= -(\lambda - 2)^3 + 12(\lambda - 2) + 16$$

$$= -\left(\lambda^3 - 6\lambda^2 + 12\lambda - 8 - 12\lambda + 24 - 16\right)$$

$$= -\left(\lambda^3 - 6\lambda^2\right)$$

$$= -\lambda^2(\lambda - 6),$$

whereupon

$$\det(A - \lambda \mathscr{I}) = 0$$

yields

$$\lambda_1 = 0, \quad \lambda_2 = 0, \quad \text{and} \quad \lambda_3 = 6.$$

Note that A is β-regressive if $6 \in \mathscr{R}_\beta$. Consider the following IVPs

$$D_\beta r_1(t) = 0, \quad r_1(t_0) = 1,$$
$$D_\beta r_2(t) = r_1(t), \quad r_2(t_0) = 0,$$
$$D_\beta r_3(t) = r_2(t) + 6r_3(t), \quad r_3(t_0) = 0.$$

Then we have

$$r_1(t) = 1, \quad t \in I,$$
$$D_\beta r_2(t) = 1, \quad r_2(t_0) = 0, \quad t \in I.$$

Thus

$$r_2(t) = t - t_0, \quad t \in I,$$

and

$$D_\beta r_3(t) = t - t_0 + 6r_3(t), \quad r_3(t_0) = 0, \quad t \in I.$$

This gives

$$r_3(t) = \int_{s_0}^{t} e_{6,\beta}(\beta(\tau))(\tau - t_0) d_\beta \tau, \quad t \in I.$$

Next, we have

$$P_0 = \begin{pmatrix} 1 & 0 & 0 \\ 0 & 1 & 0 \\ 0 & 0 & 1 \end{pmatrix},$$

$$P_1 = (A - \lambda_1 \mathscr{I}) P_0$$
$$= A P_0$$
$$= A$$
$$= \begin{pmatrix} 2 & 1 & 2 \\ 4 & 2 & 4 \\ 2 & 1 & 2 \end{pmatrix},$$

$$P_2 = (A - \lambda_1 I)(A - \lambda_2 I)$$
$$= A^2 I$$
$$= A^2$$
$$= \begin{pmatrix} 2 & 1 & 2 \\ 4 & 2 & 4 \\ 2 & 1 & 2 \end{pmatrix} \begin{pmatrix} 2 & 1 & 2 \\ 4 & 2 & 4 \\ 2 & 1 & 2 \end{pmatrix}$$
$$= \begin{pmatrix} 12 & 6 & 12 \\ 24 & 12 & 24 \\ 12 & 6 & 12 \end{pmatrix}, \quad \text{and}$$

$$P_3 = 0.$$

Hence, in view of the Putzer algorithm (Theorem 7.4.10), we obtain

$$e_{A,\beta}(t) = r_1(t) P_0 + r_2(t) P_1 + r_3(t) P_2$$
$$= \begin{pmatrix} 1 & 0 & 0 \\ 0 & 1 & 0 \\ 0 & 0 & 1 \end{pmatrix} + (t - t_0) \begin{pmatrix} 2 & 1 & 2 \\ 4 & 2 & 4 \\ 2 & 1 & 2 \end{pmatrix} + \left(\int_{t_0}^{t} e_{6,\beta}(\beta(\tau))(\tau - t_0) d_\beta \tau \right) \begin{pmatrix} 12 & 6 & 12 \\ 24 & 12 & 24 \\ 12 & 6 & 12 \end{pmatrix},$$

and

$$\begin{pmatrix} x_1(t) \\ x_2(t) \\ x_3(t) \end{pmatrix} = e_{A,\beta}(t) \begin{pmatrix} c_1 \\ c_2 \\ c_3 \end{pmatrix}, \quad t \in I,$$

where $c_1, c_2, c_3 \in \mathbb{R}$, is a general solution of the considered β-differential system provided $6 \in \mathscr{R}_\beta$.

Exercise 7.4.12. Using the β-Putzer algorithm, find $e_{A,\beta}(t)$, where

1. $A = \begin{pmatrix} 1 & 2 \\ -1 & 3 \end{pmatrix}$;

2. $A = \begin{pmatrix} 1 & -1 & 1 \\ 1 & 0 & 2 \\ -1 & 1 & 1 \end{pmatrix}$.

7.5 Nonlinear systems

In this section, we investigate the nonlinear system

$$D_\beta x_j = f_j(t, x), \quad j \in \{1, \ldots, n\}, \quad t \in [a, b], \tag{7.6}$$

subject to the initial condition

$$x_j(a) = x_{0j}, \quad j \in \{1, \ldots, n\}, \tag{7.7}$$

where

(A1) $[a, b] \subset I$, $s_0 \in [a, b]$,
(A2) $f = (f_1, \ldots, f_n)$, $f_j \in \mathscr{C}([a, b] \times \mathbb{R})$, $j \in \{1, \ldots, n\}$, $x_0 = (x_{01}, \ldots, x_{0n}) \in \mathbb{R}^n$, and $x = (x_1, \ldots, x_n)$ is the unknown function.

Denote $X = (\mathscr{C}([a, b]))^n$. In $\mathscr{C}([a, b] \times \mathbb{R})$, we define the norm

$$\|u\|_1 = \sup_{(t, y) \in [a, b] \times \mathbb{R}} |u(t, y)|,$$

provided that it exists. In X, we define the norm

$$\|x\| = \max_{j \in \{1, \ldots, n\}} \|x_j\|_1.$$

We will start our investigation with the following useful lemma.

Lemma 7.5.1. *Assume that* (A1) *and* (A2) *hold. Then,* $x \in \left(\mathscr{C}_\beta([a, b])\right)^n$ *is a solution to the IVP* (7.6)–(7.7) *if and only if* x_j, $j \in \{1, \ldots, n\}$, *satisfies*

$$x_j(t) = x_{0j} + \int_a^t f_j(s, x(s))d_\beta s, \quad t \in [a, b]. \tag{7.8}$$

Proof. Assume that x_j, $j \in \{1, \ldots, n\}$, satisfies (7.8). Since $f_j \in \mathscr{C}([a, b] \times \mathbb{R})$, $j \in \{1, \ldots, n\}$, we have that

$$\int_a^t f_j(s, x(s))d_\beta s \in \mathscr{C}_\beta^1([a, b]), \quad j \in \{1, \ldots, n\}.$$

Then $x_j \in \mathscr{C}_\beta^1([a, b])$, $j \in \{1, \ldots, n\}$. Now, β-differentiating (7.8), we get (7.6). Putting $t = a$ into (7.8) we obtain (7.7). Thus, x is a solution to the IVP (7.6)–(7.7). Conversely, assume that x is a solution to the IVP (7.6)–(7.7). Now, β-integrating (7.6) and using the initial condition (7.7), we obtain (7.8). This completes the proof. $\qquad\square$

In addition, suppose the following:

(A3)

$$|f_j(t, x^1) - f_j(t, x^2)| \le K \sum_{l=1}^{n} |x_l^1 - x_l^2|, \quad t \in [a, b], \quad x^1, x^2 \in \mathbb{R}^n,$$

$j \in \{1, \ldots, n\}$, for some positive constant K.

(A4) $|f_j(t, x)| \le M, t \in [a, b], x \in \mathbb{R}^n, j \in \{1, \ldots, n\}$, for some constant $M \ge 0$.

Theorem 7.5.2. *Assume that* (A1)–(A4) *hold. Then the IVP* (7.6)–(7.7) *has a unique solution* $x \in \left(\mathscr{C}_\beta([a, b])\right)^n$.

Proof. We define the sequence

$$x_{j0}(t) = x_{j0},$$

$$x_{jl}(t) = x_{j0}(t) + \int_a^t f(s, x_{l-1}(s)) d_\beta s, \quad t \in [a, b], \quad l \in \mathbb{N}.$$

Then, for $t \in [a, b], j \in \{1, \ldots, n\}$, we have

$$|x_{j1}(t) - x_{j0}(t)| = \left| \int_a^t f_j(s, x_0(s)) d_\beta s \right|$$

$$\le \int_a^t |f_j(s, x_0(s))| d_\beta s$$

$$\le M \int_a^t d_\beta s$$

$$= M h_1(t, a).$$

Thus,

$$|x_{j1}(t) - x_{j0}(t)| \le M h_1(t, a), \quad t \in [a, b], \quad j \in \{1, \ldots, n\}.$$

Assume that

$$|x_{jl-1}(t) - x_{jl-2}(t)| \le M n^{l-2} K^{l-2} h_{l-1}(t, a), \quad t \in [a, b], \quad j \in \{1, \ldots, n\}, \tag{7.9}$$

for some $l \in \mathbb{N}, l \ge 2$. We will prove that

$$|x_{jl}(t) - x_{jl-1}(t)| \le M n^{l-1} K^{l-1} h_l(t, a), \quad t \in [a, b], \quad j \in \{1, \ldots, n\}.$$

Indeed, we have

$$|x_{jl}(t) - x_{jl-1}(t)| = \left| \int_a^t \left(f_j(s, x_l(s)) - f_j(s, x_{l-1}(s)) \right) d_\beta s \right|$$

$$\le \int_a^t \left| f_j(s, x_l(s)) - f_j(s, x_{l-1}(s)) \right| d_\beta s$$

$$\leq K \sum_{m=1}^{n} \int_{a}^{t} |x_{ml}(s) - x_{ml-1}(s)| d_{\beta}s$$

$$\leq K M n^{l-2} A^{l-2} \sum_{m=1}^{n} \int_{a}^{t} h_{l-1}(s, a) d_{\beta}s$$

$$= M K^{l-1} n^{l-1} h_{l}(t, a), \quad t \in [a, b], \quad j \in \{1, \ldots, n\}.$$

Thus, (7.9) holds for all $l \in \mathbb{N}$, $l \geq 2$. Note that

$$\left| \lim_{l \to \infty} \left(x_{jl}(t) - x_{j0}(t) \right) \right| = \left| \sum_{l=1}^{\infty} \left(x_{jl}(t) - x_{jl-1}(t) \right) \right|$$

$$\leq \sum_{l=1}^{\infty} \left| x_{jl}(t) - x_{jl-1}(t) \right|$$

$$\leq M \sum_{l=1}^{\infty} K^{l-1} n^{l-1} h_{l}(t, a)$$

$$= \frac{M}{Kn} \sum_{l=1}^{\infty} K^{l} n^{l} h_{l}(t, a)$$

$$\leq \frac{M}{Kn} \sum_{l=1}^{\infty} K^{l} n^{l} h_{l}(b, a)$$

$$= \frac{M}{Kn} e_{Kn,\beta}(b, a)$$

$$< \infty, \quad j \in \{1, \ldots, n\}.$$

Thus, the series

$$\sum_{l=1}^{\infty} \left(x_{jl}(t) - x_{jl-1}(t) \right), \quad j \in \{1, \ldots, n\},$$

is uniformly convergent on $[a, b]$. Hence,

$$\lim_{l \to \infty} (x_{jl}(t) - x_{j0}(t)) = \sum_{l=1}^{\infty} (x_{jl}(t) - x_{jl-1}(t)), \quad j \in \{1, \ldots, n\},$$

$t \in [a, b]$, and then

$$x(t) = \lim_{l \to \infty} x_{l}(t), \quad t \in [a, b].$$

Further, $x_{j}(t)$, $t \in [a, b]$, $j \in \{1, \ldots, n\}$, satisfy (7.8). Now, in view of Lemma 7.5.1, we conclude that $x(t)$, $t \in [a, b]$, is a solution to the IVP (7.6)–(7.7). Assume that the IVP (7.6)–(7.7) has another solution $y(t)$, $t \in [a, b]$. Then by Lemma 7.5.1, we have

$$y_{j}(t) = x_{j0}(t) + \int_{a}^{t} f_{j}(s, y(s)) d_{\beta}s, \quad t \in [a, b], \quad j \in \{1, \ldots, n\}.$$

Note that for $t \in [a, b]$, $j \in \{1, \ldots, n\}$,

$$
\begin{aligned}
|y_j(t) - x_{j0}(t)| &= \left| \int_a^t f_j(s, y(s)) d_\beta s \right| \\
&\leq \int_a^t |f_j(s, y(s))| d_\beta s \\
&\leq M \int_a^t d_\beta s \\
&= M h_1(t, a),
\end{aligned}
$$

that is,

$$
|y_j(t) - x_{j0}(t)| \leq M h_1(t, a), \quad t \in [a, b], \quad j \in \{1, \ldots, n\}.
$$

Assume that

$$
|x_{jl-1}(t) - y_j(t)| \leq M n^{l-2} K^{l-2} h_{l-1}(t, a), \quad t \in [a, b], \quad j \in \{1, \ldots, n\}, \tag{7.10}
$$

for some $l \in \mathbb{N}$, $l \geq 2$. We will prove that

$$
|x_{jl}(t) - y_j(t)| \leq M n^{l-1} K^{l-1} h_l(t, a), \quad t \in [a, b], \quad j \in \{1, \ldots, n\}.
$$

In fact, we have

$$
\begin{aligned}
|x_{jl}(t) - y_j(t)| &= \left| \int_a^t \left(f_j(s, x_{l-1}(s)) - f_j(s, y(s)) \right) d_\beta s \right| \\
&\leq \int_a^t \left| f_j(s, y(s)) - f_j(s, x_{l-1}(s)) \right| d_\beta s \\
&\leq K \sum_{m=1}^n \int_a^t |y_m(s) - x_{ml-1}(s)| d_\beta s \\
&\leq K M n^{l-2} A^{l-2} \sum_{m=1}^n \int_a^t h_{l-1}(s, a) d_\beta s \\
&= M K^{l-1} n^{l-1} h_l(t, a), \quad t \in [a, b], \quad j \in \{1, \ldots, n\}.
\end{aligned}
$$

Thus, (7.10) holds for all $l \in \mathbb{N}$, $l \geq 2$. Because

$$
\lim_{l \to \infty} M K^l n^l h_l(t, a) = 0, \quad t \in [a, b],
$$

we conclude that

$$
\begin{aligned}
x_j(t) - y_j(t) &= \lim_{l \to \infty} (x_{jl}(t) - y_j(t)) \\
&= 0, \quad t \in [a, b], \quad j \in \{1, \ldots, n\}.
\end{aligned}
$$

This completes the proof. $\qquad\square$

7.6 Advanced practical problems

Problem 7.6.1. Let $I = \mathbb{R}$. Define $\beta(t) = 3t$ and

$$A(t) = \begin{pmatrix} \dfrac{t^2+2}{t+1} & t^2+3t \\ 4t-1 & 3t \end{pmatrix}, \quad t \in I.$$

Find $D_\beta A(t), t \in I$.

Answer.

$$D_\beta A(t) = \begin{pmatrix} \dfrac{3t^2+4t-2}{(t+1)(3t+1)} & 4t+3 \\ 4 & 3 \end{pmatrix}, \quad t \in I.$$

Problem 7.6.2. Let $I = \mathbb{R}$. Define $\beta(t) = 3t$,

$$A(t) = \begin{pmatrix} t+10 & t^2-2t+2 \\ t & t^2+t+1 \end{pmatrix}, \quad \text{and}$$

$$B(t) = \begin{pmatrix} t-2 & t+1 \\ t^2 & t^3-1 \end{pmatrix}, \quad t \in I.$$

Show that

$$D_\beta(AB)(t) = D_\beta A(t)B^\beta(t) + A(t)D_\beta B(t), \quad t \in I.$$

Problem 7.6.3. Let $I = \mathbb{R}$. Define $\beta(t) = \dfrac{1}{3}t + 7$, and

$$A(t) = \begin{pmatrix} t^2+2t+2 & t+1 \\ \dfrac{1}{t+1} & t^2+2 \end{pmatrix}, \quad t \in I.$$

Show that

$$(A^\beta)^{-1}(t) = (A^{-1})^\beta(t), \quad t \in I.$$

Problem 7.6.4. Let $I = \mathbb{R}$. Define $\beta(t) = 3t$,

$$A(t) = \begin{pmatrix} 1 & -1 \\ 0 & 1 \end{pmatrix}, \quad \text{and}$$

$$B(t) = \begin{pmatrix} 2 & 0 \\ -1 & 1 \end{pmatrix}, \quad t \in I.$$

Find

$$(A \oplus_\beta B)(t), \quad t \in I.$$

Answer.

$$\begin{pmatrix} 3(t+2\sqrt{t}+2) & -(t+2\sqrt{t}+2) \\ -(t+2\sqrt{t}+2) & t+2\sqrt{t}+3 \end{pmatrix}, \quad t \in I.$$

Problem 7.6.5. Find a general solution of the β-differential system

$$D_\beta x_1(t) = 2x_1(t) + 3x_2(t),$$
$$D_\beta x_2(t) = x_1(t) + 4x_2(t), \quad t \in I.$$

Answer.

$$x(t) = c_1 e_{1,\beta}(t) \begin{pmatrix} 3 \\ -1 \end{pmatrix} + c_2 e_{5,\beta}(t) \begin{pmatrix} 1 \\ 1 \end{pmatrix}, \quad t \in I,$$

where $c_1, c_2 \in \mathbb{R}$, provided that $1, 5 \in \mathscr{R}_\beta$.

Problem 7.6.6. Find a general solution of the β-differential system

$$D_\beta x_1(t) = -x_1(t) - x_2(t) - x_3(t),$$
$$D_\beta x_2(t) = x_1(t) - x_2(t) + 3x_3(t),$$
$$D_\beta x_3(t) = x_1(t) - x_2(t) + 4x_3(t), \quad t \in I.$$

Problem 7.6.7. Using the β-Putzer algorithm, find $e_{A,\beta}(t)$, where

1. $A = \begin{pmatrix} 2 & 3 \\ 1 & -4 \end{pmatrix}$;

2. $A = \begin{pmatrix} -1 & 2 & 3 \\ 1 & 1 & -4 \\ 1 & -1 & 2 \end{pmatrix}$.

7.7 Notes and references

In this chapter, we introduced the β-differential systems and investigated their structures. We have defined the β-exponential matrix-value function and deduced some of its properties. The constant case is also considered. We proved the β-analogue of the classical Liouville theorem. The β-Putzer algorithm for finding the β-exponential matrix-valued function in the case where the matrix is constant is presented. A good percentage of the contents of this chapter is new and is an original contribution of the authors.

Linear β-integral inequalities[*]

Let $I \subseteq \mathbb{R}$, and let $\beta : I \to \mathbb{R}$ be a first- or second-kind general quantum operator. Let $a, b \in I$ be such that $a < b$ and $\beta(a) = a$, and let $J = [a, b]$.

8.1 Gronwall- and Bellman-type inequalities

Theorem 8.1.1 (Gronwall-type Inequality). *Let $x \in \mathscr{C}_\beta(J)$ be a nonnegative function, and let c and p be nonnegative constants. Then the inequality*

$$x(t) \leq \int_a^t (px(s) + c)\,d_\beta s, \quad t \in J,$$

implies the inequality

$$x(t) \leq c \int_a^t e_{\ominus_\beta p, \beta}(\beta(s))d_\beta s, \quad t \in J.$$

Proof. Denote

$$g(t) = \int_a^t (px(s) + c)\,d_\beta s, \quad t \in J.$$

Then

$$g(a) = 0,$$
$$x(t) \leq g(t), \quad t \in J.$$

Also,

$$\begin{aligned} D_\beta g(t) &= px(t) + c \\ &\leq pg(t) + c, \quad t \in J, \end{aligned}$$

that is,

$$D_\beta g(t) - pg(t) \leq c, \quad t \in J.$$

[*] This book has a companion website hosting complementary materials. Visit this URL to access it: https://www.elsevier.com/books-and-journals/book-companion/9780443328046.

Generalized Quantum Calculus with Applications. https://doi.org/10.1016/B978-0-44-332804-6.00013-3

Now, we have

$$D_\beta(g(\cdot)e_{\ominus_\beta p,\beta}(\cdot))(t) = D_\beta g(t)e_{\ominus_\beta p,\beta}(\beta(t)) + (\ominus_\beta p)(t)g(t)e_{\ominus_\beta p,\beta}(t)$$

$$= D_\beta g(t)e_{\ominus_\beta p,\beta}(\beta(t)) + (\ominus_\beta p)(t)g(t)e_{\ominus_\beta p,\beta}(t)$$

$$= D_\beta g(t)e_{\ominus_\beta p,\beta}(\beta(t)) - \frac{pg(t)}{1 + p(\beta(t) - t)}e_{\ominus_\beta p,\beta}(t)$$

$$= D_\beta g(t)e_{\ominus_\beta p,\beta}(\beta(t)) - \frac{pg(t)}{1 + p(\beta(t) - t)}(1 + p(\beta(t) - t))e_{\ominus_\beta p,\beta}(\beta(t))$$

$$= D_\beta g(t)e_{\ominus_\beta p,\beta}(\beta(t)) - pg(t)e_{\ominus_\beta p,\beta}(\beta(t))$$

$$= (D_\beta g(t) - pg(t))e_{\ominus_\beta p,\beta}(\beta(t))$$

$$\leq ce_{\ominus_\beta p,\beta}(\beta(t)),$$

and hence

$$g(t)e_{\ominus_\beta p,\beta}(t) \leq c \int_a^t e_{\ominus_\beta p,\beta}(\beta(s))d_\beta s, \quad t \in J,$$

that is,

$$g(t) \leq \frac{c}{e_{\ominus_\beta p,\beta}(t)} \int_a^t e_{\ominus_\beta p,\beta}(\beta(s))d_\beta s, \quad t \in J,$$

This yields

$$x(t) \leq g(t)$$

$$\leq \frac{c}{e_{\ominus_\beta p,\beta}(t)} \int_a^t e_{\ominus_\beta p,\beta}(\beta(s))d_\beta s,$$

that is,

$$x(t) \leq \frac{c}{e_{\ominus_\beta p,\beta}(t)} \int_a^t e_{\ominus_\beta p,\beta}(\beta(s))d_\beta s, \quad t \in J.$$

This completes the proof. $\qquad\qquad\square$

Theorem 8.1.2 (Bellman-type Inequality). *Let $x, f \in \mathscr{C}_\beta(J)$ be nonnegative functions such that $f \in \mathscr{R}_\beta^+$, and let c be a nonnegative constant. Then the inequality*

$$x(t) \leq c + \int_a^t f(s)x(s)d_\beta s, \quad t \in J,$$

implies the inequality

$$x(t) \leq ce_{f,\beta}(t), \quad t \in J.$$

Proof. Denote

$$y(t) = c + \int_a^t f(s)x(s)d_\beta s, \quad t \in J.$$

Then

$$y(a) = c, \quad \text{and}$$
$$x(t) \le y(t), \quad t \in J.$$

Also,

$$D_\beta y(t) = D_\beta \left(c + \int_a^t f(s)x(s)d_\beta s \right)$$
$$= f(t)x(t)$$
$$\le f(t)y(t),$$

that is,

$$D_\beta y(t) \le f(t)y(t), \quad t \in J. \tag{8.1}$$

Further,

$$D_\beta(ye_{\ominus_\beta f,\beta}(\cdot))(t) = D_\beta y(t)e_{\ominus_\beta f,\beta}(\beta(t)) + y(t)\left(\ominus_\beta f\right)(t)e_{\ominus_\beta f,\beta}(t)$$
$$= D_\beta y(t)e_{\ominus_\beta f,\beta}(\beta(t)) - \frac{y(t)f(t)}{1+(\beta(t)-t)f(t)}e_{\ominus_\beta f,\beta}(t)$$
$$= D_\beta y(t)e_{\ominus_\beta f,\beta}(\beta(t)) - y(t)f(t)e_{\ominus_\beta f,\beta}(\beta(t))$$
$$= \left(D_\beta y(t) - y(t)f(t)\right)e_{\ominus_\beta f,\beta}(\beta(t)).$$

Thus,

$$D_\beta(ye_{\ominus_\beta f,\beta}(\cdot))(t) \le \left(D_\beta y(t) - y(t)f(t)\right)e_{\ominus_\beta f,\beta}(\beta(t)), \quad t \in J. \tag{8.2}$$

Since $f \in \mathscr{R}_\beta^+(J)$, we have

$$1 + (\beta(t) - t)f(t) > 0, \quad t \in J,$$

and

$$1 + (\beta(t) - t)\left(\ominus_\beta f\right)(t) = 1 - \frac{(\beta(t) - t)f(t)}{1 + (\beta(t) - t)f(t)}$$
$$= \frac{1}{1 + (\beta(t) - t)f(t)}$$
$$> 0, \quad t \in J,$$

that is, $\ominus_\beta f \in \mathscr{R}_\beta^+(J)$. Therefore,

$$e_{\ominus_\beta f,\beta}(t) > 0 \quad \text{and} \quad e_{\ominus_\beta f,\beta}(\beta(t)) > 0, \quad t \in J.$$

Now, inserting inequality (8.1) into inequality (8.2), we get

$$D_\beta(ye_{\ominus_\beta f,\beta}(\cdot))(t) \le 0, \quad t \in J,$$

whereupon, integrating from a to $t, t \in J$, we get

$$y(t)e_{\ominus_\beta f,\beta}(t) - y(a)e_{\ominus_\beta f,\beta}(a) \le 0,$$

which yields

$$y(t)e_{\ominus_\beta f,\beta}(t) \le c,$$

that is,

$$y(t) \le \frac{c}{e_{\ominus_\beta f,\beta}(t)}$$
$$= c\,e_{f,\beta}(t).$$

Thus,

$$y(t) \le c\,e_{f,\beta}(t), \quad t \in J.$$

This completes the proof. $\qquad\square$

Theorem 8.1.3. *Let* $x, f \in \mathscr{C}_\beta(J)$ *be nonnegative functions such that* $f \in \mathscr{R}_\beta^+$ *and* $g \in \mathscr{C}_\beta(J)$ *is a positive nondecreasing function. Then the inequality*

$$x(t) \le g(t) + \int_a^t x(s)f(s)d_\beta s, \quad t \in J, \tag{8.3}$$

implies the inequality

$$x(t) \le g(t)e_{f,\beta}(t), \quad t \in J.$$

Proof. Let

$$y(t) = \frac{x(t)}{g(t)}, \quad t \in J.$$

Then, using (8.3), we get

$$\frac{x(t)}{g(t)} \le 1 + \int_a^t \frac{x(s)f(s)}{g(t)}d_\beta s$$
$$\le 1 + \int_a^t \frac{x(s)f(s)}{g(s)}d_\beta s, \quad t \in J,$$

whereupon

$$y(t) \le 1 + \int_a^t f(s)y(s)d_\beta s, \quad t \in J.$$

Now, using Theorem 8.1.2, we obtain

$$y(t) \le e_{f,\beta}(t), \quad t \in J.$$

Thus,

$$\frac{x(t)}{g(t)} \le e_{f,\beta}(t), \quad t \in J,$$

that is,

$$x(t) \le g(t)e_{f,\beta}(t), \quad t \in J.$$

This completes the proof. $\qquad\square$

Theorem 8.1.4. *Let $x, f, g, h \in \mathscr{C}_\beta(J)$ be nonnegative functions. Then the inequality*

$$x(t) \geq f(z) + g(t) \int_z^t h(s)x(s)d_\beta s, \quad a \leq z \leq t \leq b,$$

implies the inequality

$$x(t) \geq f(z)\frac{e_{\ominus_\beta p, \beta}(z)}{e_{\ominus_\beta p, \beta}(t)}, \quad a \leq z \leq t \leq b,$$

where

$$p(z) = g(t)h(z) \quad and \quad \ominus_\beta\, p = -\frac{g(t)h(z)}{1 + (\beta(z) - z)g(t)h(z)}, \quad a \leq z \leq t \leq b.$$

Proof. Define

$$y(z) = x(t) - g(t) \int_z^t h(s)x(s)d_\beta s, \quad a \leq z \leq t \leq b.$$

Then

$$y(t) = x(t), \quad t \in J,$$
$$y(t) \geq f(z),$$
$$y(t) \leq x(t), \quad a \leq z \leq t \leq b,$$

and

$$D_\beta y(z) = g(t)h(z)x(z)$$
$$\geq g(t)h(z)y(z)$$
$$= p(z)y(z), \quad a \leq z \leq t \leq b.$$

Therefore,

$$D_\beta y(z) - p(z)y(z) \geq 0, \quad a \leq z \leq t \leq b.$$

Now, we have

$$D_\beta(y(\cdot)e_{\ominus_\beta p, \beta}(\cdot)) = D_\beta y(z)e_{\ominus_\beta p, \beta}(\beta(z)) + y(z)\left(\ominus_\beta p\right)(z)e_{\ominus_\beta p, \beta}(z)$$
$$= D_\beta y(z)e_{\ominus_\beta p, \beta}(\beta(z)) - y(z)p(z)e_{\ominus_\beta p, \beta}(\beta(z))$$
$$= \left(D_\beta y(z) - p(z)y(z)\right)e_{\ominus_\beta p, \beta}(\beta(z))$$
$$\geq 0, \quad a \leq z \leq t \leq b,$$

which yields

$$y(t)e_{\ominus_\beta p, \beta}(t) \geq y(z)e_{\ominus_\beta p, \beta}(z)$$
$$\geq f(z)e_{\ominus_\beta p, \beta}(z), \quad a \leq z \leq t \leq b,$$

that is,

$$x(t) \geq f(z)\frac{e_{\ominus\beta p,\beta}(z)}{e_{\ominus\beta p,\beta}(t)}, \quad a \leq z \leq t \leq b.$$

This completes the proof. $\qquad\qquad\square$

Theorem 8.1.5. *Let* $x, f, g, h \in \mathscr{C}_\beta(I)$ *be nonnegative functions such that* $gh \in \mathscr{R}_\beta^+$. *Then the inequality*

$$x(t) \leq f(t) + g(t)\int_a^t h(s)x(s)d_\beta s, \quad t \in I, \tag{8.4}$$

implies the inequality

$$x(t) \leq f(t) + \frac{g(t)}{e_{\ominus\beta(gh),\beta}(t)}\int_a^t h(s)f(s)e_{\ominus\beta(gh),\beta}(\beta(s))d_\beta s, \quad t \in I.$$

Proof. Let

$$y(t) = \int_a^t h(s)x(s)d_\beta s, \quad t \in I.$$

Then, using (8.4), we get

$$x(t) \leq f(t) + g(t)y(t), \quad t \in I. \tag{8.5}$$

We have

$$D_\beta y(t) = h(t)x(t)$$
$$\leq h(t)\left(f(t) + g(t)y(t)\right), \quad t \in I,$$

or

$$D_\beta y(t) - y(t)g(t)h(t) \leq h(t)f(t), \quad t \in I,$$

and

$$y(a) = 0,$$

$$D_\beta\left(y(\cdot)e_{\ominus\beta(gh),\beta}(\cdot)\right)(t) = D_\beta y(t)e_{\ominus\beta(gh),\beta}(\beta(t)) + y(t)\left(\ominus_\beta(gh)\right)(t)e_{\ominus\beta(hg),\beta}(t)$$
$$= D_\beta y(t)e_{\ominus\beta(gh),\beta}(\beta(t)) - \frac{y(t)g(t)h(t)}{1 + (\beta(t) - t)g(t)h(t)}e_{\ominus\beta(gh),\beta}(t)$$
$$= D_\beta y(t)e_{\ominus\beta(gh),\beta}(\beta(t)) - y(t)h(t)g(t)e_{\ominus\beta(gh),\beta}(\beta(t))$$
$$= \left(D_\beta y(t) - y(t)h(t)g(t)\right)e_{\ominus\beta(gh),\beta}(\beta(t))$$
$$\leq h(t)f(t)e_{\ominus\beta(gh),\beta}(\beta(t)), \quad t \in I.$$

By β-integration of the last inequality from a to t, we get

$$y(t)e_{\ominus\beta(gh),\beta}(t) \leq \int_a^t h(s)f(s)e_{\ominus\beta(gh),\beta}(\beta(s))d_\beta s, \quad t \in I.$$

Since $gh \in \mathscr{R}_\beta^+(I)$, we have

$$e_{\ominus_\beta(gh),\beta}(t) > 0, \quad t \in I.$$

Therefore,

$$y(t) \le \frac{1}{e_{\ominus_\beta(gh),\beta}(t)} \int_a^t h(s)f(s)e_{\ominus_\beta(gh),\beta}(\beta(s))d_\beta s, \quad t \in I.$$

Now, in view of (8.5), we obtain

$$x(t) \le f(t) + g(t)y(t)$$
$$\le f(t) + \frac{g(t)}{e_{\ominus_\beta(gh),\beta}(t)} \int_a^t h(s)f(s)e_{\ominus_\beta(gh),\beta}(\beta(s))d_\beta s, \quad t \in I.$$

This completes the proof. $\qquad\qquad\qquad\qquad\qquad\qquad\qquad\qquad\qquad\qquad$ $\square$

8.2 Volterra-type β-integral inequalities

Theorem 8.2.1. *Let x, $f \in \mathscr{C}(J)$ and $k \in \mathscr{C}(J \times J)$ be nonnegative on $J \times J$ such that*

$$x(t) \le f(t) + \int_a^t k(t,s)x(s)d_\beta s, \quad t \in J. \tag{8.6}$$

Also, let

$$k_1(t,s) = k(t,s) \quad and$$
$$k_l(t,s) = \int_a^t k(t,s_1)k_{l-1}(s_1,s)d_\beta s_1, \quad l \in \mathbb{N}, \quad l \ge 2, \quad a \le s \le t \le b.$$

Suppose that the series

$$H(t,s) = \sum_{n=1}^\infty k_n(t,s)$$

is uniformly convergent on $a \le s \le t \le b$. Then

$$x(t) \le f(t) + \int_a^t H(t,s)f(s)d_\beta s, \quad t \in J. \tag{8.7}$$

Proof. First, we will prove that

$$x(t) \le f(t) + \sum_{l=1}^n \int_a^t k_l(t,s)f(s)d_\beta s + \int_a^t k_{n+1}(t,s)x(s)d_\beta s, \quad t \in J, \tag{8.8}$$

for all $n \in \mathbb{N}$. We have

$$x(t) \le f(t) + \int_a^t k(t,s)x(s)d_\beta s$$

$$\leq f(t) + \int_a^t k(t,s)\left(f(s) + \int_a^s k(s,s_1)x(s_1)d\beta s_1\right)d\beta s$$

$$= f(t) + \int_a^t k(t,s)f(s)d\beta s + \int_a^t\int_a^s k(t,s)k(s,s_1)x(s_1)d\beta s_1 d\beta s$$

$$= f(t) + \int_a^t k(t,s)f(s)d\beta s + \int_a^t\left(\int_a^{s_1} k(t,s)k(s,s_1)d\beta s\right)x(s_1)d\beta s_1$$

$$= f(t) + \int_a^t k_1(t,s)f(s)d\beta s + \int_a^t k_2(t,s)x(s)d\beta s, \quad t\in J.$$

Hence, for $t\in J$, we get

$$x(t) \leq f(t) + \int_a^t k(t,s)x(s)d\beta s$$

$$\leq f(t) + \int_a^t k(t,s)\left(f(s) + \int_a^s k_1(s,s_1)f(s_1)d\beta s_1 + \int_a^s k_2(s,s_1)x(s_1)d\beta s_1\right)$$

$$= f(t) + \int_a^t k(t,s)f(s)d\beta s + \int_a^t\int_a^s k(t,s)k_1(s,s_1)f(s_1)d\beta s_1 d\beta s$$

$$+ \int_a^t\int_a^s k(t,s)k_2(s,s_1)x(s_1)d\beta s_1 d\beta s$$

$$= f(t) + \int_a^t k_1(t,s)f(s)d\beta s + \int_a^t\left(\int_a^{s_1} k(t,s)k_1(s,s_1)d\beta s\right)f(s_1)d\beta s_1$$

$$+ \int_a^t\left(\int_a^{s_1} k(t,s)k_2(s,s_1)d\beta s\right)x(s_1)d\beta s_1$$

$$= f(t) + \int_a^t k_1(t,s)f(s)d\beta s + \int_a^t k_2(t,s)f(s)d\beta s + \int_a^t k_3(t,s)x(s)d\beta s$$

$$= f(t) + \int_a^t (k_1(t,s)+k_2(t,s))f(s)d\beta s + \int_a^t k_3(t,s)x(s)d\beta s$$

i.e.,

$$x(t) \leq f(t) + \int_a^t (k_1(t,s)+k_2(t,s))f(s)d\beta s + \int_a^t k_3(t,s)x(s)d\beta s, \quad t\in J.$$

Thus, (8.8) holds for $n=1$. Assume that

$$x(t) \leq f(t) + \sum_{l=1}^n \int_a^t k_l(t,s)f(s)d\beta s + \int_a^t k_{n+1}(t,s)x(s)d\beta s, \quad t\in J,$$

for some $n\in\mathbb{N}$. We will prove that

$$x(t) \leq f(t) + \sum_{l=1}^{n+1} \int_a^t k_l(t,s)f(s)d\beta s + \int_a^t k_{n+2}(t,s)x(s)d\beta s, \quad t\in J.$$

Indeed, for $t \in J$, we have

$$x(t) \le f(t) + \int_a^t k(t,s)x(s)d_\beta s$$

$$\le f(t) + \int_a^t k(t,s)\left(f(s) + \sum_{l=1}^n \int_a^s k_l(s,s_1)f(s_1)d_\beta s_1 + \int_a^s k_{n+1}(s,s_1)x(s_1)d_\beta s_1 \right) d_\beta s$$

$$= f(t) + \int_a^t k(t,s)f(s)d_\beta s + \sum_{l=1}^n \int_a^t \int_a^s k(t,s)k_l(s,s_1)f(s_1)d_\beta s_1 d_\beta s$$

$$+ \int_a^t \int_a^s k(t,s)k_{n+1}(s,s_1)x(s_1)d_\beta s_1 d_\beta s$$

$$= f(t) + \int_a^t k(t,s)f(s)d_\beta s + \sum_{l=1}^n \int_a^t \left(\int_a^{s_1} k(t,s)k_l(s,s_1)d_\beta s \right) f(s_1)d_\beta s_1$$

$$+ \int_a^t \left(\int_a^{s_1} k(t,s)k_{n+1}(s,s_1)d_\beta s \right) x(s_1)d_\beta s_1$$

$$= f(t) + \int_a^t k_1(t,s)f(s)d_\beta s + \sum_{l=1}^n \int_a^t k_{l+1}(t,s)f(s)d_\beta s + \int_a^t k_{n+2}(t,s)x(s)d_\beta s$$

$$= f(t) + \int_a^t k_1(t,s)f(s)d_\beta s + \sum_{l=2}^{n+1} \int_a^t k_l(t,s)f(s)d_\beta s + \int_a^t k_{n+2}(t,s)x(s)d_\beta s$$

$$= f(t) + \sum_{l=1}^{n+1} \int_a^t k_l(t,s)f(s)d_\beta s + \int_a^t k_{n+2}(t,s)x(s)d_\beta s.$$

Therefore, (8.8) holds for all $n \in \mathbb{N}$. Now, letting $n \to \infty$ in (8.8) and using the fact that $H(t,s) = \sum_{n=1}^{\infty} k_n(t,s)$ is a uniformly convergent series on $a \le s \le t \le b$, we obtain (8.7). This completes the proof. $\square$

Theorem 8.2.2. *Let $x \in \mathscr{C}(J)$ be a nonnegative function, and let $k \in \mathscr{C}(J \times J)$ be a nonnegative function such that $k(t,s)$ is nondecreasing in t for each $s \in J$. If*

$$x(t) \le c + \int_a^t k(t,s)x(s)d_\beta s, \quad t \in J, \tag{8.9}$$

for a nonnegative constant c, then

$$x(t) \le c e_{k(t,\cdot),\beta}(t), \quad t \in J. \tag{8.10}$$

Proof. Fix $T \in J$. Then using (8.9), for $a \le t \le T$, we have

$$x(t) \le c + \int_a^t k(T,s)x(s)d_\beta s.$$

Now applying Theorem 8.1.2, for $t \in [a, T]$, we obtain

$$x(t) \le c e_{k(T,\cdot),\beta}(t), \quad t \in [a, T].$$

In particular,

$$x(T) \le c e_{k(T,\cdot),\beta}(T).$$

Since $T \in J$ was arbitrarily chosen, we get (8.10). This completes the proof. $\qquad\square$

Theorem 8.2.3. *Let $x \in \mathscr{C}(J)$ be a nonnegative function, and let $k \in \mathscr{C}(J \times J)$ be a nonnegative function such that $k(t,s)$ is nondecreasing in t for each $s \in J$. Also, let $g \in \mathscr{C}(J)$ be a nondecreasing positive function. If*

$$x(t) \le g(t) + \int_a^t k(t,s)x(s)d_\beta s, \quad t \in J, \tag{8.11}$$

then

$$x(t) \le g(t)e_{k(t,\cdot),\beta}(t), \quad t \in J. \tag{8.12}$$

Proof. Let $T \in J$ be arbitrarily chosen and fixed. Then using (8.11), we obtain

$$x(t) \le g(t) + \int_a^t k(T,s)x(s)d_\beta s, \quad a \le t \le T.$$

Now applying Theorem 8.1.3, for $a \le t \le T$, we obtain

$$x(t) \le g(t)e_{k(T,\cdot),\beta}(t).$$

In particular, when $t = T$,

$$x(T) \le g(T)e_{k(T,\cdot),\beta}(T).$$

Since $T \in J$ was arbitrarily chosen, we obtain (8.12). This completes the proof. $\qquad\square$

8.3 The inequalities of Gamidov

Theorem 8.3.1 (Gamidov Inequality I). *Let x, f, g_i, $h_i \in \mathscr{C}_\beta(I)$, $i \in \{1, \dots, n\}$, be nonnegative functions, and*

$$g(t) = \sup_{i \in \{1,\dots,n\}} g_i(t) \quad \text{and}$$

$$h(t) = \sum_{i=1}^n h_i(t), \quad t \in I.$$

Then the inequality

$$x(t) \le f(t) + \sum_{i=1}^n g_i(t) \int_a^t h_i(s)x(s)d_\beta s, \quad t \in I, \tag{8.13}$$

implies the inequality

$$x(t) \le f(t) + \frac{g(t)}{e_{\ominus\beta(gh),\beta}(t)} \int_a^t h(s)f(s)e_{\ominus\beta(gh),\beta}(\beta(s))d_\beta s, \quad t \in I. \tag{8.14}$$

Proof. Using (8.13), we obtain

$$x(t) \le f(t) + g(t) \int_a^t \sum_{i=1}^n h_i(s)x(s)d_\beta s$$

$$= f(t) + g(t) \int_a^t h(s)x(s)d_\beta s, \quad t \in I.$$

Hence, in view of Theorem 8.1.5, we get (8.14). This completes the proof.   □

Theorem 8.3.2 (Gamidov Inequality II). *Let $n \in \mathbb{N}$, $n \ge 2$, let $x, f, g_1, g_2, h_i \in \mathscr{C}_\beta(I)$, $i \in \{1, \ldots, n\}$, be nonnegative functions, let*

$$a = t_1 \le t_2 \le \cdots \le t_n = b,$$

and let c_i, $i \in \{1, \ldots, n\}$, be nonnegative constants. Denote

$$m_i = c_i \int_{t_1}^{t_i} h_i(s)x(s)d_\beta s, \quad i \in \{2, \ldots, n\},$$

$$F(t) = f(t) + g_2(t) \sum_{i=2}^n m_i,$$

$$p_1(t) = f(t) + \frac{g_1(t)}{e_{\ominus\beta(Fh_1),\beta}(t)} \int_a^t h_1(s)f(s)e_{\ominus\beta(Fh_1)}(\beta(s))d_\beta s, \quad \text{and}$$

$$p_2(t) = g_2(t) + \frac{g_1(t)}{e_{\ominus\beta(Fh_1),\beta}(t)} \int_a^t h_1(s)g_2(s)e_{\ominus\beta(Fh_1)}(\beta(s))d_\beta s.$$

Suppose that

$$M^{-1} = 1 - \sum_{j=2}^n c_j \int_{t_1}^{t_j} h_j(s)p_2(s)d_\beta s > 0.$$

Then the inequality

$$x(t) \le f(t) + \frac{g_1(t)}{e_{\ominus\beta(Fh_1),\beta}(t)} \int_{t_1}^t h_1(s)x(s)d_\beta s, \quad t \in I, \tag{8.15}$$

implies the inequality

$$x(t) \le p_1(t) + \frac{Mp_2(t)}{e_{\ominus\beta(Fh_1),\beta}(t)} \sum_{i=2}^n c_i \int_{t_1}^{t_i} h_i(s)p_1(s)d_\beta s, \quad t \in I. \tag{8.16}$$

Proof. Using inequality (8.15), we obtain

$$x(t) \le \left(f(t) + g_2(t) \sum_{i=2}^{n} m_i \right) + g_1(t) \int_{t_1}^{t} h_1(s)x(s)d_\beta s$$

$$= F(t) + g_1(t) \int_{t_1}^{t} h_1(s)x(s)d_\beta s, \quad t \in I.$$

Hence, in view of Theorem 8.1.5, we obtain

$$x(t) \le F(t) + \frac{g_1(t)}{e_{\ominus\beta(Fh_1),\beta}(t)} \int_{t_1}^{t} h_1(s)F(s)e_{\ominus\beta(Fh_1),\beta}(\beta(s))d_\beta s$$

$$= f(t) + g_2(t) \sum_{i=2}^{n} m_i + \frac{g_1(t)}{e_{\ominus\beta(Fh_1),\beta}(t)} \int_{t_1}^{t} h_1(s) \left(f(s) + g_2(s) \sum_{i=2}^{n} m_i \right) e_{\ominus\beta(Fh_1),\beta}(\beta(s))d_\beta s$$

$$= f(t) + \frac{g_1(t)}{e_{\ominus\beta(Fh_1),\beta}(t)} \int_{t_1}^{t} h_1(s)f(s)e_{\ominus\beta(Fh_1),\beta}(\beta(s))d_\beta s$$

$$+ \sum_{i=2}^{n} m_i \left(g_2(t) + \frac{g_1(t)}{e_{\ominus\beta(Fh_1),\beta}(t)} \int_{t_1}^{t} h_1(s)g_2(s)e_{\ominus\beta(Fh_1),\beta}(\beta(s))d_\beta s \right)$$

$$= p_1(t) + p_2(t) \sum_{i=2}^{n} m_i,$$

that is,

$$x(t) \le p_1(t) + p_2(t) \sum_{i=2}^{n} m_i, \quad t \in I. \tag{8.17}$$

Note that

$$\sum_{i=2}^{n} m_i = \sum_{i=2}^{n} c_i \int_{t_1}^{t_i} h_i(s)x(s)d_\beta s$$

$$= \sum_{i=2}^{n} c_i \int_{t_1}^{t_i} h_i(s) \left(p_1(s) + p_2(s) \sum_{i=2}^{n} m_i \right) d_\beta s$$

$$= \sum_{i=2}^{n} c_i \int_{t_1}^{t_i} h_i(s)p_1(s)d_\beta s + \sum_{i=2}^{n} m_i \left(\sum_{j=2}^{n} c_j \int_{t_1}^{t_j} h_j(s)p_2(s)d_\beta s \right),$$

that is,

$$\sum_{i=2}^{n} m_i \left(1 - \sum_{j=2}^{n} c_j \int_{t_1}^{t_j} h_j(s)p_2(s)d_\beta s \right) \le \sum_{i=2}^{n} c_i \int_{t_1}^{t_i} h_i(s)p_1(s)d_\beta s.$$

Therefore,

$$\sum_{i=2}^{n} m_i \leq M \sum_{i=2}^{n} c_i \int_{t_1}^{t_i} h_i(s) p_1(s) d_\beta s.$$

Now, in view of (8.17), we get (8.16). This completes the proof. $\qquad\square$

Henceforth, we denote $\mathbb{R}_+ = [0, \infty)$ and suppose that $\mathbb{R}_+ \cap I \neq \emptyset$ and $\sup I = \infty$.

Theorem 8.3.3. *Let* $x, f \in \mathscr{C}_\beta(\mathbb{R}_+ \cap I)$ *be nonnegative functions such that*

$$1 - (\beta(t) - t) f(t) > 0, \quad t \in \mathbb{R}_+ \cap I.$$

Also, let c be a nonnegative constant. Then the inequality

$$x(t) \leq c + \int_t^\infty f(s) x(s) d_\beta s, \quad t \in \mathbb{R}_+ \cap I, \tag{8.18}$$

implies the inequality

$$x(t) \leq c e_{\ominus \beta g, \beta}(t), \quad t \in \mathbb{R}_+ \cap I,$$

where

$$g(t) = \frac{f(t)}{1 - (\beta(t) - t) f(t)}, \quad t \in \mathbb{R}_+ \cap I.$$

Proof. Let

$$y(t) = c + \int_t^\infty f(s) x(s) d_\beta s, \quad t \in \mathbb{R}_+ \cap I.$$

Then

$$\lim_{t \to \infty} y(t) = c,$$

and using (8.18), we have

$$x(t) \leq y(t), \quad t \in \mathbb{R}_+ \cap I. \tag{8.19}$$

Also, we have

$$D_\beta y(t) = -f(t) x(t)$$
$$\geq -f(t) y(t),$$

that is,

$$D_\beta y(t) + f(t) y(t) \geq 0, \quad t \in \mathbb{R}_+ \cap I.$$

Using the last inequality, we get

$$D_\beta \left(y(t) e_{g,\beta}(t) \right) = D_\beta y(t) e_{g,\beta}(\beta(t)) + y(t) g(t) e_{g,\beta}(t)$$

$$
\begin{aligned}
&= D_\beta y(t) e_{g,\beta}(\beta(t)) + \frac{y(t)g(t)}{1+(\beta(t)-t)g(t)} \left(1+(\beta(t)-t)g(t)\right) e_{g,\beta}(t) \\
&= D_\beta y(t) e_{g,\beta}(\beta(t)) + \frac{y(t)g(t)}{1+(\beta(t)-t)g(t)} e_{g,\beta}(\beta(t)) \\
&= \left(D_\beta y(t) + \frac{g(t)}{1+(\beta(t)-t)g(t)} y(t) \right) e_{g,\beta}(\beta(t)) \\
&= \left(D_\beta y(t) + \frac{\frac{f(t)}{1-(\beta(t)-t)f(t)}}{1+\frac{(\beta(t)-t)f(t)}{1-(\beta(t)-t)f(t)}} y(t) \right) e_{g,\beta}(\beta(t)) \\
&= \left(D_\beta y(t) + f(t)y(t) \right) e_{g,\beta}(\beta(t)).
\end{aligned}
\tag{8.20}
$$

Since $g \geq 0$, we observe that

$$
e_{g,\beta}(\beta(t)) \geq 0, \quad t \in \mathbb{R}_+ \cap I.
$$

Hence, using (8.20), we obtain

$$
D_\beta \left(y(t) e_{g,\beta}(t) \right) \geq 0, \quad t \in \mathbb{R}_+ \cap I.
$$

Now β-integrating the last inequality from t to ∞, we get

$$
\lim_{\tau \to \infty} \left(y(\tau) e_{g,\beta}(\tau) \right) - y(t) e_{g,\beta}(t) \geq 0,
$$

that is,

$$
y(t) e_{g,\beta}(t) \leq c,
$$

and thus

$$
y(t) \leq c e_{\ominus_\beta g, \beta}(t), \quad t \in \mathbb{R}_+ \cap I.
$$

From (8.19) and the last inequality, we obtain

$$
x(t) \leq c e_{\ominus_\beta g, \beta}(t), \quad t \in \mathbb{R}_+ \cap I.
$$

This completes the proof. $\qquad\square$

Theorem 8.3.4. *Let $x, g \in \mathscr{C}_\beta(\mathbb{R}_+ \cap I)$ be nonnegative functions, and let $f \in \mathscr{C}_\beta(\mathbb{R}_+ \cap I)$ be a positive decreasing function such that*

$$
f(t) - (\beta(t)-t)g(t) > 0, \quad t \in \mathbb{R}_+ \cap I.
$$

Then the inequality

$$
x(t) \leq f(t) + \int_t^\infty g(s)x(s)d_\beta s, \quad t \in \mathbb{R}_+ \cap I,
$$

implies the inequality

$$
x(t) \leq f(t) e_{\ominus_\beta h, \beta}(t), \quad t \in \mathbb{R}_+ \cap I,
$$

where

$$h(t) = \frac{g(t)}{f(t) - (\beta(t) - t)g(t)}, \qquad t \in \mathbb{R}_+ \cap I.$$

Proof. Using the fact that f is decreasing and nonnegative on $\mathbb{R}_+ \cap I$, we have

$$\frac{x(t)}{f(t)} \leq 1 + \int_t^\infty \frac{g(s)}{f(t)} x(s) d_\beta s$$
$$\leq 1 + \int_t^\infty \frac{g(s)}{f(s)} x(s) d_\beta s, \qquad t \in \mathbb{R}_+ \cap I.$$

Now in view of Theorem 8.3.3, we obtain

$$\frac{x(t)}{f(t)} \leq e_{\ominus \beta h, \beta}(t),$$

that is,

$$x(t) \leq f(t) e_{\ominus \beta h, \beta}(t), \qquad t \in \mathbb{R}_+ \cap I.$$

This completes the proof. $\qquad\qquad\qquad\qquad\qquad\qquad\qquad\qquad\qquad\qquad\qquad\qquad\quad$ $\square$

Theorem 8.3.5. *Let $x, f, g \in \mathscr{C}_\beta(\mathbb{R}_+ \cap I)$ be nonnegative functions, let $\gamma \in \mathscr{C}_\beta(\mathbb{R}_+ \cap I)$ be a positive decreasing function, and let c be a nonnegative constant such that*

$$1 - (\beta(t) - t)g(t) > 0,$$
$$\gamma(t) - c(\beta(t) - t)g(t) > 0,$$

and γx is a bounded function on $\mathbb{R}_+ \cap I$. Suppose there exists a sufficiently large $a \in \mathbb{R}_+ \cap I$ such that

$$\eta = \int_a^\infty f(s) d_\beta s + \int_a^\infty g(s) d_\beta s < 1.$$

Then the inequality

$$x(t) \leq c + \int_a^t f(s) x(s) d_\beta s + \frac{1}{\gamma(t)} \int_t^\infty \gamma(s) g(s) x(s) d_\beta s, \qquad t \in \mathbb{R}_+ \cap I,$$

implies the inequalities:

1.

$$x(t) \leq \frac{c}{(1-\eta)} e_{\ominus \beta h, \beta}(t), \qquad t \in \mathbb{R}_+ \cap I, \quad t \geq a,$$

where

$$h(t) = \frac{g(t)}{c\gamma(t) - (1-\eta)(\beta(t) - t)g(t)}, \qquad t \in \mathbb{R}_+ \cap I, \quad t \geq a,$$

2.

$$x(t) \le c e_{\ominus \beta h_1, \beta}(t), \quad t \in \mathbb{R}_+ \cap I, \quad t \le a,$$

where

$$h_1(t) = \frac{g(t)}{1 - (\beta(t) - t)g(t)}, \quad t \in \mathbb{R}_+ \cap I, \quad t \le a.$$

Proof. **1.** Let $t \ge a, t \in \mathbb{R}_+ \cap I$. We set

$$y(t) = \max_{a \le s \le t} x(s).$$

Then $y \in \mathscr{C}_\beta([a, \infty) \cap I)$ is an increasing function such that

$$x(t) \le y(t), \quad t \in \mathbb{R}_+ \cap I, \quad t \ge a.$$

Also, y is a bounded function on $\mathbb{R}_+ \cap I \cap [a, \infty)$. For a given $t \ge a$, there exists $t_1 \in [a, t]$ such that

$$y(t) = x(t_1).$$

Then using the given inequality, we have

$$y(t) = x(t_1)$$

$$\le c + \int_a^{t_1} f(s)x(s)d_\beta s + \frac{1}{\gamma(t_1)} \int_{t_1}^\infty \gamma(s)g(s)x(s)d_\beta s$$

$$\le c + \int_a^{t_1} f(s)y(s)d_\beta s + \frac{1}{\gamma(t_1)} \int_{t_1}^\infty \gamma(s)g(s)y(s)d_\beta s$$

$$= c + \int_a^{t_1} f(s)y(s)d_\beta s + \frac{1}{\gamma(t_1)} \int_{t_1}^t \gamma(s)g(s)y(s)d_\beta s + \frac{1}{\gamma(t_1)} \int_t^\infty \gamma(s)g(s)y(s)d_\beta s$$

$$\le c + \int_a^{t_1} f(s)y(s)d_\beta s + \frac{1}{\gamma(t_1)} \int_{t_1}^t \gamma(t_1)g(s)y(s)d_\beta s + \frac{1}{\gamma(t_1)} \int_t^\infty \gamma(s)g(s)y(s)d_\beta s$$

$$= c + \int_a^{t_1} f(s)y(s)d_\beta s + \int_{t_1}^t g(s)y(s)d_\beta s + \frac{1}{\gamma(t)} \int_t^\infty \gamma(s)g(s)y(s)d_\beta s$$

$$= c + \int_a^{t_1} f(s)y(s)d_\beta s + y(t) \int_{t_1}^t g(s)d_\beta s + \frac{1}{\gamma(t)} \int_t^\infty \gamma(s)g(s)y(s)d_\beta s$$

$$\le c + \int_a^{t_1} f(s)y(s)d_\beta s + y(t) \int_a^\infty g(s)d_\beta s + \frac{1}{\gamma(t)} \int_t^\infty \gamma(s)g(s)y(s)d_\beta s$$

$$\le c + y(t) \int_a^{t_1} f(s)d_\beta s + y(t) \int_a^\infty g(s)d_\beta s + \frac{1}{\gamma(t)} \int_t^\infty \gamma(s)g(s)y(s)d_\beta s$$

$$\le c + y(t) \left(\int_a^\infty f(s)d_\beta s + \int_a^\infty g(s)d_\beta s \right) + \frac{1}{\gamma(t)} \int_t^\infty \gamma(s)g(s)y(s)d_\beta s$$

$$= c + \eta y(t) + \frac{1}{\gamma(t)} \int_t^\infty \gamma(s)g(s)y(s)d_\beta s, \quad t \in \mathbb{R}_+ \cap I, \quad t \ge a,$$

that is,

$$(1 - \eta)y(t) \leq c + \frac{1}{\gamma(t)} \int_t^\infty \gamma(s)g(s)y(s)d_\beta s.$$

Hence,

$$y(t)\gamma(t) \leq \frac{c}{1-\eta}\gamma(t) + \frac{1}{1-\eta} \int_t^\infty \gamma(s)g(s)y(s)d_\beta s, \quad t \in \mathbb{R}_+ \cap I, \quad t \geq a.$$

Now, in view of Theorem 8.3.4, it follows that

$$y(t)\gamma(t) \leq \frac{c\gamma(t)}{1-\eta}e_{\ominus\beta h,\beta}(t),$$

that is,

$$y(t) \leq \frac{c}{(1-\eta)}e_{\ominus\beta h,\beta}(t), \quad t \in \mathbb{R}_+ \cap I, \quad t \geq a.$$

Thus,

$$x(t) \leq \frac{c}{(1-\eta)}e_{\ominus\beta h,\beta}(t), \quad t \in \mathbb{R}_+ \cap I, \quad t \geq a.$$

2. Let $t \in \mathbb{R}_+ \cap I, t \leq a$. Then

$$\int_a^t f(s)x(s)d_\beta s \leq 0,$$

and

$$x(t) \leq c + \frac{1}{\gamma(t)} \int_t^\infty \gamma(s)g(s)x(s)d_\beta s$$

$$\leq c + \int_t^\infty g(s)x(s)d_\beta s, \quad t \in \mathbb{R}_+ \cap I, \quad t \leq a.$$

Now in view of Theorem 8.3.3, it follows that

$$x(t) \leq ce_{\ominus\beta h_1,\beta}(t), \quad t \in \mathbb{R}_+ \cap I, \quad t \leq a.$$

This completes the proof. $\qquad\qquad\square$

8.4 Simultaneous inequalities

In this section, we assume that $0 \in I$ and $\beta(0) = 0$.

Theorem 8.4.1. *Let $x, y, h_1, h_2, h_3, h_4 \in \mathscr{C}_\beta(\mathbb{R}_+ \cap I)$ be nonnegative functions, and let λ, k_1, and k_2 be nonnegative constants such that*

$$x(t) \le k_1 + \int_0^t h_1(s)x(s)d_\beta s + \int_0^t e_{\lambda,\beta}(s)h_2(s)y(s)d_\beta s \quad \text{and}$$

$$y(t) \le k_2 + \int_0^t e_{\ominus\beta\lambda,\beta}(s)h_3(s)x(s)d_\beta s + \int_0^t h_4(s)y(s)d_\beta s, \quad t \in \mathbb{R}_+ \cap I. \tag{8.21}$$

Then

$$x(t) \le (k_1 M + k_2)e_{\lambda \oplus \beta h,\beta}(t), \quad \text{and}$$

$$y(t) \le (k_1 M + k_2)e_{h,\beta}(t), \quad t \in \mathbb{R}_+ \cap I,$$

where

$$h(t) = \max\{h_1(t) + h_3(t),\ h_2(t) + h_4(t)\}, \quad \text{and}$$

$$M = \max_{t \ge 0} e_{\ominus\beta\lambda,\beta}(t), \quad t \in \mathbb{R}_+ \cap I.$$

Proof. Since $\lambda \ge 0$, we have

$$e_{\lambda,\beta}(s) \le e_{\lambda,\beta}(t), \quad t \ge s, \quad s, t \in \mathbb{R}_+ \cap I,$$

and

$$e_{\ominus\beta\lambda,\beta}(t) \le e_{\ominus\beta\lambda,\beta}(s), \quad t \ge s, \quad s, t \in \mathbb{R}_+ \cap I.$$

From the first inequality of (8.21), we find

$$x(t) \le k_1 + \int_0^t h_1(s)x(s)d_\beta s + \int_0^t e_{\lambda,\beta}(s)h_2(s)y(s)d_\beta s$$

$$= k_1 + \int_0^t h_1(s)x(s)d_\beta s + e_{\lambda,\beta}(t)\int_0^t h_2(s)y(s)d_\beta s,$$

whereupon

$$e_{\ominus\beta\lambda,\beta}(t)x(t) \le k_1 e_{\ominus\beta\lambda,\beta}(t) + e_{\ominus\beta\lambda,\beta}(t)\int_0^t h_1(s)x(s)d_\beta s + \int_0^t h_2(s)y(s)d_\beta s$$

$$\le k_1 e_{\ominus\beta\lambda,\beta}(t) + \int_0^t e_{\ominus\beta\lambda,\beta}(s)h_1(s)x(s)d_\beta s + \int_0^t h_2(s)y(s)d_\beta s$$

$$\le M k_1 + \int_0^t e_{\ominus\beta\lambda,\beta}(s)h_1(s)x(s)d_\beta s + \int_0^t h_2(s)y(s)d_\beta s,$$

that is,

$$e_{\ominus\beta\lambda,\beta}(t)x(t) \le M k_1 + \int_0^t e_{\ominus\beta\lambda,\beta}(s)h_1(s)x(s)d_\beta s + \int_0^t h_2(s)y(s)d_\beta s, \quad t \in \mathbb{R}_+ \cap I. \tag{8.22}$$

Denote

$$F(t) = e_{\ominus\beta\lambda,\beta}(t)x(t) + y(t), \quad t \in \mathbb{R}_+ \cap I.$$

Now using (8.22) and the second inequality of (8.21), we obtain

$$F(t) \leq (k_1 M + k_2) + \int_0^t e_{\ominus\beta\lambda,\beta}(s)\,(h_1(s) + h_3(s))\,d_\beta s + \int_0^t (h_2(s) + h_4(s))\,y(s)d_\beta s$$

$$\leq (k_1 M + k_2) + \int_0^t e_{\ominus\beta\lambda,\beta}(s)x(s)h(s)d_\beta s + \int_0^t h(s)y(s)d_\beta s$$

$$= (k_1 M + k_2) + \int_0^t F(s)h(s)d_\beta s, \quad t \in \mathbb{R}_+ \cap I.$$

In view of Theorem 8.1.2, the last inequality yields

$$F(t) \leq (k_1 M + k_2)e_{h,\beta}(t), \quad t \in \mathbb{R}_+ \cap I,$$

that is,

$$e_{\ominus\beta\lambda,\beta}(t)x(t) \leq (k_1 M + k_2)e_{h,\beta}(t).$$

Thus

$$x(t) \leq (k_1 M + k_2)e_{\lambda,\beta}(t)e_{h,\beta}(t)$$
$$= (k_1 M + k_2)e_{\lambda\oplus\beta h,\beta}(t), \quad t \in \mathbb{R}_+ \cap I,$$

and

$$y(t) \leq (k_1 M + k_2)e_{h,\beta}(t), \quad t \in \mathbb{R}_+ \cap I.$$

This completes the proof. $\qquad\square$

Theorem 8.4.2. *Let $x, y, p, a, b, h_1, h_2, h_3, h_4 \in \mathscr{C}_\beta(\mathbb{R}_+ \cap I)$ be nonnegative functions, and let λ be a nonnegative constant such that*

$$x(t) \leq a(t) + p(t)\int_0^t h_1(s)x(s)d_\beta s + p(t)\int_0^t e_{\lambda,\beta}(s)h_2(s)y(s)d_\beta s \quad and$$

$$y(t) \leq b(t) + p(t)\int_0^t e_{\ominus\beta\lambda,\beta}(s)h_3(s)x(s)d_\beta s + p(t)\int_0^t h_4(s)y(s)d_\beta s, \quad t \in \mathbb{R}_+ \cap I. \tag{8.23}$$

Then

$$x(t) \leq (a(t) + b(t)e_{\lambda,\beta}(t)) + \frac{p(t)e_{\lambda,\beta}(t)}{e_{\ominus\beta(hp),\beta}(t)}\int_0^t h(s)\Big(a(s)e_{\ominus\beta\lambda,\beta}(s) + b(s)\Big)e_{\ominus\beta(hp),\beta}(\beta(s))d_\beta s$$

and

$$y(t) \leq (a(t)e_{\ominus\beta\lambda,\beta}(t) + b(t)) + \frac{p(t)}{e_{\ominus\beta(hp),\beta}(t)}\int_0^t h(s)\Big(a(s)e_{\ominus\beta\lambda,\beta}(s) + b(s)\Big)e_{\ominus\beta(hp),\beta}(\beta(s))d_\beta s,$$

$t \in \mathbb{R}_+ \cap I$, *where*

$$h(t) = \max\{h_1(t) + h_3(t), \ h_2(t) + h_4(t)\}, \quad t \in \mathbb{R}_+ \cap I.$$

Proof. Since $\lambda \geq 0$, we have

$$e_{\lambda,\beta}(s) \leq e_{\lambda,\beta}(t), \quad t \geq s, \quad s, t \in \mathbb{R}_+ \cap I,$$

and

$$e_{\ominus\beta\lambda,\beta}(t) \leq e_{\ominus\beta\lambda,\beta}(s), \quad t \geq s, \quad s, t \in \mathbb{R}_+ \cap I.$$

From the first inequality of (8.23), we find

$$x(t) \leq a(t) + p(t) \int_0^t h_1(s)x(s)d_\beta s + p(t) \int_0^t e_{\lambda,\beta}(t)h_2(s)y(s)d_\beta s$$

$$= a(t) + p(t) \int_0^t h_1(s)x(s)d_\beta s + e_{\lambda,\beta}(t)p(t) \int_0^t h_2(s)y(s)d_\beta s,$$

whereupon

$$e_{\ominus\beta\lambda,\beta}(t)x(t) \leq a(t)e_{\ominus\beta\lambda,\beta}(t) + e_{\ominus\beta\lambda,\beta}(t)p(t) \int_0^t h_1(s)x(s)d_\beta s + p(t) \int_0^t h_2(s)y(s)d_\beta s$$

$$\leq a(t)e_{\ominus\beta\lambda}(t) + p(t) \int_0^t e_{\ominus\beta\lambda}(s)h_1(s)x(s)d_\beta s + p(t) \int_0^t h_2(s)y(s)d_\beta s, \quad t \in \mathbb{R}_+ \cap I,$$

that is,

$$e_{\ominus\beta\lambda,\beta}(t)x(t) \leq a(t)e_{\ominus\beta\lambda}(t) + p(t) \int_0^t e_{\ominus\beta\lambda}(s)h_1(s)x(s)d_\beta s$$

$$+ p(t) \int_0^t h_2(s)y(s)d_\beta s, \quad t \in \mathbb{R}_+ \cap I. \tag{8.24}$$

Denote

$$F(t) = e_{\ominus\beta\lambda,\beta}(t)x(t) + y(t), \quad t \in \mathbb{R}_+ \cap I.$$

Now using (8.24) and the second inequality of (8.23), we obtain

$$F(t) \leq (a(t)e_{\ominus\beta\lambda,\beta}(t) + b(t)) + p(t) \int_0^t e_{\ominus\beta\lambda,\beta}(s)\,(h_1(s) + h_3(s))\,x(s)d_\beta s$$

$$+ p(t) \int_0^t (h_2(s) + h_4(s))\,y(s)d_\beta s$$

$$\leq (a(t)e_{\ominus\beta\lambda,\beta}(t) + b(t)) + p(t) \int_0^t e_{\ominus\beta\lambda,\beta}(s)x(s)h(s)d_\beta s + p(t) \int_0^t h(s)y(s)d_\beta s$$

$$= (a(t)e_{\ominus\beta\lambda,\beta}(t) + b(t)) + p(t) \int_0^t F(s)h(s)d_\beta s, \quad t \in \mathbb{R}_+ \cap I,$$

that is,

$$F(t) \le (a(t)e_{\ominus\beta\lambda,\beta}(t) + b(t)) + p(t)\int_0^t F(s)h(s)d_\beta s, \quad t \in \mathbb{R}_+ \cap I.$$

In view of Theorem 8.1.5, the last inequality yields

$$F(t) \le (a(t)e_{\ominus\beta\lambda,\beta}(t) + b(t)) + \frac{p(t)}{e_{\ominus\beta(hp),\beta}(t)}\int_0^t h(s)\Big(a(s)e_{\ominus\beta\lambda,\beta}(s) + b(s)\Big)e_{\ominus\beta(hp),\beta}(\beta(s))d_\beta s,$$

$t \in \mathbb{R}_+ \cap I$, that is,

$$e_{\ominus\beta\lambda,\beta}(t)x(t) + y(t) \le (a(t)e_{\ominus\beta\lambda,\beta}(t) + b(t))$$
$$+ \frac{p(t)}{e_{\ominus\beta(hp),\beta}(t)}\int_0^t h(s)\Big(a(s)e_{\ominus\beta\lambda,\beta}(s) + b(s)\Big)e_{\ominus\beta(hp),\beta}(\beta(s))d_\beta s.$$

Thus

$$x(t) \le (a(t) + b(t)e_{\lambda,\beta}(t)) + \frac{p(t)e_{\lambda,\beta}(t)}{e_{\ominus\beta(hp),\beta}(t)}\int_0^t h(s)\Big(a(s)e_{\ominus\beta\lambda,\beta}(s) + b(s)\Big)e_{\ominus\beta(hp),\beta}(\beta(s))d_\beta s$$

and

$$y(t) \le (a(t)e_{\ominus\beta\lambda,\beta}(t) + b(t))$$
$$+ \frac{p(t)}{e_{\ominus\beta(hp),\beta}(t)}\int_0^t h(s)\Big(a(s)e_{\ominus\beta\lambda,\beta}(s) + b(s)\Big)e_{\ominus\beta(hp),\beta}(\beta(s))d_\beta s, \quad t \in \mathbb{R}_+ \cap I.$$

This completes the proof. $\qquad\square$

8.5 **The Pachpatte inequalities**

In this section, we assume that $0 \in I$ and $\beta(0) = 0$.

Theorem 8.5.1 (Pachpatte Inequality I). *Let $x, f, g \in \mathscr{C}_\beta(\mathbb{R}_+ \cap I)$ be nonnegative functions such that*

$$x(t) \le c + \int_0^t f(s)x(s)d_\beta s + \int_0^t f(s)\left(\int_0^s g(y)x(y)d_\beta y\right)d_\beta s,$$

$t \in \mathbb{R}_+ \cap I$, *where c is a nonnegative constant. Then*

$$x(t) \le ce_{f+g,\beta}(t),$$

and

$$x(t) \le c\left(1 + \int_0^t f(s)e_{f+g,\beta}(s)d_\beta s\right), \quad t \in \mathbb{R}_+ \cap I.$$

Proof. Let

$$z(t) = c + \int_0^t f(s)x(s)d_\beta s + \int_0^t f(s)\left(\int_0^s g(y)x(y)d_\beta y\right)d_\beta s, \quad t \in \mathbb{R}_+ \cap I.$$

Then

$$z(0) = c,$$
$$x(t) \le z(t), \quad t \in \mathbb{R}_+ \cap I, \quad \text{and}$$
$$D_\beta z(t) = f(t)x(t) + f(t)\int_0^t g(y)x(y)d_\beta y$$
$$= f(t)\left(x(t) + \int_0^t g(y)x(y)d_\beta y\right)$$
$$\le f(t)\left(z(t) + \int_0^t g(y)z(y)d_\beta y\right), \quad t \in \mathbb{R}_+ \cap I.$$

$$(8.25)$$

Let

$$m(t) = z(t) + \int_0^t g(y)z(y)d_\beta y, \quad t \in \mathbb{R}_+ \cap I.$$

Then we have

$$z(t) \le m(t) \quad \text{and}$$
$$D_\beta z(t) \le f(t)m(t), \quad t \in \mathbb{R}_+ \cap I.$$

Further,

$$m(0) = z(0)$$
$$= c, \quad \text{and}$$
$$D_\beta m(t) = D_\beta z(t) + g(t)z(t)$$
$$\le f(t)m(t) + g(t)m(t)$$
$$= (f(t) + g(t))m(t), \quad t \in \mathbb{R}_+ \cap I.$$

Hence,

$$m(t) \le m(0) + \int_0^t (f(y) + g(y))m(y)d_\beta y$$
$$= c + \int_0^t (f(y) + g(y))m(y)d_\beta y, \quad t \in \mathbb{R}_+ \cap I.$$

In view of Theorem 8.1.2, the last inequality yields

$$m(t) \le c e_{f+g,\beta}(t), \quad t \in \mathbb{R}_+ \cap I.$$

Also,

$$x(t) \le z(t)$$

$$\leq m(t)$$
$$\leq ce_{f+g,\beta}(t), \quad t \in \mathbb{R}_+ \cap I.$$

Now using (8.25), we get

$$D_\beta z(t) \leq cf(t)e_{f+g,\beta}(t), \quad t \in \mathbb{R}_+ \cap I.$$

Then

$$z(t) - z(0) \leq c \int_0^t f(s)e_{f+g,\beta}(s)d_\beta s,$$

$$z(t) \leq c \left(1 + \int_0^t f(s)e_{f+g,\beta}(s)d_\beta s \right), \quad t \in \mathbb{R}_+ \cap I.$$

Therefore,

$$x(t) \leq c \left(1 + \int_0^t f(s)e_{f+g,\beta}(s)d_\beta s \right), \quad t \in \mathbb{R}_+ \cap I.$$

This completes the proof. $\qquad\qquad\Box$

Theorem 8.5.2 (Pachpatte Inequality II). *Let $x, f, p, g \in \mathscr{C}_\beta(\mathbb{R}_+ \cap I)$ be nonnegative functions, and let c be a nonnegative constant such that*

$$x(t) \leq c + \int_0^t (f(s)x(s) + p(s))\, d_\beta s + \int_0^t f(s) \left(\int_0^s g(y)x(y)d_\beta y \right) d_\beta s, \quad t \in \mathbb{R}_+ \cap I.$$

Then

$$x(t) \leq c + \int_0^t (p(s) + q(s)f(s))\, d_\beta s$$
$$+ \int_0^t \frac{f(s)}{e_{\ominus\beta(f+g),\beta}(s)} \int_0^s (f(y) + g(y))\, q(y)e_{\ominus\beta(f+g),\beta}(\beta(y))d_\beta y d_\beta s, \quad t \in \mathbb{R}_+ \cap I,$$

where

$$q(t) = c + \int_0^t p(s)d_\beta s, \quad t \in \mathbb{R}_+ \cap I.$$

Proof. Let

$$z(t) = c + \int_0^t (f(s)x(s) + p(s))\, d_\beta s + \int_0^t f(s) \left(\int_0^s g(y)x(y)d_\beta y \right) d_\beta s, \quad t \in \mathbb{R}_+ \cap I.$$

Then

$$z(0) = c,$$
$$x(t) \leq z(t), \quad t \in \mathbb{R}_+ \cap I,$$

and

$$D_\beta z(t) = f(t)x(t) + p(t) + f(t)\int_0^t g(y)x(y)d_\beta y$$

$$\leq z(t)f(t) + p(t) + f(t)\int_0^t g(y)z(y)d_\beta y$$

$$= p(t) + f(t)\left(z(t) + \int_0^t g(y)z(y)d_\beta y\right), \quad t \in \mathbb{R}_+ \cap I.$$

Let

$$v(t) = z(t) + \int_0^t g(y)z(y)d_\beta y, \quad t \in \mathbb{R}_+ \cap I.$$

Then we have

$$z(t) \leq v(t) \quad \text{and}$$
$$D_\beta z(t) \leq p(t) + f(t)v(t), \quad t \in \mathbb{R}_+ \cap I.$$

Also,

$$v(0) = z(0)$$
$$= c, \quad \text{and}$$
$$D_\beta v(t) = D_\beta z(t) + g(t)z(t)$$
$$\leq p(t) + f(t)v(t) + g(t)v(t)$$
$$\leq p(t) + (f(t) + g(t))\, v(t), \quad t \in \mathbb{R}_+ \cap I.$$

Hence,

$$v(t) \leq c + \int_0^t p(s)d_\beta s + \int_0^t (f(s) + g(s))\, v(s)d_\beta s$$
$$= q(t) + \int_0^t (f(s) + g(s))\, v(s)d_\beta s, \quad t \in \mathbb{R}_+ \cap I.$$

In view of Theorem 8.1.5, the last inequality yields

$$v(t) \leq q(t) + \frac{1}{e_{\ominus_\beta(f+g),\beta}(t)}\int_0^t (f(s) + g(s))q(s)e_{\ominus_\beta(f+g),\beta}(\beta(s))d_\beta s, \quad t \in \mathbb{R}_+ \cap I.$$

Thus

$$D_\beta z(t) \leq p(t) + q(t)f(t) + \frac{f(t)}{e_{\ominus_\beta(f+g),\beta}(t)}\int_0^t (f(s) + g(s))q(s)e_{\ominus_\beta(f+g),\beta}(\beta(s))d_\beta s,$$

and

$$z(t) \leq c + \int_0^t (p(s) + q(s)f(s))\, d_\beta s$$

$$+ \int_0^t \frac{f(s)}{e_{\ominus\beta(f+g),\beta}(s)} \int_0^s (f(y)+g(y))\, q(y) e_{\ominus\beta(f+g),\beta}(\beta(y))\, d_\beta y\, d_\beta s, \quad t \in \mathbb{R}_+ \cap I.$$

Consequently,

$$x(t) \le c + \int_0^t (p(s)+q(s)f(s))\, d_\beta s$$

$$+ \int_0^t \frac{f(s)}{e_{\ominus\beta(f+g),\beta}(s)} \int_0^s (f(y)+g(y))\, q(y) e_{\ominus\beta(f+g),\beta}(\beta(y))\, d_\beta y\, d_\beta s, \quad t \in \mathbb{R}_+ \cap I.$$

This completes the proof. $\qquad\square$

Corollary 8.5.3. *Let $x, f, g, p \in \mathscr{C}_\beta(\mathbb{R}_+ \cap I)$ be nonnegative functions, and let c be a nonnegative constant such that*

$$x(t) \le c + \int_0^t f(s)x(s)\, d_\beta s + \int_0^t f(s)\left(\int_0^s (g(y)x(y)+p(y))\, d_\beta y \right) d_\beta s, \quad t \in \mathbb{R}_+ \cap I.$$

Then

$$x(t) \le c + \int_0^t (q(s)+l(s)f(s))\, d_\beta s$$

$$+ \int_0^t \frac{f(s)}{e_{\ominus\beta(f+g),\beta}(s)} \int_0^s (f(y)+g(y))\, l(y) e_{\ominus\beta(f+g),\beta}(\beta(y))\, d_\beta y\, d_\beta s, \quad t \in \mathbb{R}_+ \cap I,$$

where

$$q(t) = f(t) \int_0^t p(y)\, d_\beta y, \quad and$$

$$l(t) = c + \int_0^t q(s)\, d_\beta s, \quad t \in \mathbb{R}_+ \cap I.$$

Proof. We have

$$x(t) \le c + \int_0^t f(s)x(s)\, d_\beta s + \int_0^t f(s) \int_0^s p(y)\, d_\beta y\, d_\beta s + \int_0^t f(s)\left(\int_0^s g(y)x(y)\, d_\beta y \right) d_\beta s$$

$$= c + \int_0^t (f(s)x(s)+q(s))\, d_\beta s + \int_0^t f(s)\left(\int_0^s g(y)x(y)\, d_\beta y \right) d_\beta s, \quad t \in \mathbb{R}_+ \cap I.$$

Now using Theorem 8.5.2, we get the desired result. This completes the proof. $\qquad\square$

Corollary 8.5.4. *Let $x, f, g, h \in \mathscr{C}_\beta(\mathbb{R}_+ \cap I)$ be nonnegative functions, and let c be a nonnegative constant such that*

$$x(t) \le c + \int_0^t f(s)x(s)\, d_\beta s + \int_0^t g(s)\left(x(s) + \int_0^s h(y)x(y)\, d_\beta y \right) d_\beta s, \quad t \in \mathbb{R}_+ \cap I.$$

Then

$$x(t) \leq c + c \int_0^t (f(s) + g(s))\, d_\beta s$$

$$+ c \int_0^t \frac{(f(s) + g(s))}{e_{\ominus\beta(f+g+h),\beta}(s)} \int_0^s (f(y) + g(y) + h(y))\, e_{\ominus\beta(f+g+h),\beta}(\beta(y))\, d_\beta y\, d_\beta s, \quad t \in \mathbb{R}_+ \cap I.$$

Proof. We have

$$x(t) \leq c + \int_0^t (f(s) + g(s))\, x(s)\, d_\beta s + \int_0^t g(s) \left(\int_0^s h(y) x(y)\, d_\beta y \right) d_\beta s$$

$$\leq c + \int_0^t (f(s) + g(s))\, x(s)\, d_\beta s + \int_0^t (f(s) + g(s)) \left(\int_0^s h(y) x(y)\, d_\beta y \right) d_\beta s, \quad t \in \mathbb{R}_+ \cap I.$$

Now using Theorem 8.5.2, we get the desired result. This completes the proof. $\qquad\qquad\square$

Theorem 8.5.5 (Pachpatte Inequality III). *Let* $x, f, g, h, p \in \mathscr{C}_\beta(\mathbb{R}_+ \cap I)$ *be nonnegative functions such that*

$$x(t) \leq h(t) + p(t) \int_0^t f(s) x(s)\, d_\beta s + p(t) \int_0^t f(s) p(s) \left(\int_0^s g(y) x(y)\, d_\beta y \right) d_\beta s, \quad t \in \mathbb{R}_+ \cap I.$$

Then

$$x(t) \leq h(t) + p(t) l(t) + \frac{p(t)}{e_{\ominus\beta q,\beta}(t)} \int_0^t q(s) h(s) e_{\ominus\beta q,\beta}(\beta(s))\, d_\beta s, \quad t \in \mathbb{R}_+ \cap I,$$

where

$$l(t) = \int_0^t (f(s) + g(s))\, h(s)\, d_\beta s, \quad and$$

$$q(t) = p(t)\,(f(t) + g(t)), \quad t \in \mathbb{R}_+ \cap I.$$

Proof. Let

$$z(t) = \int_0^t f(s) x(s)\, d_\beta s + \int_0^t f(s) p(s) \left(\int_0^s g(y) x(y)\, d_\beta y \right) d_\beta s, \quad t \in \mathbb{R}_+ \cap I.$$

Then

$$z(0) = 0,$$
$$x(t) \leq h(t) + p(t) z(t), \quad t \in \mathbb{R}_+ \cap I,$$

and

$$D_\beta z(t) = f(t) x(t) + f(t) p(t) \int_0^t g(s) x(s)\, d_\beta s$$

$$\leq f(t) h(t) + f(t) p(t) z(t) + f(t) p(t) \int_0^t g(s)\,(h(s) + p(s) z(s))\, d_\beta s$$

$$= f(t)\left(h(t) + p(t)\left(z(t) + \int_0^t g(s)\,(h(s) + p(s)z(s))\,d_\beta s\right)\right), \quad t \in \mathbb{R}_+ \cap I.$$

We set

$$m(t) = z(t) + \int_0^t g(s)\,(h(s) + p(s)z(s))\,d_\beta s, \quad t \in \mathbb{R}_+ \cap I.$$

Then

$$z(t) \le m(t),$$
$$m(0) = z(0)$$
$$= 0,$$

and

$$D_\beta z(t) \le f(t)\,(h(t) + p(t)m(t)), \quad t \in \mathbb{R}_+ \cap I.$$

Also,

$$\begin{aligned}
D_\beta m(t) &= D_\beta z(t) + g(t)h(t) + g(t)p(t)z(t) \\
&\le f(t)\,(h(t) + p(t)m(t)) + g(t)h(t) + g(t)p(t)m(t) \\
&= (f(t) + g(t))\,h(t) + p(t)\,(f(t) + g(t))\,m(t), \quad t \in \mathbb{R}_+ \cap I.
\end{aligned}$$

Hence,

$$\begin{aligned}
m(t) &\le \int_0^t (f(s) + g(s))\,h(s)d_\beta s + \int_0^t p(s)\,(f(s) + g(s))\,m(s)d_\beta s \\
&= l(t) + \int_0^t q(s)m(s)d_\beta s, \quad t \in \mathbb{R}_+ \cap I.
\end{aligned}$$

In view of Theorem 8.1.5, the last inequality yields

$$m(t) \le l(t) + \frac{1}{e_{\ominus\beta q,\beta}(t)} \int_0^t q(s)l(s)e_{\ominus\beta q,\beta}(\beta(s))d_\beta s, \quad t \in \mathbb{R}_+ \cap I.$$

Therefore,

$$\begin{aligned}
x(t) &\le h(t) + p(t)z(t) \\
&\le h(t) + p(t)m(t) \\
&\le h(t) + p(t)l(t) + \frac{p(t)}{e_{\ominus\beta q,\beta}(t)} \int_0^t q(s)l(s)e_{\ominus\beta q,\beta}(\beta(s))d_\beta s, \quad t \in \mathbb{R}_+ \cap I.
\end{aligned}$$

This completes the proof. $\qquad\square$

Theorem 8.5.6 (Pachpatte Inequality IV). *Let $x, f, g, h \in \mathscr{C}_\beta(\mathbb{R}_+ \cap I)$ be nonnegative functions, and let c be a nonnegative constant such that*

$$x(t) \le c + \int_0^t f(s)x(s)d_\beta s + \int_0^t f(s)\left(\int_0^s g(y)x(y)d_\beta y\right)d_\beta s$$

$$+ \int_0^t f(s) \left(\int_0^s g(y) \left(\int_0^y h(\tau) x(\tau) d_\beta \tau \right) d_\beta y \right) d_\beta s, \quad t \in \mathbb{R}_+ \cap I.$$

Then

$$x(t) \le c \left(1 + \int_0^t (f(s) + g(s)) \, e_{f+g+h,\beta}(s) d_\beta s \right), \quad t \in \mathbb{R}_+ \cap I.$$

Proof. Let

$$z(t) = c + \int_0^t f(s) x(s) d_\beta s + \int_0^t f(s) \left(\int_0^s g(y) x(y) d_\beta y \right) d_\beta s$$
$$+ \int_0^t f(s) \left(\int_0^s g(y) \left(\int_0^y h(\tau) x(\tau) d_\beta \tau \right) d_\beta y \right) d_\beta s, \quad t \in \mathbb{R}_+ \cap I.$$

Then

$$x(t) \le z(t),$$
$$z(0) = c, \quad \text{and}$$
$$D_\beta z(t) = f(t) x(t) + f(t) \int_0^t g(y) x(y) d_\beta y + f(t) \int_0^t g(y) \left(\int_0^y h(s) x(s) d_\beta s \right) d_\beta y$$
$$= f(t) \left(x(t) + \int_0^t g(y) x(y) d_\beta y + \int_0^t g(y) \left(\int_0^y h(s) x(s) d_\beta s \right) d_\beta y \right)$$
$$\le f(t) \left(z(t) + \int_0^t g(y) z(y) d_\beta y + \int_0^t g(y) \left(\int_0^y h(s) z(s) d_\beta s \right) d_\beta y \right), \quad t \in \mathbb{R}_+ \cap I.$$

Let

$$v(t) = z(t) + \int_0^t g(y) z(y) d_\beta y + \int_0^t g(y) \left(\int_0^y h(s) z(s) d_\beta s \right) d_\beta y, \quad t \in \mathbb{R}_+ \cap I.$$

Then

$$v(0) = z(0)$$
$$= c,$$
$$D_\beta z(t) \le f(t) v(t),$$
$$z(t) \le v(t), \quad \text{and}$$
$$D_\beta v(t) = D_\beta z(t) + g(t) z(t) + g(t) \int_0^t h(s) z(s) d_\beta s$$
$$\le f(t) v(t) + g(t) v(t) + g(t) \int_0^t h(s) v(s) d_\beta s$$
$$= (f(t) + g(t)) v(t) + g(t) \int_0^t h(s) v(s) d_\beta s$$
$$\le (f(t) + g(t)) v(t) + (f(t) + g(t)) \int_0^t h(s) x(s) d_\beta s, \quad t \in \mathbb{R}_+ \cap I.$$

Hence,

$$v(t) \le c + \int_0^t (f(s) + g(s))\, v(s) d_\beta s + \int_0^t (f(s) + g(s)) \left(\int_0^s h(y) v(y) d_\beta y \right) d_\beta s, \quad t \in \mathbb{R}_+ \cap I.$$

In view of Theorem 8.5.1, the last inequality yields

$$v(t) \le c \left(1 + \int_0^t (f(s) + g(s)) e_{f+g+h,\beta}(s) d_\beta s \right), \quad t \in \mathbb{R}_+ \cap I.$$

Consequently, we obtain

$$x(t) \le c \left(1 + \int_0^t (f(s) + g(s)) e_{f+g+h,\beta}(s) d_\beta s \right), \quad t \in \mathbb{R}_+ \cap I.$$

This completes the proof. $\qquad\qquad\qquad\qquad\qquad\qquad\qquad\qquad\qquad\qquad\qquad\qquad\qquad\qquad\quad$ $\square$

Theorem 8.5.7. *Let $x, k, p, f, g, h \in \mathscr{C}_\beta(\mathbb{R}_+ \cap I)$ be nonnegative functions, and let c be a nonnegative constant such that $f(t) \le g(t)$, $t \in \mathbb{R}_+ \cap I$, and*

$$x(t) \le k(t) + p(t) \left(\int_0^t f(s) x(s) d_\beta s + \int_0^t f(s) p(s) \left(\int_0^s g(\tau) x(\tau) d_\beta \tau \right) d_\beta s \right.$$
$$\left. + \int_0^t f(s) p(s) \left(\int_0^s g(\tau) p(\tau) \left(\int_0^\tau h(y) x(y) d_\beta y \right) d_\beta \tau \right) d_\beta s \right), \quad t \in \mathbb{R}_+ \cap I.$$

Then

$$x(t) \le k(t) + p(t) l(t) + 2 p(t) r(t) + \frac{2 p(t)}{e_{\ominus \beta c, \beta}(t)} \int_0^t c(s) b(s) e_{\ominus \beta c, \beta}(\beta(s)) d_\beta s, \quad t \in \mathbb{R}_+ \cap I,$$

where

$$q(t) = f(t) k(t) + f(t) p(t) \int_0^t g(s) k(s) d_\beta s + f(t) p(t) \int_0^t g(\tau) p(\tau) \left(\int_0^\tau h(y) k(y) d_\beta y \right) d_\beta \tau,$$

$$r(t) = \int_0^t (p(s) g(s) + h(s) p(s)) l(s) d_\beta s,$$

$$c(t) = 2 (p(t) g(t) + h(t) p(t)), \quad and$$

$$l(t) = \int_0^t q(s) d_\beta s, \quad t \in \mathbb{R}_+ \cap I.$$

Proof. Let

$$z(t) = \int_0^t f(s) x(s) d_\beta s + \int_0^t f(s) p(s) \left(\int_0^s g(\tau) x(\tau) d_\beta \tau \right) d_\beta s$$
$$+ \int_0^t f(s) p(s) \left(\int_0^s g(\tau) p(\tau) \left(\int_0^\tau h(y) x(y) d_\beta y \right) d_\beta \tau \right) d_\beta s, \quad t \in \mathbb{R}_+ \cap I.$$

Then

$$z(0) = 0,$$
$$x(t) \le k(t) + p(t)z(t), \quad \text{and}$$

$$D_\beta z(t) = f(t)x(t) + f(t)p(t)\int_0^t g(s)x(s)d_\beta s + f(t)p(t)\int_0^t g(\tau)p(\tau)\left(\int_0^\tau h(y)x(y)d_\beta y\right)d_\beta \tau$$

$$\le f(t)k(t) + f(t)p(t)z(t) + f(t)p(t)\int_0^t g(s)k(s)d_\beta s + f(t)p(t)\int_0^t p(s)g(s)z(s)d_\beta s$$

$$+ f(t)p(t)\int_0^t g(\tau)p(\tau)\left(\int_0^\tau h(y)k(y)d_\beta y\right)d_\beta \tau$$

$$+ f(t)p(t)\int_0^t g(\tau)p(\tau)\left(\int_0^\tau h(y)p(y)z(y)d_\beta y\right)d_\beta \tau$$

$$= q(t) + f(t)p(t)z(t) + f(t)p(t)\int_0^t p(s)g(s)z(s)d_\beta s$$

$$+ f(t)p(t)\int_0^t g(\tau)p(\tau)\left(\int_0^\tau h(y)p(y)z(y)d_\beta y\right)d_\beta \tau$$

$$= q(t) + f(t)p(t)\left(z(t) + \int_0^t p(s)g(s)z(s)d_\beta s + \int_0^t g(\tau)p(\tau)\left(\int_0^\tau h(y)p(y)z(y)d_\beta y\right)d_\beta \tau\right).$$

Thus,

$$D_\beta z(t) \le q(t) + f(t)p(t)\left(z(t) + \int_0^t p(s)g(s)z(s)d_\beta s\right.$$

$$\left. + \int_0^t g(\tau)p(\tau)\left(\int_0^\tau h(y)p(y)z(y)d_\beta y\right)d_\beta \tau\right), \quad t \in \mathbb{R}_+ \cap I.$$

Let

$$v(t) = z(t) + \int_0^t p(s)g(s)z(s)d_\beta s + \int_0^t g(\tau)p(\tau)\left(\int_0^\tau h(y)p(y)z(y)d_\beta y\right)d_\beta \tau, \quad t \in \mathbb{R}_+ \cap I.$$

Then

$$v(0) = z(0)$$
$$= 0,$$
$$D_\beta z(t) \le q(t) + f(t)p(t)v(t),$$
$$z(t) \le v(t), \quad \text{and}$$

$$D_\beta v(t) = D_\beta z(t) + p(t)g(t)z(t) + p(t)g(t)\int_0^t h(y)p(y)z(y)d_\beta y$$

$$\le q(t) + f(t)p(t)v(t) + p(t)g(t)v(t) + p(t)g(t)\int_0^t h(y)p(y)v(y)d_\beta y$$

$$\le q(t) + g(t)p(t)\left(2v(t) + \int_0^t h(y)p(y)v(y)d_\beta y\right), \quad t \in \mathbb{R}_+ \cap I.$$

Hence,

$$v(t) \le \int_0^t q(s)d_\beta s + 2\int_0^t p(s)g(s)v(s)d_\beta s + \int_0^t g(s)p(s)\int_0^s h(y)p(y)v(y)d_\beta y d_\beta s$$

$$\le l(t) + 2\int_0^t p(s)g(s)v(s)d_\beta s + 4\int_0^t g(s)p(s)\int_0^s h(y)p(y)v(y)d_\beta y d_\beta s, \quad t \in \mathbb{R}_+ \cap I.$$

In view of Theorem 8.5.5, the last inequality yields

$$v(t) \le l(t) + 2r(t) + \frac{2}{e_{\ominus\beta c,\beta}(t)}\int_0^t c(s)b(s)e_{\ominus\beta c,\beta}(\beta(s))d_\beta s, \quad t \in \mathbb{R}_+ \cap I.$$

Therefore,

$$x(t) \le k(t) + p(t)z(t)$$
$$\le k(t) + p(t)v(t)$$
$$\le k(t) + p(t)l(t) + 2p(t)r(t) + \frac{2p(t)}{e_{\ominus\beta c,\beta}(t)}\int_0^t c(s)b(s)e_{\ominus\beta c,\beta}(\beta(s))d_\beta s, \quad t \in \mathbb{R}_+ \cap I.$$

This completes the proof. $\qquad\square$

Theorem 8.5.8. *Let x, f, $g \in \mathscr{C}_\beta(\mathbb{R}_+ \cap I)$ be nonnegative functions, and let $h \in \mathscr{C}_\beta(\mathbb{R}_+ \cap I)$ be a positive nondecreasing function such that*

$$x(t) \le h(t) + \int_0^t f(s)x(s)d_\beta s + \int_0^t f(s)\left(\int_0^s g(y)x(y)d_\beta y\right)d_\beta s, \quad t \in \mathbb{R}_+ \cap I.$$

Then

$$x(t) \le h(t)\left(1 + \int_0^t f(s)e_{f+g,\beta}(s)d_\beta s\right), \quad t \in \mathbb{R}_+ \cap I.$$

Proof. Since h is positive and nondecreasing on $\mathbb{R}_+ \cap I$, we get

$$\frac{x(t)}{h(t)} \le 1 + \frac{1}{h(t)}\int_0^t f(s)x(s)d_\beta s + \frac{1}{h(t)}\int_0^t f(s)\left(\int_0^s g(y)x(y)d_\beta y\right)d_\beta s$$

$$\le 1 + \int_0^t f(s)\frac{x(s)}{h(s)}d_\beta s + \int_0^t f(s)\left(\int_0^s g(y)\frac{x(y)}{h(y)}d_\beta y\right)d_\beta s, \quad t \in \mathbb{R}_+ \cap I.$$

Using Theorem 8.5.1, we obtain

$$\frac{x(t)}{h(t)} \le 1 + \int_0^t f(s)e_{f+g,\beta}(s)d_\beta s, \quad t \subset \mathbb{R}_+ \cap I,$$

that is,

$$x(t) \le h(t)\left(1 + \int_0^t f(s)e_{f+g,\beta}(s)d_\beta s\right), \quad t \in \mathbb{R}_+ \cap I.$$

This completes the proof. $\qquad\square$

Theorem 8.5.9. *Let* $x, h, p \in \mathscr{C}_\beta(\mathbb{R}_+ \cap I)$ *be nonnegative functions, let* $f \in \mathscr{C}_\beta^2(\mathbb{R}_+ \cap I)$ *and* $g \in \mathscr{C}_\beta^1(\mathbb{R}_+ \cap I)$ *be positive functions, and let* c *be a positive constant such that*

$$x(t) \le c + \int_0^t f(s)\left(h(s) + \int_0^s g(\tau)\left(\int_0^\tau p(y)x(y)d_\beta y\right)d_\beta\tau\right)d_\beta s, \quad t \in \mathbb{R}_+ \cap I.$$

Then

$$x(t) \le \left(c + \int_0^t f(s)h(s)d_\beta s\right)\left(1 + \frac{1}{e_{\ominus\beta q,\beta}(t)}\int_0^t q(s)e_{\ominus\beta q,\beta}(\beta(s))d_\beta s\right), \quad t \in \mathbb{R}_+ \cap I,$$

where

$$q(t) = f(t)\int_0^t g(y)\int_0^y p(\tau)d_\beta\tau\, d_\beta y, \quad t \in \mathbb{R}_+ \cap I.$$

Proof. Let

$$l(t) = c + \int_0^t f(s)h(s)d_\beta s, \quad t \in \mathbb{R}_+ \cap I.$$

Then

$$x(t) \le l(t) + \int_0^t f(s)\int_0^s g(\tau)\left(\int_0^\tau p(y)x(y)d_\beta y\right)d_\beta\tau\, d_\beta s, \quad t \in \mathbb{R}_+ \cap I.$$

Since l is a positive and nondecreasing function on $\mathbb{R}_+ \cap I$, we have

$$\frac{x(t)}{l(t)} \le 1 + \frac{1}{l(t)}\int_0^t f(s)\int_0^s g(\tau)\left(\int_0^\tau p(y)x(y)d_\beta y\right)d_\beta d_\beta s$$

$$\le 1 + \int_0^t f(s)\int_0^s g(\tau)\left(\int_0^\tau p(y)\frac{x(y)}{l(y)}d_\beta y\right)d_\beta\tau\, d_\beta s, \quad t \in \mathbb{R}_+ \cap I.$$

Let

$$z(t) = 1 + \int_0^t f(s)\int_0^s g(\tau)\left(\int_0^\tau p(y)\frac{x(y)}{l(y)}d_\beta y\right)d_\beta\tau\, d_\beta s, \quad t \in \mathbb{R}_+ \cap I.$$

Then

$$\frac{x(t)}{l(t)} \le z(t),$$

$$z(0) = 1, \quad \text{and}$$

$$D_\beta z(t) = f(t)\int_0^t g(\tau)\left(\int_0^\tau p(y)\frac{x(y)}{l(y)}d_\beta y\right)d_\beta\tau, \quad t \in \mathbb{R}_+ \cap I.$$

This yields

$$\frac{D_\beta z(t)}{f(t)} = \int_0^t g(\tau)\left(\int_0^\tau p(y)\frac{x(y)}{l(y)}d_\beta y\right)d_\beta\tau,$$

$$D_\beta\left(\frac{D_\beta z(t)}{f(t)}\right) = g(t)\int_0^t p(y)\frac{x(y)}{l(y)}d_\beta y,$$

$$\frac{1}{g(t)} D_\beta \left(\frac{D_\beta z(t)}{f(t)} \right) = \int_0^t p(y) \frac{x(y)}{l(y)} d_\beta y,$$

$$D_\beta \left(\frac{1}{g(t)} D_\beta \left(\frac{D_\beta z(t)}{f(t)} \right) \right) = p(t) \frac{x(t)}{l(t)}$$

$$\leq p(t) z(t), \quad t \in \mathbb{R}_+ \cap I.$$

Therefore,

$$\frac{D_\beta \left(\frac{1}{g(t)} D_\beta \left(\frac{D_\beta z(t)}{f(t)} \right) \right)}{z(t)} \leq p(t), \quad t \in \mathbb{R}_+ \cap I.$$

Note that

$$D_\beta z(t) \geq 0, \quad z(t) \geq 0 \quad \text{and} \quad \frac{1}{g(t)} D_\beta \left(\frac{D_\beta z(t)}{f(t)} \right) \geq 0.$$

Therefore,

$$\frac{D_\beta \left(\frac{1}{g(t)} D_\beta \left(\frac{D_\beta z(t)}{f(t)} \right) \right)}{z(t)} \leq p(t) + \frac{\frac{1}{g(t)} D_\beta \left(\frac{D_\beta z(t)}{f(t)} \right) D_\beta z(t)}{z(t) z(\beta(t))}, \quad t \in \mathbb{R}_+ \cap I.$$

Since z is nondecreasing on $\mathbb{R}_+ \cap I$, we have

$$\frac{D_\beta \left(\frac{1}{g(t)} D_\beta \left(\frac{D_\beta z(t)}{f(t)} \right) \right) z(t)}{(z(t))^2} \geq \frac{D_\beta \left(\frac{1}{g(t)} D_\beta \left(\frac{D_\beta z(t)}{f(t)} \right) \right) z(t)}{z(t) z(\beta(t))}, \quad t \in \mathbb{R}_+ \cap I.$$

Therefore

$$\frac{D_\beta \left(\frac{1}{g(t)} D_\beta \left(\frac{D_\beta z(t)}{f(t)} \right) \right) z(t)}{z(t) z(\beta(t))} \leq p(t) + \frac{\frac{1}{g(t)} D_\beta \left(\frac{D_\beta z(t)}{f(t)} \right) D_\beta z(t)}{z(t) z(\beta(t))},$$

that is,

$$D_\beta \left(\frac{\frac{1}{g(t)} D_\beta \left(\frac{D_\beta z(t)}{f(t)} \right)}{z(t)} \right) \leq p(t), \quad t \in \mathbb{R}_+ \cap I.$$

Then

$$\frac{\frac{1}{g(t)} D_\beta \left(\frac{D_\beta z(t)}{f(t)} \right)}{z(t)} \leq \int_0^t p(s) d_\beta s,$$

that is,

$$\frac{D_\beta \left(\frac{D_\beta z(t)}{f(t)} \right)}{z(t)} \leq g(t) \int_0^t p(s) d_\beta s,$$

and thus

$$\frac{D_\beta \left(\frac{D_\beta z(t)}{f(t)} \right) z(t)}{(z(t))^2} \leq g(t) \int_0^t p(s) d_\beta s, \quad t \in \mathbb{R}_+ \cap I.$$

Since z is nondecreasing, we obtain

$$\frac{D_\beta\left(\frac{D_\beta z(t)}{f(t)}\right)z(t)}{z(t)z(\beta(t))} \le \frac{D_\beta\left(\frac{D_\beta z(t)}{f(t)}\right)z(t)}{(z(t))^2}$$

$$\le g(t)\int_0^t p(s)d_\beta s + \frac{\frac{D_\beta z(t)}{f(t)}D_\beta z(t)}{z(t)z(\beta(t))},$$

that is,

$$D_\beta\left(\frac{\frac{D_\beta z(t)}{f(t)}}{z(t)}\right) \le g(t)\int_0^t p(s)d_\beta s, \quad t \in \mathbb{R}_+ \cap I.$$

Then

$$\frac{\frac{D_\beta z(t)}{f(t)}}{z(t)} \le \int_0^t g(s)\int_0^s p(y)d_\beta y\, d_\beta s,$$

that is,

$$\frac{D_\beta z(t)}{f(t)} \le z(t)\int_0^t g(s)\int_0^s p(y)d_\beta y\, d_\beta s.$$

Thus

$$D_\beta z(t) \le z(t)f(t)\int_0^t g(s)\int_0^s p(y)d_\beta y\, d_\beta s,$$

and, finally,

$$z(t) \le 1 + \int_0^t\left(f(s)\int_0^s g(y)\int_0^y p(\tau)d_\beta\tau d_\beta y\right)z(s)d_\beta s, \quad t \in \mathbb{R}_+ \cap I.$$

In view of Theorem 8.1.5, the last inequality yields

$$z(t) \le 1 + \frac{1}{e_{\ominus\beta q,\beta}(t)}\int_0^t q(s)e_{\ominus\beta q,\beta}(\beta(s))d_\beta s, \quad t \in \mathbb{R}_+ \cap I.$$

Consequently,

$$x(t) \le l(t)\left(1 + \frac{1}{e_{\ominus\beta q,\beta}(t)}\int_0^t q(s)e_{\ominus\beta q,\beta}(\beta(s))d_\beta s\right)$$

$$= \left(c + \int_0^t f(s)h(s)d_\beta s\right)\left(1 + \frac{1}{e_{\ominus\beta q,\beta}(t)}\int_0^t q(s)e_{\ominus\beta q,\beta}(\beta(s))d_\beta s\right), \quad t \in \mathbb{R}_+ \cap I.$$

This completes the proof. $\qquad\square$

Theorem 8.5.10. *Let* $x, h, p \in \mathscr{C}_\beta(\mathbb{R}_+ \cap I)$ *be nonnegative functions, let* $f \in \mathscr{C}_\beta^1(\mathbb{R}_+ \cap I)$ *be a positive function, and let* c *be a positive constant such that*

$$x(t) \leq c + \int_0^t f(s)\left(h(s) + \int_0^s p(y)x(y)d_\beta y\right)d_\beta s, \quad t \in \mathbb{R}_+ \cap I.$$

Then

$$x(t) \leq g(t) + \frac{1}{e_{\ominus\beta l,\beta}(t)}\int_0^t l(s)g(s)e_{\ominus\beta l,\beta}(\beta(s))d_\beta s, \quad t \in \mathbb{R}_+ \cap I,$$

where

$$g(t) = c + \int_0^t f(s)h(s)d_\beta s, \quad and$$

$$l(t) = f(t)\int_0^t p(y)d_\beta y, \quad t \in \mathbb{R}_+ \cap I.$$

Proof. Let

$$z(t) = c + \int_0^t f(s)\left(h(s) + \int_0^s p(y)x(y)d_\beta y\right)d_\beta s, \quad t \in \mathbb{R}_+ \cap I.$$

Then

$$x(t) \leq z(t),$$
$$z(0) = c, \quad \text{and}$$
$$D_\beta z(t) = f(t)h(t) + f(t)\int_0^t p(y)x(y)d_\beta y, \quad t \in \mathbb{R}_+ \cap I.$$

Hence

$$D_\beta z(t) \geq 0,$$
$$\frac{D_\beta z(t) - f(t)h(t)}{f(t)} = \int_0^t p(y)x(y)d_\beta y$$
$$\geq 0, \quad t \in \mathbb{R}_+ \cap I,$$

and

$$D_\beta\left(\frac{D_\beta z(t) - f(t)h(t)}{f(t)}\right) = p(t)x(t)$$
$$\leq p(t)z(t), \quad t \in \mathbb{R}_+ \cap I.$$

Therefore,

$$\frac{D_\beta\left(\frac{D_\beta z(t) - f(t)h(t)}{f(t)}\right)}{z(t)} \leq p(t)$$

$$\leq p(t) + \frac{\frac{D_\beta z(t) - f(t)h(t)}{f(t)} D_\beta z(t)}{z(t)z(\beta(t))}, \qquad t \in \mathbb{R}_+ \cap I.$$

Consequently,

$$\frac{D_\beta\left(\frac{D_\beta z(t) - f(t)h(t)}{f(t)}\right) z(t)}{z(t)z(\beta(t))} \leq \frac{D_\beta\left(\frac{D_\beta z(t) - f(t)h(t)}{f(t)}\right) z(t)}{(z(t))^2}$$

$$\leq p(t) + \frac{\frac{D_\beta z(t) - f(t)h(t)}{f(t)} D_\beta z(t)}{z(t)z(\beta(t))}, \qquad t \in \mathbb{R}_+ \cap I,$$

which yields

$$D_\beta\left(\frac{\frac{D_\beta z(t) - f(t)h(t)}{f(t)}}{z(t)}\right) \leq p(t)$$

and

$$\frac{D_\beta z(t) - f(t)h(t)}{f(t)z(t)} \leq \int_0^t p(s)d_\beta s,$$

that is,

$$\frac{D_\beta z(t) - f(t)h(t)}{z(t)} \leq f(t)\int_0^t p(s)d_\beta s,$$

and thus

$$D_\beta z(t) \leq f(t)h(t) + z(t)f(t)\int_0^t p(s)d_\beta s, \qquad t \in \mathbb{R}_+ \cap I.$$

Therefore,

$$z(t) \leq c + \int_0^t f(s)h(s)d_\beta s + \int_0^t f(s)\left(\int_0^s p(y)d_\beta y\right) z(s)d_\beta s$$

$$= g(t) + \int_0^t l(s)z(s)d_\beta s, \qquad t \in \mathbb{R}_+ \cap I.$$

In view of Theorem 8.1.5, the last inequality yields

$$z(t) \leq g(t) + \frac{1}{e_{\ominus \beta l, \beta}(t)}\int_0^t l(s)g(s)e_{\ominus \beta l, \beta}(\beta(s))d_\beta s, \qquad t \in \mathbb{R}_+ \cap I,$$

whereupon

$$x(t) \leq g(t) + \frac{1}{e_{\ominus \beta l, \beta}(t)}\int_0^t l(s)g(s)e_{\ominus \beta l, \beta}(\beta(s))d_\beta s, \qquad t \in \mathbb{R}_+ \cap I.$$

This completes the proof. $\qquad\qquad\qquad\qquad\qquad\qquad\qquad\qquad\qquad\quad\Box$

Theorem 8.5.11. *Let* $x, v, p, f, g \in \mathscr{C}_\beta(I)$ *be nonnegative functions, and let* p *be a decreasing function on* I *such that*

$$x(t) \geq v(s) - p(t)\left(\int_s^t f(y)v(y)d_\beta y + \int_s^t f(y)\left(\int_y^t g(\tau)v(\tau)d_\beta\tau\right)d_\beta y\right), \quad a \leq s \leq t \leq b.$$

Then

$$x(t) \geq v(s)\left(1 + p(t)e_{k,\beta}(t)\int_s^t \frac{f(y)}{e_{k,\beta}(y)}d_\beta y\right)^{-1}, \quad a \leq s \leq t \leq b,$$

where

$$k(s) = \frac{h(s)}{1 - (\beta(s) - s)h(s)}, \quad a \leq s \leq t \leq b,$$

and

$$h(s) = p(s)f(s) + g(s), \quad a \leq s \leq t \leq b,$$

satisfying $1 - (\beta(s) - s)h(s) > 0$ *for* $a \leq s \leq t$.

Proof. Let $t \in I$ be fixed and denote

$$z(s) = x(t) + p(t)\left(\int_s^t f(y)v(y)d_\beta y + \int_s^t f(y)\left(\int_y^t g(\tau)v(\tau)d_\beta\tau\right)d_\beta y\right),$$

$a \leq s \leq t \leq b$. Then

$$z(t) = x(t),$$
$$v(s) \leq z(s), \quad \text{and}$$
$$D_\beta z(s) = -p(t)f(s)\left(v(s) + \int_s^t g(y)v(y)d_\beta y\right)$$
$$\geq -p(t)f(s)\left(z(s) + \int_s^t g(y)z(y)d_\beta y\right), \quad a \leq s \leq t \leq b.$$

Let

$$r(s) = z(s) + \int_s^t g(y)z(y)d_\beta y, \quad a \leq s \leq t \leq b.$$

Then

$$z(s) \leq r(s), \quad \text{and}$$
$$D_\beta r(s) = D_\beta z(s) - g(s)z(s)$$
$$\geq -p(s)f(s)r(s) - g(s)z(s)$$
$$\geq -p(s)f(s)r(s) - g(s)r(s),$$

which yields

$$D_\beta r(s) + (p(s)f(s) + g(s))r(s) \geq 0,$$

that is,

$$D_\beta r(s) + h(s)r(s) \geq 0, \quad a \leq s \leq t \leq b.$$

Note that

$$\begin{aligned}
D_\beta \left(r(s)e_{k,\beta}(s) \right) &= D_\beta r(s)e_{k,\beta}(\beta(s)) + r(s)k(s)e_{k,\beta}(s) \\
&= D_\beta r(s)\left(1 + (\beta(s) - s)k(s) \right) e_{k,\beta}(s) + r(s)k(s)e_{k,\beta}(s) \\
&= \frac{k(s)}{h(s)} \left(D_\beta r(s) + h(s)r(s) \right) e_{k,\beta}(s) \\
&\geq 0, \quad a \leq s \leq t \leq b.
\end{aligned}$$

Then

$$r(t)e_{k,\beta}(t) \geq r(s)e_{k,\beta}(s),$$

that is,

$$r(t) \geq r(s)\frac{e_{k,\beta}(s)}{e_{k,\beta}(t)},$$

and thus

$$\begin{aligned}
r(s) &\leq r(t)\frac{e_{k,\beta}(t)}{e_{k,\beta}(s)} \\
&= x(t)\frac{e_{k,\beta}(t)}{e_{k,\beta}(s)}, \quad a \leq s \leq t \leq b.
\end{aligned}$$

Therefore

$$D_\beta z(s) \geq -p(t)f(s)x(t)\frac{e_{k,\beta}(t)}{e_{k,\beta}(s)},$$

and

$$z(t) - z(s) \geq -p(t)x(t)e_{k,\beta}(t) \int_s^t \frac{f(y)}{e_{k,\beta}(y)} d_\beta y,$$

which gives

$$\begin{aligned}
z(t) &\geq z(s) - p(t)x(t)e_{,\beta}(t) \int_s^t \frac{f(y)}{e_{,\beta}(y)} d_\beta y \\
&\geq v(s) - p(t)x(t)e_{,\beta}(t) \int_s^t \frac{f(y)}{e_{,\beta}(y)} d_\beta y.
\end{aligned}$$

Then

$$x(t) \geq v(s) - p(t)x(t)e_{k,\beta}(t) \int_s^t \frac{f(y)}{e_{k,\beta}(y)} d_\beta y,$$

that is,

$$\left(1 + p(t)e_{k,\beta}(t)\int_{s}^{t}\frac{f(y)}{e_{k,\beta}(y)}d_{\beta}y\right)x(t) \geq v(s),$$

and thus

$$x(t) \geq v(s)\left(1 + p(t)e_{k,\beta}(t)\int_{s}^{t}\frac{f(y)}{e_{k,\beta}(y)}d_{\beta}y\right)^{-1}, \quad a \leq s \leq t \leq b.$$

This completes the proof. $\square$

8.6 **Notes and references**

In this chapter, we deduct some linear β-integral inequalities such as Gronwall-type, Bellman-type, Volterra-type, Gamidov-type, and Pachpatte-type β-integral inequalities. The materials in Sections 8.2, 8.3, 8.4, and 8.5 are new and authors' original contributions. Most of the results of this chapter are new, and some of the results can be found in [12].

Bibliography

[1] B. Ahmad, S. Ntouyas, J. Tariboon, Quantum Calculus. New Concepts, Impulsive IVPs and BVPs, Inequalities, World Scientific Publishing Co., Singapore, 2016.

[2] M. Bohner, A.C. Peterson, Dynamic Equations on Time Scales. An Introduction with Applications, Birkhäuser Boston, MA, 2001.

[3] M. Bohner, A.C. Peterson, Advances in Dynamic Equations on Time Scales, Birkhäuser Boston, MA, 2002.

[4] M. Bohner, S.G. Georgiev, Multivariable Dynamic Calculus on Time Scales, Springer International Publishing, Switzerland, 2016.

[5] J.L. Cardoso, A β-Strum–Liouville problem associated with the general quantum operator, Journal of Difference Equations and Applications 27 (4) (2021) 579–595.

[6] J.L. Cardoso, Variations around a general quantum operator, The Ramanujan Journal 54 (2021) 555–569.

[7] A.M.C.B. da Cruz, N. Martins, General quantum variational calculus, Statistics, Optimization & Information Computing 6 (1) (2018) 22–41.

[8] S. Elaydi, An Introduction to Difference Equations, third edition, Springer International Publishing, New York, 2008.

[9] N. Faried, E.M. Shehata, R.M. El Zafarani, Theory of nth-order linear general quantum difference equations, Advances in Difference Equations 264 (2018) 2018.

[10] S.G. Georgiev, S. Tikare, Taylor's formula for general quantum calculus, Journal of Mathematical Modeling 11 (3) (2023) 491–505.

[11] A.E. Hamza, A-S.M. Sarhan, E. Shehata, K.A. Aldwoah, A general quantum calculus, Advances in Difference Equations 2015 (2015) 182.

[12] A.E. Hamza, E. Shehata, Some inequalities based on a general quantum difference operator, Journal of Inequalities and Applications 2015 (2015) 38.

[13] A.E. Hamza, E.M. Shehata, Existence and uniqueness of solutions of general quantum difference equations, Advances in Dynamical Systems and Applications 11 (1) (2016) 45–58.

[14] A.E. Hamza, A-S.M. Sarhan, E. Shehata, Exponential, trigonometric and hyperbolic functions associated with a general quantum difference operator, Advances in Dynamical Systems and Applications 12 (1) (2017) 25–38.

[15] A.E. Hamza, E.M. Shehata, P. Agarwal, Leibniz's rule and Fubini's theorem associated with a general quantum difference operator, in: Computational Mathematics and Variational Analysis, Springer International Publishing, Cham, 2020, pp. 121–134.

[16] V. Kac, P. Cheung, Quantum Calculus, Springer, 2002.

[17] A.O. Karim, E.M. Shehata, J.L. Cardoso, The directional derivative in general quantum calculus, Symmetry 14 (2022) 1766.

[18] W.G. Kelley, A.C. Peterson, Difference Equations: An Introduction with Applications, Academic Press, San Diego, 2001.

[19] A.B. Malinowska, Delfim F.M. Torres, Quantum Variational Calculus, Springer Cham, 2014.

[20] R.E. Mickens, Difference Equations: Theory, Applications and Advanced Topics, third edition, CRC Press, Boca Raton, 2015.

[21] A-S.M. Sarhan, E.M. Shehata, On the fixed points of certain types of functions for constructing associated calculi, Journal of Fixed Point Theory and Applications 20 (2018) 124.

[22] E. Shehata, N. Faried, R.M. El Zafarani, A general quantum Laplace transform, Advances in Difference Equations 2020 (2020) 613.

[23] E. Shehata, R.M. El Zafarani, A β-convolution theorem associated with the general quantum difference operator, Journal of Function Spaces 2022 (2022) 1581362.

Index